Ecological Studies, Vol. 113

Analysis and Synthesis

Edited by

O.L. Lange, Würzburg, FRG
H.A. Mooney, Stanford, USA

Ecological Studies

Volumes published since 1989 are listed at the end of this book

F. Stuart Chapin III   Christian Körner   (Eds.)

Arctic and Alpine Biodiversity: Patterns, Causes and Ecosystem Consequences

With 68 Figures

Springer-Verlag
Berlin Heidelberg New York
London Paris Tokyo
Hong Kong Barcelona
Budapest

Prof. Dr. F. Stuart Chapin III
Department of Integrative Biology
University of California
Berkeley, CA 94720-3140
USA

Prof. Dr. Christian Körner
Department of Botany
University of Basel
Schönbeinstraße 6
4056 Basel
Switzerland

Design of the cover illustration by Mark W. Chapin

ISBN 3-540-57948-6 Springer-Verlag Berlin Heidelberg New York

Library of Congress Cataloging-in-Publication Data. Arctic and alpine biodiversity: patterns, causes, and ecosystem consequences/F. Stuart Chapin III and Christian Körner/eds.) p. cm. – (Ecological studies; vol. 113) Based on papers from a workshop held in Kongsvold, Norway, Aug. 17–20, 1993. Includes bibliographical references and index. ISBN 3-540-57948-6 (Springer-Verlag Berlin Heidelberg New York). 1. Biotic communities – Arctic regions – Congresses. 2. Mountain ecology – Congresses. 3. Biological diversity – Congresses. 4. Plant communities – Arctic regions – Ecology – Congresses. 5. Mountain plants – Ecology – Congresses. I. Chapin, F. Stuart (F. Stuart), III. II. Körner, Christian. III. Series: Ecological studies; v. 113. QH84. 1.A7 1995 574.5' 2621 – dc20 94-40221

Typesetting: Thomson Press (I) Ltd., Madras

SPIN: 10427000 31/3130/SPS – 5 4 3 2 1 0 – Printed on acid-free paper

Preface

As human populations expand and have increasing access to technology, two general environmental concerns have arisen. First, human populations are having increasing impact on the earth system, such that we are altering the biospheric carbon pools, basic processes of elemental cycling and the climate system of the earth. Because of time lags and feedbacks, these processes are not easily reversed. These alterations are occurring now more rapidly than at any time in the last several million years. Secondly, human activities are causing changes in the earth's biota that lead to species extinctions at a rate and magnitude rivaling those of past geologic extinction events. Although environmental change is potentially reversible at some time scales, the loss of species is irrevocable. Changes in diversity at other scales are also cause for concern. Habitat fragmentation and declines in population sizes alter genetic diversity. Loss or introduction of new functional groups, such as nitrogen fixers or rodents onto islands can strongly alter ecosystem processes. Changes in landscape diversity through habitat modification and fragmentation alter the nature of processes within and among vegetation patches.

Although both ecological changes altering the earth system and the loss of biotic diversity have been major sources of concern in recent years, these concerns have been largely independent, with little concern for the environmental causes the ecosystem consequences of changes in biodiversity. These two processes are clearly interrelated. Changes in ecological systems cause changes in diversity. Unfortunately, we know much less about the converse. What types and magnitudes of change in diversity alter the way in which ecosystems and the earth system function? What are the processes and circumstances under which this occurs?

The Scientific Committee on Problems in the Environment (SCOPE) and the United Nations Environment Programme (UNEP) have recently initiated a Global Biodiversity Assessment. An important component of this assessment is a series of studies on the causes and consequences of changes in biodiversity. An initial meeting sought to define the general issues and principles required to predict the nature of interaction between global change and biodiversity (Schulze and Mooney 1993, Ecol. Stud. vol. 99). Because it was clear that the nature of this interaction differs strikingly among ecosystem types, an additional series of meetings was planned to assess the causes and consequences of changes in biodiversity in 14 major ecosystems. This book addresses this issue in arctic and alpine ecosystems, covering about 8% of the global land area, an area similar to that occupied by boreal forests or by all the crop land on earth. Arctic and alpine ecosystems are critical because:

1. high latitudes are predicted to undergo more pronounced warming than other regions of the globe;
2. cold regions are the areas where climatic warming would have the greatest ecological consequences;
3. high altitudes, due to reduced pressure, are regions where CO_2 should be particularly limiting and where rising CO_2 might strongly stimulate plant growth;
4. arctic ecosystems with their large frozen pools of carbon and methane may exert strong feedbacks to global climate; and
5. due to their relative simplicity, these ecosystems may show clear effects of species on ecosystem processes and may, therefore, be strongly affected by loss or gain of species.

Hence, arctic and alpine ecosystems provide unique insights into causes and consequences of diversity in general. Furthermore, arctic and alpine ecosystems are the only biome with a global distribution, making them ideal for global monitoring of environmental change.

The book is subdivided into three sections, whose objectives are to:

1. describe the patterns of arctic and alpine diversity and suggest causes for these patterns;
2. develop a framework for predicting how biodiversity may change; and
3. discuss the ecosystem consequences of changes in biodiverisity.

Most of the chapters of this book are the product of a workshop held 17–20 August 1993 in Kongsvold, Norway. Kongsvold is situated in Dovre Fjell National Park, where arctic and alpine biomes merge. The workshop was sponsored by the US National Science Foundation, the Swiss IGPB National Committee, and various personal travel grants. The Kongsvold field station, which is one of Europe's oldest ecological field stations, provided a blend of a long tradition of nature conservation and a natural diversity of landscapes that inspired many long discussions of the future of biodiversity and its consequences in cold-dominated ecosystems.

Berkeley, USA F.S. Chapin III
Basel, Switzerland Ch. Körner
December 1994

Contents

List of Contributors

O. Agakhanjanz

Department of Physical Geography, Pedagogic University of Belarus, Sovetskaja 18, 220809 Minsk, Belarus/ CIS; and DFG-Guest Professor at the Department of Ecology in Bielefeld, Germany

B. Ammann

Institute for Geobotany, University of Bern, Altenbergrain 21, 3013 Bern, Switzerland

P.M. Anderson

Quaternary Research Center, University of Washington, Seattle, WA 98195, USA

G. Bonan

National Center for Atmospheric Research, P.O. Box 3000, Boulder, CO 80307-3000, USA

S.-W. Breckle

Department of Ecology, Bielefeld University, P.O. Box 100310, 33501 Bielefeld, Germany

M.S. Bret-Harte

Department of Integrative Biology, University of California, Berkeley, CA 94720-3140, USA

L.B. Brubaker

College of Forest Resources, University of Washington, Seattle, WA 98195, USA

J.P. Bryant

Institute for Arctic Biology, University of Alaska, Fairbanks, AK 99775-0180, USA

T.V. Callaghan

Centre for Arctic Biology, School of Biological Sciences, University of Manchester, Manchester M13 9PL, UK

F.S. Chapin III

Department of Integrative Biology, University of California, Berkeley, CA 94720-3140, USA

M.C. Chapin

Department of Integrative Biology, University of California, Berkeley, CA 94720-3140 USA

Yu.I. Chernov

Institute of Animal Evolutionary Morphology and Ecology, Leninski Prospect, 3, Moscow 117071, Russia

V.I. Chuprynin

Pacific Institute of Geography, Far-East Branch, Russian Academy of Sciences, 7 Radio Street, 690 041 Vladivostok, Russia

M. Gottfried

Institute of Plant Physiology, Department of Vegetation Ecology and Conservation Biology, University of Vienna, Althanstrasse 14, 1090 Vienna, Austria

G. Grabherr

Institute of Plant Physiology, Department of Vegetation Ecology and Conservation Biology, University of Vienna, Althanstrasse 14, 1090 Vienna, Austria

A. Gruber

Institute of Plant Physiology, Department of Vegetation Ecology and Conservation Biology, University of Vienna, Althanstrasse 14, 1090 Vienna, Austria

A.E. Hershey

Department of Biology, University of Minnesota – Duluth, Duluth, MN 55812, USA

S.E. Hobbie

Department of Integrative Biology, University of California, Berkeley, CA 94720-3140, USA

F.S. Hu

College of Forest Resources, University of Washington, Seattle, WA 98195, USA

R.L. Jefferies

Department of Botany, University of Toronto, 25 Willcocks Street, Toronto, Ontario M5S 3B2, Canada

S. Jonasson

Botanical Institute, Department of Plant Ecology, University of Copenhagen, Oster Farimagsgade 2D, 1353 Copenhagen K, Denmark

G.W. Kling

Department of Biology, University of Michigan, Ann Arbor, MI 48109-1048, USA

Ch. Körner

Institute of Botany, University of Basel, Schönbeinstrasse 6, 4056 Basel, Switzerland

J.B. McGraw

Department of Biology, West Virginia University, Morgantown, WV 26506-6057, USA

R.W. Merritt

Department of Entomology, Michigan State University, East Lansing, MI 48823, USA

E. Meyer

Institut für Zoologie, Technikerstrasse 25, 6020 Innsbruck, Austria

M.C. Miller

Department of Biological Sciences, University of Cincinnati, Cincinnati, OH 45221, USA

D.F. Murray

Herbarium of the University of Alaska Museum, University of Alaska Fairbanks, AK 99775-6960, USA

A.P. Oreshko

Pacific Institute of Geography, Far-East Branch,
Russian Academy of Sciences, 7 Radio Street,
690 041 Vladivostok, Russia

J. Pastor

Natural Resources Research Institute, University of Minnesota,
Duluth, MN 55811, USA

H. Pauli

Institute of Plant Physiology, Department of Vegetation Ecology
and Conservation Biology, University of Vienna, Althanstrasse 14,
1090 Vienna, Austria

J.F. Reynolds

Department of Botany, Duke University, Darham, NC 27708,
USA

J. Schimel

Institute of Arctic Biology, POB 757000, University of Alaska,
Fairbanks, AK 99775, USA

G.R. Shaver

The Ecosystems Center, Marine Biological Laboratory,
Woods Hole, MA 02543, USA

K. Thaler

Institut für Zoologie, Technikerstrasse 25, 6020 Innsbruck, Austria

M.D. Walker

Joint Facility for Regional Ecosystem Analysis,
Institute of Arctic and Alpine Research, University of Colorado,
Boulder, CO 80309-0450, USA

O.R. Young

Dickey Center Institute of Arctic Studies, 6193 Murdough Center,
Dartmouth College, Hanover, NH 03755, USA

S.A. Zimov

North-East Scientific Station, Pacific Institute of Geography,
Far-East Branch, Russian Academy of Sciences,
Republic of Sakha, Yakutia, Cherskii, Russia

Part I
Patterns and Causes of Diversity

1 Patterns and Causes of Arctic Plant Community Diversity

M.D. Walker

Attempts to explain and quantify community diversity have been a major paradigm in the development of modern ecology (e.g. Pielou 1975; MacArthur 1960; Whittaker 1965). Questions of the functional significance of diversity, and indeed whether species diversity alone has any functional significance, are also abundant (e.g. Hurlbert 1971; May 1973; Schulze and Mooney 1993). The taxonomic and genetic diversity within a community, the diversity among communities, and the diversity of communities on a landscape all contribute to regional diversity and are all aspects of community diversity. Genetic and species diversity are the building blocks of communities, and define the set of potential plant communities for a given region, but the plant communities into which these taxonomic and genetic units are organized are the most direct and easily measurable indication of overall ecosystem diversity, becuase they represent the integration of species and landscape. Understanding how the diversity of communities may change following a directional change in climate requires an understanding of the processes that control diversity at different levels.

1.1 Background and Definitions

Whittaker (1972) recognized and defined four aspects of community diversity, each of which have an implicit scale associated with them: (1) alpha diversity, the species diversity within a stand or community, (2) beta diversity, the degree of differentiation among communities within a landscape, (3) gamma diversity, the total species diversity within a landscape, and (4) delta diversity, the change in floras from one landscape or area to another.

1.1.1 Alpha Diversity

One of the key ecological debates of the past century has concerned the ecological importance, or lack thereof, of alpha diversity. This debate has centered on whether diversity acts as a stabilizing influence in a community or ecosystem, with more species resulting in a greater degree of ecological redundancy (May 1973). This debate became mired in problems with defining stability and was

Joint Facility for Regional Ecosystem Analysis, Institute of Arctic and Alpine Research, University of Colorado, Boulder, CO 80309-0450, USA

Ecological Studies, Vol. 113
Chapin/Körner (eds) Arctic and Alpine Biodiversity

further complicated by the idea that complexity, rather than simple diversity, was necessary for stability. More central to the present discussion is the large body of work on the mechanisms that control alpha diversity, much of which comes from the field of theoretical population biology (e.g. May 1975). Understanding and explaining these mechanisms are critical to predicting changes in alpha diversity with a changing climate, as climate and diversity are not likely to be simply linked.

Alpha diversity has two main subcomponents: richness (the total number of species) and evenness (the uniformity of their distribution). Many different methods of calculating richness and evenness have been published (Magurran 1988). Studies of the species diversity of natural communities have often found that richness alone gives equally useful information as richness and evenness (e.g. Fridriksson 1989).

1.1.2 Beta Diversity

Beta diversity is a measure of community diversity independent of any classification system. There has been relatively little published on the ecological importance of and underlying controls on beta diversity; most of the contributions are from outside the field of plant ecology (e.g. Cody 1970). Beta diversity can be represented as a ratio between alpha and gamma diversity (Whittaker 1972):

$$D_\beta = \frac{D_\gamma}{\overline{D_\alpha} - 1},$$

where D_β is beta diversity, D_γ is gamma diversity, and $\overline{D_\alpha}$ is the average sample alpha diversity. As average sample alpha diversity decreases relative to gamma diversity, beta diversity increases. Another quantitative measure of beta diversity is the length of the first axis of a detrended correspondence analysis ordination, which is in "SD units" (Gauch and Whittaker 1972). One SD unit represents the average width of a species dispersion along the floristic gradient constructed by the ordination, with the underlying assumption that species have a unimodal, Gaussian response to the gradient. For most data sets, one SD unit represents approximately 50% dissimilarity, and four SD units represent approximately 100% dissimilarity. The richness and evenness of community types (associations) can also be considered as a measure of beta diversity, although not strictly within Whittaker's original definition.

1.1.3 Gamma Diversity

Gamma diversity, or total regional diversity, represents the sum of alpha diversities and a fundamental constraint on those alpha diversities. It can also be considered as essentially equivalent to alpha diversity at the next level of organization, and the same theoretical and quantitative considerations apply.

A problem with quantitative data on species richness of large areas is the log-linear relationship between species and area, which makes simple comparisons of differently sized areas difficult, because small areas will have disproportionately high numbers of species per unit area compared to larger areas (Gleason 1922; Cailleux 1961; Billings 1992). Given that this log-linear relationship has been shown to hold true for most empirical data sets (Preston 1962 a,b), the most useful index of floristic richness should be a logarithm of species number per some set area. For example, Cailleux (1961) demonstrated a linear relationship betwen mean annual temperature and the logarithm of areally adjusted vascular species richness on a global scale.

Most data on gamma diversity are not based on equally sized areas but loosely defined regions, referred to as regional or local floras, which are usually some unit representing an area that can be reached conveniently by foot or vehicle from a central location, such as the Toolik Lake region, the Prudhoe Bay region, etc. containing all of the major habitats present in that region. Tolmachev (1931) formalized this concept as a concrete or local flora, which is defined as the flora of the minimal area necessary to include the most characteristic habitats for a given macroclimate. Abundant data on local floras of the Russian Arctic exist, but most have not yet been published or analyzed (V.Yu. Raszhivin and N.V. Matveyeva, pers. comm.). The Russian data represent an untapped source for testing specific theories about controls on arctic species diversity (e.g. Chernov, this Vol.). In practice, the Russian and North American approaches are not that different, as in all cases the investigator is seeking to visit all habitats within an "ecologically homogeneous area", and the species-area curve will tend to flatten in any case unless the size of the study or search increases geometrically.

1.1.4 Delta Diversity

Delta diversity is the functional equivalent of beta diversity at the next level of organization; it is the degree to which local floras are differentiated. Although there are many qualitative comparisons of specific local arctic floras (e.g. Young 1971; Murray 1978), I know of no quantitative analysis of delta diversity in the Arctic, and indeed there are few published analyses of delta diversity of plants for any system (e.g. Kruger and Taylor 1979). Delta diversity is an important component in predicting response to global change, because the degree of differentiation among local floras is one indicator of the potential for invasion of new species.

1.2 Arctic Species Diversity: The First Filter

Community diversity is fundamentally and inextricably linked to species diversity. Körner (this Vol.) explains the causes of diversity of a given site, community, etc. as a function of a series of "filters" with differently sized pores. The first filter is the presence of the species somewhere in the Arctic. Only about 0.4% of the earth's known vascular plant species occur in the Arctic (Billings 1992). The first filter is so fine that relative size differences among additional pores are small, and species that get through the first filter have a high probability of getting through additional ones.

The Arctic is one end of a global gradient of summer temperature, biologically useful heat, and taxonomic diversity. Temperature has strong effects on almost all aspects of arctic ecosystems, including soil stability, moisture, and nutrient availability, and any one or many of these may effectively limit an individual species' presence in the ecosystem. Even habitat diversity, considered an important control on species diversity in all systems, can be strongly related to thermal energy in the Arctic, as periglacial land and surface forms result from the interaction between climate and local geological conditions (Washburn 1973). I refer to the gradient of summer warmth (usually measured in degree-days above 0 °C) and summer temperature (usually measured as mean July temperature) synonymously as a gradient of thermal energy, recognizing that they have different effects and importance. Temperature has direct biological effects that may control species presence and therefore diversity, for example, Q_{10} and key thresholds such as frost resistance (e.g. Körner and Larcher 1988). In other cases, the total seasonal heat energy may be critical, such as for the development of reproductive structures. Callaghan and Jonasson (this Vol.) report an increase in seed set and viability of high arctic populations of *Dryas octopetala* following experimentally increased temperatures. Both temperature and thermal energy are a consequence of latitudinal differences in solar input, and it is their combined influence that ultimately affects species diversity. Biological or physical processes related to either temperature or heat energy may limit individual species at almost any stage in the life cycle and are not expected to be the same among species.

1.3 Biogeographical Patterns Within the Arctic: The Second Set of Filters

Most recent estimates of the size of the arctic flora, which vary depending upon the classification of subspecific taxa and species groups, include about 1500 vascular, 750 bryophyte, and 1200 lichen species (Murray 1992; B. Murray, pers. comm.). Approximately 60% of the vascular flora is in common throughout, increasing to about 90% in common in the most northerly areas (Polunin 1959; Hultén 1962; Young 1971; Yurtsev et al. 1978; Billings 1992). If one begins at some arbitrary point in the southern Arctic, there will be a gradient of decreasing diversity to the north, with species being lost and only a few new truly arctic species being gained (i.e., the rate of loss is much greater than the rate of gain), and a gradient of increasing total diversity to the east or west, with new species coming into the flora and others being lost locally (i.e., the rates of gains and losses are approximately equal).

1.3.1 Climatic Gradients

The south to north gradient of decreasing diversity is essentially a localized version of the equatorial-polar gradient of diversity and productivity, and therefore can be related primarily to changes in thermal energy. The importance of thermal energy to arctic ecosystems has been a major theme of every comprehensive treatment of these ecosystems (Billings 1974; Ives and Barry 1974; Yurtsev et al. 1978; Aleksandrova 1980, 1988; Chernov 1985; Bliss and Matveyeva 1992). Young (1971) noted that within the Arctic: (1) there is a decrease in overall vascular plant species diversity from south to north, (2) the overall flora is similar throughout relative to other biomes, (3) the vegetation structure is similar throughout, and (4) there are few species endemic to the most northerly regions, although there are some truly arctic species-(see, for example, Beschel 1970; Murray 1992). Young also noted that the northerly floras are primarily depauperate subsets of the southerly floras, indicating that competition is an unlikely factor explaining the lack of species in northerly areas. Young proposed that latitudinally decreasing species diversity could be explained solely by the sum of mean monthly temperatures for all months in which the mean is above 0 °C. This measure is similar to but less precise than thawing-degree days, which represent an annual sum of biologically useful heat (thawing-degree days are the sum of mean daily temperatures for all days when the mean is >0 °C). Mean annual temperature is not as well correlated with species numbers, because low winter temperatures do not necessarily exclude species, and annual means are heavily weighted by winter values.

Young's hypothesis, which was a statement of correlation rather than mechanism or cause, has yet to be refuted, and further studies have supported it. Rannie (1986) made a direct test of Young's supposition that summer temperature is the only variable needed to predict vascular species diversity. Using published floristic data from 38 localities in the Canadian Arctic, she showed a remarkably high correlation between July mean temperature, July degree-days >0 °C, and July degree-days >5 °C, with these various measures explaining 94–95% of the variance in species diversity (Fig. 1.). Temperature had a slightly better correlation with diversity than did degree-days, but the strong correlation between temperature and degree-days makes it impossible to isolate one or the other as a more important casual agent. Although the surveys that went into Rannie's analysis were based on different areas, they were all large enough that the species-area relationship was of minor importance. This is an example of why the

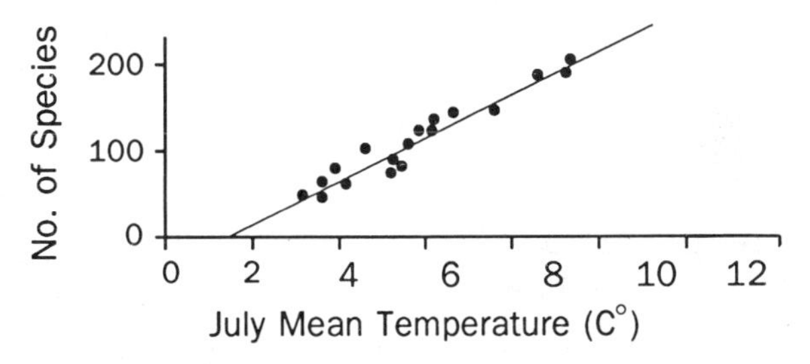

Fig. 1. Correlation between July mean temperature and number of species for local floras of the Canadian Arctic Archipelago. (Redrawn from Rannie 1986 with permission of the Arctic Institute of North America)

local flora concept may be more useful for comparative purposes than strict equal area comparisons.

At a finer scale, D.A. Walker (1985) divided the Prudhoe Bay region into three zones following a steep coastal-inland temperature gradient of decreasing temperature, and examined the size of the vascular flora and its distribution according to a zonation system developed by Young (1971). Young divided the Arctic into four zones based on northernmost limits of species distribution, so that a species classified as zone 4 (the southernmost of the four zones) has its northern limit somewhere in zone 4. The result of Walker's analysis was an increase in total number of species from 115 in the coastal zone to 188 in the inland zone, a decrease in zone 1 and 2 species from the coast inland, and an increase in zone 3 and 4 species in the same manner (Fig. 2).

The Taymyr Peninsula in Siberia offers almost an ideal gradient of summer temperature from tree line to polar desert, with little change in relief, no difference in winter temperature, and continuous continental landmass (Matveyeva, 1994). The Taymyr is the only place on the globe that offers such a gradient without complications such as islands, complex geology, and strong oceanic climate influences. The diversity of vascular species in local floras of the Taymyr decreases from about 250 in the south to about 50 in the north (Chernov and Matveyeva 1979).

Other climatic gradients also influence diversity. The polar deserts of the Canadian Arctic Archipelago and those of Siberia are both to the north of the 2 °C July isotherm (Aleksandrova 1988). Both areas are mostly barren, with total plant cover of less than 5%.The Canadian communities, on the Queen Elizabeth Islands, are "dominated" by vascular plants, whereas in the Siberian communities,

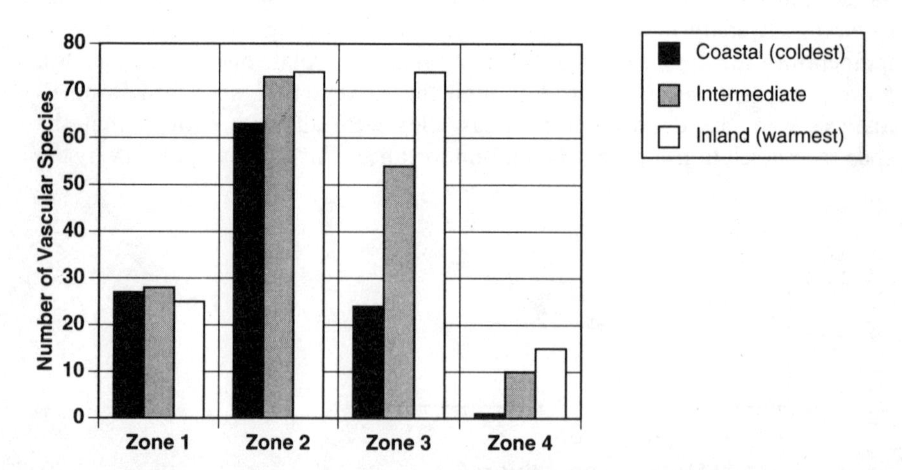

Fig. 2. Number of vascular species in three climatic zones (coastal, intermediate, and inland) at Prudhoe Bay, Alaska, according to Young's (1971) zones of northern limit. Zone 1 species have their northern limit in the polar deserts, and zone 4 species have northern limits in the Low Arctic. Data from D.A. Walker (1985)

on Cape Cheluyskin of the Taymyr Peninsula, lichens and mosses dominate (Matveyeva 1979; Bliss and Svoboda 1984; Aleksandrova 1988; Bliss and Matveyeva 1992). Differences in precipitation and possibly summer fogginess are the most likely causes of these differences; mean annual precipitation is 100–250 mm in the Asian sites versus only 75–150 mm in the North American sites (Bliss and Matveyeva 1992), although a comprehensive comparative treatment of the two systems has not yet been completed.

1.3.2 Gradients of Geological History

Perhaps the most significant historical event in the Arctic has been the series of glaciations that have occurred there over the past million years (Péwé 1975). Although most land areas were covered by large continental glaciers, with only small localized nonglaciated areas, such as nunataks, available as refugia, most of Beringia, covering what is now northern Alaska and the Chukotka Peninsula of eastern Russia, was an extensive, arctic, ice-free area (Hopkins 1967). Lowered sea levels exposed much of the continental shelf between Asia and North America, making a large, contiguous ice-free zone. The modern flora of Beringia is disproportionately high, with a greater number of endemics relative to other arctic areas.

The Beringian species, defined as species that are either endemic to some region of what was the Pleistocene ice-free zone or those that have North American-Asian distributions, and which add to the richness of the Beringian flora, do not dominate the modern communities, but instead are mostly found in specialized habitats with good drainage and often southern exposures (Yurtsev 1982; D.A. Walker 1985; M.D. Walker 1990). M.D. Walker (1990) compared the continental distributions of species that occurred only on pingos[1], a relatively rare habitat which offers good drainage on the otherwise flat wet coastal plain of Alaska, with those species that occurred on pingos but were also found in other more common local habitats. Seventy-one percent of the species that occurred only on pingos had distributions restricted to either Asia and North America, only North America, or only Alaska and the Yukon, with the remaining 29% being circumpolar or nearly so in distribution, whereas the widespread species had 54 and 46% of their species in these respective categories.

Geological history also influences finer scale patterns. When two southern Arctic Foothills, Alaska, landscapes were compared, which differed in their time since release from glaciation but which had the same macroclimate and degree of topographic relief, there was less richness and evenness of vegetation community types on older landscapes (D.A. Walker et al. unpubl. data; Table 1). The causes of the observed differences in diversity are likely related to those observed in other

[1] Pingos are ice-cored, dome-shaped hills ranging in height to as much as 50m, although normally less than 15 m, and in basal diameter to as much as 1000 m, although normally less than 300 m. They have constant initial substrate, although zoogenic, pedogenic, and cryogenic processes tend to increase microsite substrate differences over time (D.A. Walker et al. 1985).

Table 1. Richness and evenness of community types on three differently-aged glacial drift surfaces, southern Arctic Foothills, Alaska. The Sagavanirktok drift is broadly mid-Pleistocene (300–500 Ka) in age, the Itkillik I drift is approximately 50 ka, and the Itikillik II drift is approximately 10 ka (Hamilton 1986). Vegetation data are from D. A. Walker et. al (1989 and unpub.)

	Total richness (number of vegetation types)	Areally adjusted richness (number of types per 10 m^2)	Shannon-Wiener index[a]	Evenness index[b]	Total area (ha)	Total vegetated area (ha)
Imnavait Creek map area (1:6 000 scale)						
Sagavanirktok	29	0.187	0.784	0.536	155 640	154 359
Itkillik I	17	1.81	0.832	0.676	9 405	9 390
Toolik Lake map area (1:5 000 scale)						
Itkillik I	20	0.292	0.686	0.527	68 576	67 951
Itkillik II	27	0.373	0.933	0.652	72 288	72 136

[a] $H' = -\sum_{i=1}^{s} p_i \log_{10} P_i$, where H′ is the Shannon-Weiner diversity index, s is the number of vegetation types, and p_i is the proportional abundance of the ith vegetation type.

[b] $J = \frac{H'}{\log_{10} S}$, where J is the evenness index (Pielou 1969) H′ is the Shannon-Weiner diversity index, and s is the number of vegetation types. The denominator is the maximum value for H′; as the maximum approaches the actual value, the types are more evenly distributed.

glacial landscapes over shorter time sequences – that diversity initially increases and then declines as dominance patterns develop (Zollitsch 1969; Reiners et al. 1971; Matthews 1992). *Eriophorum vaginatum* tussock tundra is the dominant community type in these landscapes. Although shifts in species dominance related to moisture and nutrient conditions cause visual variation among tussock tundra stands, its among-sample compositional similarity is consistently high, and its species diversity is low (M.D. Walker et al. 1994). The development of tussock tundra is dependent upon the process of paludification, with development of a *Sphagnum* moss layer resulting in cold soils, decreased depth of thaw, decreased nutrient availability, and increased soil moisture (Jorgenson 1984). A dense canopy of shrubs and tussock grasses forms within this moss layer. The Arctic Foothills pattern is not a simple successional sequence, however, since even the oldest sites would have experienced glacial climates during subsequent advances. The older landscapes were probably sparsely vegetated during the glacial intervals, and the presence of at least some vegetation and soil there, coupled with the existence of an established hydrologic pattern, one of the first geomorphic patterns to appear following deglaciation (Matthews 1992), may have given the older landscapes an "edge" in development.

1.4 Distribution of Species in Communities: The Third Set of Filters

The final set of filters through which a species passes are those within the local landscape. Once a species has found its way into the local flora or landscape, its distribution and abundance there will depend on its internal constraints and their interaction with the environment and its interaction with other species.

1.4.1 Gradients of Biological Interaction

A recurrent theme in the literature of arctic vegetation is an apparent lack of well-structured communities in the most severe arctic environments, which early work explained as reflecting a lack of competition in those environments (Griggs 1934; Savile 1960). These visual impressions were based on visits to high arctic and polar desert areas and were not backed by data or analyses, and the actual role of competition in shaping high arctic communities remains unknown. Even with very few species, competition for safe seedling establishment sites may be intense, and may effectively control what species are present in a region. Rather than invoking competition, or lack thereof, as an important mechanism, it is probably more fruitful to consider the gradient from the southern to northern Arctic as a gradient of potential biological interaction. As both the number of species and the ecosystem productivity decrease, the potential for biotic interaction should also decrease, making physical factors increasingly important. In the north, thermal energy, an important limiting factor not subject to competition, takes on a greater and greater role, until it finally overwhelms all other factors. Although any locality will have some degree of microclimatic variability that will control species distributions, as temperature becomes increasingly important, the role of secondary environmental gradients in shaping communities should decrease.

Because community diversity is a function of the total number of species available, we expect community diversity at the landscape level to decline as the number of species declines. The number of possible combinations of *n* species in a single space is $2^n - 1$. Assuming a local flora of only 100 species, there would be 1.27×10^{30} possible combinations of species. In practice, the concrete members of an association have an average similarity of about 50% (Gauch 1982), and the actual number of community types will be far less than this theoretical maximum. Nonetheless, this mathematical relationship indicates that as the total flora decreases, the number of potential community types should also decrease, and the degree of similarity between stands should increase. This effect should occur independently of biological or physical factors that may effect species and community distribution. This also leads to a specific testable hypothesis, i.e., the presence of any other pattern indicates that factors other than simple flora size are influencing community distributions.

The importance of competition versus individual species' response to gradients in shaping communities is difficult to quantify, and there has been little experimental work on this question for the Arctic. I used north- and south-facing slopes of pingos in the Prudhoe Bay and surrounding regions as microcosms of larger climatic gradients (M.D. Walker 1990). Temperature differences on these opposing surfaces are equivalent at least to a gradient from Low to High Arctic. From a sample of 41 pingos, overall (gamma) vascular species diversity was lower on north-facing slopes (92 total species on north-facing slopes versus 140 species on south-facing slopes), as was average alpha diversity (mean of 14 versus 18 species) and beta diversity (sum of SD units for four DCA ordination axes 9.0 versus 14.3). Ordinations of north-facing slopes also had lower eigenvalues, indicating noisier patterns in the floristic gradients, and there were fewer and weaker relationships between environmental variables and ordination axes. The first axis of the north-facing slope ordination was significantly rank correlated with five environmental variables, with a mean rank correlation coefficient of 0.378; the south-facing slope axis was correlated with six environmental variables, with a mean rank correlation coefficient of 0.443. Although it is clear from the ordinations that factors other than temperature are affecting community composition on both north- and south-facing slopes, it is also clear that there is an apparent increase in "randomness" to the distribution of species on north-facing slopes, which might be explained by a decrease in the importance of competition on these slopes.

1.4.2 Habitat Diversity

The mosaic of habitats present on a landscape at a point in time are a result of disturbances at all spatial and temporal scales, including those of geological proportion (Table 2). Although disturbance is a primary factor in shaping any landscape and therefore the diversity of that landscape (During et al. 1988), the Arctic differs from other biomes in having a great diversity of mesoscale patterns

Table 2. Hierarchy of natural disturbances, northern Alaska. (Walker and Walker 1991)

Hierarchical domain (Delcourt and Delcourt 1988)	Disturbance	Geomorphic effect	Vegetation effect	Spatial scale (m^2)	Event frequency (years)
Megascale (continental to global)	Continental drift and uplift of Brooks Range	Formation of physiographic provinces and bedrock types, establishment of regional drainage patterns	Evolution of arctic Flora	$>10^{12}$	$>10^{6}$
Macroscale (regional)	Climate fluctuations associated with ice ages	Formation of glacial surfaces, marine terraces, sand sea, loess deposits, alluvial deposits, regional paludification, sediment deposition, development of of second-order and higher drainages	Speciation and extinction; development, movement, and displacement of plant communities	10^{4}–10^{10}	10^{4}–10^{6}
Mesoscale (regional)	Climatic fluctuations during the Holocene	Development and alteration of permafrost, colluviation of steep slopes, development and alteration of many periglacial features (e.g. icewedges, pingos), thaw-lake cycle, development of water tracks and first-order drainages, soil formation	Species migration, ecotene displacement, changes in landscape mosaic, movement of tree line	10^{2}–10^{10}	10^{3}–10^{5}
	Loess deposition	Alkaline silty soils, dilution of organic matter	Addition of nutrients, calciphilous flora	10^{6}–10^{10}	10^{1}–10^{2}
Microscale (site)					
Macrosite	Tundra fires	Local fluvial and thermal erosion	Burning, recycling and loss of nutrients, opening of canopy	10^{4}–10^{8}	10^{3}–10^{4}
Mesosite	Growth and erosion of ice wedges	Formation of ice-wedge-polygon microtopography, thermokarst pits	Changes in vegetation mosaic by alteration of microenvironment	10^{0}–10^{3}	10^{0}–10^{4}

Table 2. (*contd.*)

Hierarchical domain (Delcourt and Delcourt 1988)	Disturbance	Geomorphic effect	Vegetation effect	Spatial scale (m^2)	Event frequency (years)
Mesosite	Major storms and storm surges	Debris flows, large rock falls, floodplain alteration, coastal erosion, rill formation	Burial and removal of vegetation, salt kill in coastal areas	10^{-1}–10^6	10^0–10^2
	Annual snow and runoff cycle:	Springs, icings, solifluction features	Summer-long water source, alteration of soil temperature and growing season length	10^4–10^6	10°
	Groundwater discharge, melting of active layer Snowback formation and melting	Annual erosion and sedimentation, formation of nivation hollows[a],	Unstable substrate, short growing season, summer-long water source, winter protection	10^2–10^5	10^0
	Spring flood	Fluvial erosion and sedimentation	Addition of nutrients, burial, removal of plants		
	Oil seeps	Deposition of hydrocarbons	Killing of vegetation	10^2–10^4	10^{-2}–10^1
Microsite	Animal disturbances	Caribou trails, animal dens, craters, local erosion, krotovinas	Removal of plants, addition and recycling of nutrients	10^{-2}–10^2	10^{-1}–10^2
	Daily and annual freeze-thaw cycle	Needle ice formation, frost scars	Physical disturbance of plant roots and seedings	10^{-4}–10^4	10^{-2}–10^0

[a]Nivation hollows are amphitheatre-like depressions caused by mass wasting and freeze and thaw cycles. They are considered to be the first stage in cirque formation but are not caused by glaciation (Lewis 1939).

related to periglacial activity. Features such as frost boils, high- and low-centered ice wedge polygons, thaw lake basins, and pingos all have major influences on vegetation patterning in arctic landscapes, and many treatments of arctic vegetation have used these landscape features as a primary organizing theme (e.g. D.A. Walker 1985; Matveyeva 1988; D.A. Walker et al. 1989; M.D. Walker et al. 1991).

Changes in substrate or topography at almost any spatial scale will lead to a change in species composition. Matveyeva (1988) considered the vegetation of frost boils (small sorted circles 1–2 m diameter with active frost churning and often mostly bare soil in the center) and the surrounding matrix in upland mesic tundra to be a single plant community, and based on this definition, found alpha diversities as high as 130 to 160 in an area of 100 m^2, up to 50 species on only 1 m^2, and up to 25 species on 0.1 m^2, but only 40, 30, and 20 species, respectively, on similarly sized areas in mires. D.A. Walker (1985), on the other hand, treated the vegetation of different zones (rim, trough, and center) within ice-wedge polygon tundra as a series of separate communities, and he described less diverse communities with almost complete turnover of species with only a few centimeters of change in topographic position.

Frequency and intensity of disturbance also affect diversity. D.A. Walker et al. (1989) compared the community type diversity of five major terrain unit types in a watershed of the southern Arctic Foothills in northern Alaska. They reported the highest diversity on floodplains (Shannon index 0.857), which are frequently disturbed, and the lowest diversity on retransported hillslope deposits (Shannon index 0.471),which are stable sites and the primary areas of tussock tundra development. Bedrock outcrops, glacial till, and basin colluvium had intermediate diversities (Shannon indices 0.624, 0.599, and 0.583, respectively).

1.4.3 Diversity Hot Spots: Focal Points and Oases

Species and community diversity are unevenly distributed in most landscapes, and much of the diversity may be concentrated in relatively small areas. The evenness of community diversity is largely a function of topography, because of its overwhelming importance in controlling the steepness of environmental gradients. In mountainous areas, species turnover is high and relatively continuous. There are many arctic mountain ranges, but the majority of the arctic landscapes consist of gently rolling hills or mostly flat plains. In these less topographically diverse landscapes, turnover tends to be gradual, with community diversity concentrated in areas of unusual resources. These areas of unusual resources represent focal points for the landscape, which I define as an environmental resource patch characterized by high species richness and interaction. An obvious example would be a desert oasis or water hole, where animals interact extensively and plant cover is high.

What determines a focal point depends on what is limiting in a particular landscape. "Oases" have been described from polar deserts, associated with springs or in permanently moist situations below snowbeds (Edlund and Alt

1989; Svoboda and Freedman 1994). These oases have plant cover many orders of magnitude higher than the surrounding barrens, contain most of the local plant species diversity, and may be locally very important to wildlife. Pingos are important focal points in certain arctic regions (D.A. Walker et al. 1985; M.D. Walker et al. 1991; M.D. Walker 1990). On the flat Alaskan coastal plain, topographic relief is in short supply, so the slopes of pingos are rare habitats and landscape focal points. Pingos have very high species diversity in a small area because: (1) their dome shape results in a high degree of habitat diversity in a small area, including north- and south-facing slopes, ablation areas, and snow-beds, (2) their steep slopes discourage peat formation, resulting in warmer soils with higher nutrient availability, (3) their gravel and sand substrates make excellent denning areas for squirrel and fox, creating a moderate level of disturbance and also bringing nutrients into the system. South-facing slopes of pingos are particularly species rich, as the total solar input is much higher on these sites. The "equivalent latitude" of a 15° slope at 70 °N actual latitude is 55° (based on the equation of Lee 1962). Also, pingos may represent relict Beringian habitats which support taxa that were common on the cold, dry Beringian plains during the Pleistocene glaciations and which have now been replaced by mesic taxa following the development of a peat layer at the beginning of the Holocene; the floristic affinities of the pingo species support this concept (Young 1982; M.D. Walker et al. 1991). A few range disjunctions have been described from pingos, mostly lichens to the north of their known ranges, but also some polar desert species that are common to the north but rather rare at that latitude, such as *Draba subcapitata*.

Rivers, and particularly steep, south-facing river bluffs, may also serve as landscape focal points. The steep slopes of rivers are warm, well-drained habitats, and combined with the possibility for migration along the river corridor are therefore natural concentrations of diversity. Zanokha (1989) has described rare extrazonal communities on steep, south-facing river bluffs of the Rogozinka River on the Taymyr Peninsula. These communities are dominated by boreal taxa in an area that is mostly dominated by *Dryas punctata* mesic tundra. These sites have deep, well-developed soils that rest directly on bedrock. In northern Alaska there are occasional hot springs in the northern foothills of the Brooks Range that support stands of *Populus balsamifera* well to the north of the tree line, which sits on the southern slope of the range, a major topographic barrier (Murray 1980). These are warm-water springs that create localized warm microclimates, and many of these stands support extreme range disjunctions of boreal taxa.

In almost all cases, these hot spots of plant diversity are also locally or regionally important to wildlife. Relative to their spatial extent, the intensity of animal activity on pingos is orders of magnitude higher than in the surrounding landscape. Ground squirrels, arctic fox, various birds of prey, collared lemmings, and grizzly bear are drawn to these points. Rivers are extremely important to wildlife, with steep banks or bluffs offering nesting, denning, and hunting areas, and the floodplains offering forage and cover for many species. Even the small stands of *Populus balsamifera* are locally important to moose. These intense concentrations of wildlife create disturbances, increase nutrient input, and therefore help perpetuate the high plant species diversity.

1.5 Global Change and Community Dynamics

Existing models of global climate change predict various degrees of temperature shifts in different parts of the Arctic, with increases predicted in some areas and decreases in others (Mitchell et al. 1991). If temperature and thermal energy are indeed the most important ultimate causes of patterns in species diversity, then shifts in temperature should have major consequences for these patterns. It is shifts in summer temperature and degree-days, either positive or negative, that will potentially have the strongest effects on diversity. Changes in mean annual temperature cannot be directly related to diversity, since winter climate will dominate such an average.

Given this overwhelming importance of temperature, and the hypothesized gradient of biological interaction that may exist in the Arctic, a reasonable hypothesis may be that colder areas, which now have fewer species and a stronger direct tie to climate, may be the areas that will experience the most rapid change. Also, the closer summer temperatures are to 0 °C, then the more a small change will mean in the total heat energy available throughout the growing season. Alternatively, the lack of carbon and nutrient stores in the soils of these regions could effectively prevent any quick response to a warming. The first species to migrate into the northern areas following a warming should be those which have been lacking due to a threshold response to summer temperature or heat energy but which are able to grow and reproduce under low nutrient conditions. A combination of simulation modeling and field experimentation could be used to test the feasibility of these hypotheses, and long-term monitoring sites should be placed along major gradients of climate.

Focal points may also be critical during a changing climate. In a warming climate, these sites may be important as seed sources. In a cooling climate, these rare areas, which may be of critical importance to certain wildlife species, may lose much of their plant species diversity.

1.6 Conclusions

1. The principles that organize communities do not vary among biomes, but the relative importance of various physical and biological controls may shift substantially, with physical factors being of primary importance in arctic ecosystems.
2. Thermal energy is the main factor correlated with trends in arctic species diversity, both within the Arctic and relative to the rest of the globe.
3. Other major biogeographical gradients within the Arctic include moisture and geological history.

4. Because thermal energy is not subject to biological competition, the gradient of thermal energy may also represent a gradient of biological interaction, with northernmost communities having the strongest physical controls.
5. Habitat diversity, which is a function of disturbance at all spatial and temporal scales, is the most important factor controlling the distribution of species among communities. Mesoscale periglacial features such as frost boils, ice-wedge polygons, thaw lakes, and pingos are primary controls on the diversity and composition of arctic plant communities.
6. Landscape focal points are uncommon habitats that support high species diversity. Springs, seeps, pingos, and steep river bluffs may all serve this function in certain arctic landscapes.
7. Landscape focal points are usually of great local importance to wildlife.
8. Because of the importance of thermal energy in shaping patterns of arctic community diversity at many scales, either positive or negative shifts in summer temperature should have a dramatic impact on diversity.

References

Aleksandrova VD (1980) The Arctic and Antarctic: their division into geobotanical areas. Cambridge University Press, Cambridge

Aleksandrova VD (1988) Vegetation of the Soviet polar deserts. Cambridge University Press, New York

Beschel RE (1970) The diversity of tundra vegetation. In: Fuller AD, Kevan PG (eds) Proc. Conf. on Productivity and conservation in northern circumpolar lands. IUCN Publications new series no 16. International union for the conservation of Nature, Morges, Switzerland

Billings WD (1974) Arctic and alpine vegetation: plant adaptations to cold summer climates. In: Barry RG, Ives JD (eds) Arctic and alpine environments. Methuen, London, pp 403–443

Billings WD (1992) Phytogeographic and evolutionary potential of the arctic flora and vegetation in a changing climate. In: Chapin FS III, Jefferies RL, Reynolds JF, Shaver GS, Svoboda J (eds) Arctic ecosystems in a changing climate: an ecophysiological perspective. Academic Press, San Diego, pp 91–109

Bliss LC, Matveyeva NV (1992) Circumpolar arctic vegetation. In: Chapin FS III, Jefferies RL, Reynolds, JF, Shaver GR, Svoboda J (eds) Arctic ecosystems in a changing climate: an ecophysiological perspective. Academic Press, San Diego, pp 59–89

Bliss LC, Svoboda J (1984) Plant communities and plant production in the western Queen Elizabeth Islands. Holarct Ecol 7: 325–344

Cailleux A (1961) Biogéographie Mondiale. Presses Universitaire de France, Paris

Chernov Yu, Matveyeva NV (1979) Zakonomernosti zonal'nogo raspredeleniya soobshchestv na Taimyre (The regularities of zonal distribution of communities on Taymyr). Academy of Sciences USSR, Leningrad, pp 166–200

Chernov Yu (1985) The living tundra. Cambridge University Press, Cambridge

Cody ML (1970) Chilean bird distributions. Ecology 51:455–464

Delcourt HR, Delcourt PA (1988) Quarternary landscape ecology: relevant scales in space and time. Landscape Ecology 2:23–44

During HJ, Werger MJA, Willems HJ (eds) (1988) Diversity and pattern in plant communities. SPB Academic Publishing, The Hague

Edlund SA, Alt BT (1989) Regional congruence of vegetation and summer climate patterns in the Queen Elizabeth Islands, Northwest Territories, Canada. Arctic 42:3–23

Fridriksson S (1989) The volcanic island of Surtsey, Iceland, a quarter century after it 'rose from the sea'. Environ Conserv 16:157–162

Gauch HG (1982) Multivariate analysis in community ecology. Cambridge University Press, Cambridge
Gauch HG, Whittaker RH (1972) Coenocline simulation. Ecology 53:446–451
Gleason HA (1922) On the relation between species and area. Ecology 3:158–162
Griggs RF (1934) The problem of arctic vegetation. J Wash Acad Sci 24:153–175
Hamilton TD (1986) Late Cenozoic glaciation of the Central Brooks Range. In: Hamilton TD, Reed KM, Thorson RM (eds) Glaciation in Alaska: the geologic record. Fairbanks, Alaska Geological Society, pp 9–49
Hopkins DM (ed) (1967) The Bering Land Bridge. Stanford University Press, Stanford
Hurlbert SH (1971) The non–concept of species diversity. Ecology 52:67–77
Hultén E (1962) The circumpolar plants. Sven Vetensk Akad. Handl 5:1–275
Ives JD, Barry RG (1974) Arctic and alpine environments. Methuen, London
Jorgenson MT (1984) Controls of the geographic variability of soil heat flux near Toolik Lake, Alaska. PhD Thesis, University of Alaska, Fairbanks
Körner CH, Larcher W (1988) Plant life in cold climates. In: Long SP, Woodward FI (eds) Plants and temperature. Symposia of the Society for Experimental Biology, No 22. Company of Biologists, Cambridge, pp 25–57
Kruger FJ, Taylor HC (1979) Plant species diversity in Cape Fynbos: gamma and delta diversity. Vegetatio 41: 85–93
Lee R (1962) Theory of the equivalent slope. Mon Weather Rev 90: 165–166
Lewis WV (1939) Snow-patch erosion on Iceland. Geogr J 94: 153–161
MacArthur RH (1960) On the relative abundance of species. Am Nat 94:25–36
Magurran AE (1988) Ecological diversity and its measurement. Princeton University Press, Princeton
Matthews JA (1992) The ecology of recently-deglaciated terrain: a geoecological approach to glacier forelands and primary succession. Cambridge University Press, Cambridge, 386 pp
Matveyeva NV (1979) Struktura rasttel'nogo pokrova polyarnykh pustyn' polyostrova Taymyr (mys Chelyuskin). [The structure of the plant cover in the polar desert of the Taymyr Peninsula (Cape Chelyuskin)] Arkticheskiye tundry i polyarnyye pustyni Taymyra [The arctic tundras and polar deserts of Taynyr]. Academy of Sciences USSR, Leningrad, pp 5–27
Matveyeva NV (1988) The horizontal structure of tundra communities. In: During HJ, Werger MJA, Willems JH (eds) Diversity and pattern in plant communities. SPB Academic Publishing, The Hague, pp 59–65
Matveyeva NV (1994) Floristic classification and ecology of tundra vegetation of the Taymyr Peninsula, northern Siberia. J Veg Sci 5:813–828
May RM (1973) Stability and complexity in model ecosystems. Princeton University Press, Princeton
May RM (1975) Patterns of species abundance and diversity. In: Cody ML, Diamond JM (eds) Ecology and evolution of communities. Belknap Press, Cambridge, pp 81–120
Mitchell JFB, Manabe S, Meleshko V, Tokioka T (1991) Equilibrium climate change and its implications for the future. In: Houghton JT, Jenkins GJ, Ephraums JJ (eds) Climate change: the IPCC scientific assessment. Cambridge University Press, Cambridge, pp 134–167
Murray DF (1978) Vegetation, floristics, and phytogeography of northern Alaska. In: Tieszen LL (ed) Vegetation and production ecology of an Alaskan arctic tundra. Springer, Berlin, Heidelberg, New York, pp 19–36
Murray DF (1980) Balsam poplar in arctic Alaska. In: Rutter NW, Schweger CE (eds) Proc of the 5th Biennial Conf of the American Quarternary Association, Special AMQUA issue – The ice-free corridor and the peopling of the new world. Can J Anthropol 1, Edmonton, Alberta, pp 29–32
Murray DF (1992) Vascular plant diversity in Alaskan arctic tundra. Northwest Environ 8:29–52
Péwé TL (1975) Quaternary geology of Alaska. US Geol Surv Pro Pap 835, US Geological Survey, Washington
Pielou EC (1969) An introduction to mathematical ecology. Wiley, New York.
Pielou EC (1975) Ecological diversity. John Wiley, New York
Polunin N (1959) Circumpolar arctic flora. Clarendon Press, Oxford

Preston FW (1962a) The canonical distribution of commonness and rarity: part I. Ecology 43:185–215
Preston FW (1962b) The canonical distribution of commonness and rarity: part II. Ecology 43:410–432
Rannie WF (1986) Summer air temperature and number of vascular species in arctic Canada. Arctic 39:133–137
Reiners WA, Worley IA, Lawrence DB (1971) Plant diversity on a chronosequence at Glacier Bay, Alaska. Ecology 52:55–69
Savile DBO (1960) Limitations of the competitive exclusion principle. Science 132:1761
Schulze ED, Mooney HA (eds) (1993) Biodiversity and ecosystem function. Springer Berlin Heidelberg New York
Svoboda J, Freedman B (eds) (1994) Ecology of a polar oasis. Alexandra Fiord, Ellesmere Island, Canada. Captus University Publications, Toronto
Tolmachev AI (1931) Materialy dlya flory evropeiskikh arktichekhikh ostrovov. (Data for flora of the European arctic islands). Zh Russk Bot Obshchestva 31(5–5):459–472
Walker DA (1985) Vegetation and environmental gradients of the Prudhoe Bay region, Alaska. US Army Cold Reg Res Eng Lab, Rep 85–14, Hanover
Walker DA, Walker MD (1991) History and Pattern of disturbance in Alaskan arctic terrestrial ecosystems: a hierarchical approach to analysing landscape change. J Appl Ecol 28:244–276
Walker DA, Walker MD, Everett KR, Webber PJ (1985) Pingos of the Prudhoe Bay Region, Alaska. Arc Alp Res 17:321–336
Walker DA, Binnian E, Evans BM, Lederer ND, Nordstrand E, Webber PJ (1989) Terrain, vegetation and landscape evolution of the R4D research site, Brooks Range Foothills, Alaska. Holarct Ecol 12:238–261
Walker MD (1990) Vegetation and floristics of pingos, central Arctic Coastal Plain, Alaska. Diss Bot 149, J Cramer, Stuttgart
Walker MD, Walker DA, Everett KR, Short SK (1991) Steppe vegetation on south–facing slopes of pingos, central Arctic Coastal Plain, Alaska. Arc Alp Res 23:170–188
Walker MD, Walker DA, Auerbach NA (1994) Plant Communities of a tussock tundra landscape in the Brooks Range Foothills, Alaska. J Veg Sci 5:843–866
Washburn AL (1973) Periglacial processes and environments. Edward Arnold, London
Whittaker RH (1965) Dominance and diversity in land plant communities. Science 147: 250–260
Whittaker RH (1972) Evolution and measurement of species diversity. Taxon 21:213–251
Young SB (1971) The vascular flora of St. Lawrence Island with special reference to floristic zonation in the arctic regions. Contrib Gray Herb 201:11–115
Young SB (1982) The vegetation of land–bridge Beringia. In: Hopkins DM, Matthews JV Jr, Schweger CE, Young SB (eds) Paleoecology of Beringia. Academic Press, New York, pp 179–194
Yurtsev BA (1982) Relicts of the xerophyte vegetation of Beringia. In: Hopkins DM, Matthews JV Jr, Schweger CE, Young SB (eds) Paleoecology of Beringia. Academic Press, New York, pp 157–177
Yurtsev BA, Tolmatchev AI, Rebristaya OV (1978) The floristic determination and subdivision of the Arctic. In: Yurtsev BA (eds) Arkicheskaya floristicheskaya oblast'. Izd Nauka, Leningrad, pp 9–104
Zanokha L (1989) Lugovye soobshchestva tundrovoi zony (na primere Taimyra). [Grasslands of the tundra zone (with Taymyr as an example)]. Candidate Thesis, Komarov Botanical Institute, St Petersburg
Zollitsch B (1969) Die Vegetationsentwicklung im Pasterzenvorfeld. Wiss Alpenvereinsheft (München) 21:267–290

2 Causes of Arctic Plant Diversity: Origin and Evolution

D.F. Murray

2.1 Introduction

The term biodiversity was coined rather recently as a shorthand reference to biological diversity, i.e., for the sum of all taxa of plants and animals (Wilson 1988), yet the study of biodiversity is the oldest branch of biology (Weber and Wittmann 1992). Today, frames of reference other than taxonomic ones have become important as we recognize the need for understanding diversity at various levels of organization, from inclusive to restrictive, from communities to genotypes. Whatever the level of our focus in the hierarchy, we must ultimately have a precise knowledge of the component taxa.

As we inventory the plants of the Arctic, it is impossible for me not to wonder about the origin and evolution of this flora and which of many historic and contemporary events and processes have most effectively shaped plant diversity. If we are to predict how climate change will affect taxa in both evolutionary and geographic senses, we must understand how, when, and where they were created and then brought to their present positions (cf. Smith 1969).

The reduction of plant diversity with increasing latitude naturally finds its ultimate expression in the Arctic where the abiotic environment challenges even the physiological limits of plants. Changes in dominant life forms and species composition of vegetation are the basis for schemes of the major zonation for the Arctic (Polunin 1951; Young 1971; Yurtsev et al. 1978; Rannie 1986; Edlund and Alt 1989) all of which have demonstrated a pattern of progressive loss of taxa northward and a general agreement of this loss with decreasing temperatures. It is wrong, however, to see only the loss of the boreal species at their northern limits, for there is also some replacement with a distinct arctic element that is generally not known farther south.

The total pool of vascular plants in the Arctic is about 1500 taxa. About 500 of these comprise the flora of the Alaskan Arctic Slope, and localities of roughly the same area in northern Alaska can have from 100 to 300 taxa. We do not see a greater species diversity on the ancient unglaciated surfaces as some contend (Murray 1992; Walker, this Vol.). Obviously, since plants are not evenly distributed, there are differences within the Arctic as to which taxa are actually present: (1) within the confines of one region, such as the Coastal Plain, Foothills, or

Herbarium of the University of Alaska Museum, University of Alaska Fairbanks, AK 99775-6960, USA

Ecological Studies, Vol. 113
Chapin/Körner (eds) Arctic and Alpine Biodiversity

Mountain provinces of the Arctic Slope of Alaska, (2) at localities of comparable size within the same physiographic province, as in the case of Barrow and Prudhoe Bay on the Coastal Plain and (3) even between localities only a few kilometers apart in the Brooks Range (Murray and Murray 1982). The diversity of plants of any one place is determined by the history of the flora and landscapes, present climate, and the diversity of lithology, landforms, and habitats (cf. Murray 1987, 1992).

2.2 Historical Factors of Arctic Plant Diversity

Because many life forms, genera, and even species are shared by arctic and alpine localities in the northern hemisphere, by places distant from one another, the two floras are often treated as a single entity, the arctic-alpine flora. They mix and become one, in practical terms, in subarctic mountain ranges of North America and northeastern Asia, but each of these major floristic elements has had different origins and histories. Tolmachev (1960), Hultén (1958), Weber (1965), and Hedberg (1992) have drawn our attention to the high mountain systems of Central Asia and western North America as sources of taxa for the arctic flora. Their view is supported by the knowledge that some of these mountain ranges are old enough and high enough to have supported treeless (alpine) environments, and presumably also an alpine flora, in late Tertiary (Neogene) time. Elements of these Tertiary alpine floras moved northward along their respective cordillera to the Arctic. Floristic connections today between mountain ranges such as the Altai and the Rocky Mountains are given as evidence for early migrations (Weber 1987).

During the Tertiary, the Arctic, which is now synonymous with treeless landscapes, actually supported continuous forests across Asia and America. The two continents were joined by a dry land connection for millions of years prior to the Pliocene flooding of the Bering Strait. Exchanges of biota would have been strongly filtered by the forests which dominated the Neogene lowlands. There is a fossil record from which we reconstruct an Arctic covered by forests composed of major floristic elements totally absent from the present environment. In the shift from treed to treeless landscapes at the end of the Tertiary or in earliest Pleistocene, the Arctic lost a rich flora of trees, shrubs, and herbs that are extinct or found today far to the south of their ancient arctic positions. Yet, in every list of Tertiary plant fossils from the Arctic, there are also taxa (genera and species) familiar to us in arctic tundra today, particularly bog and riparian habitats, but also herb communities of well-drained uplands, such as *Cerastium arcticum/alpinum*, *Draba* sp., *Dryas integrifolia* and *D. octopetala*, *Oxyria digyna*, *Papaver* sp., *Saxifraga oppositifolia*, *Silene* sp., and *Stellaria* sp. (Bennike and Bøcher 1990; Matthews and Ovenden 1990).

Critical to further development of the arctic flora was the exchange of plants in Beringia throughout the Pleistocene, between Asia and America (and largely from Asia to America), when, during glacial maxima and concomitant periods of lower sea levels, the shallow seafloor of the Bering Strait was again exposed and a dry land connection between the continents was reformed, several times for thousands of years each time. North America's flora became changed by each new wave of Asiatic plants, from dry grasslands (steppes) and woodlands and from arctic and alpine tundras. Beringia in the Pleistocene was a center of plant diversity from these various sources, while the remainder of the high-latitude Northern Hemisphere was glaciated. Over roughly the past 12000 years, the circumpolar Arctic has been colonized by elements from this Beringian species pool and from other refugia in North America and northeastern Asia (Hultén 1937). The arctic flora of North America has its origins in the preadapted survivors of the arctic Tertiary forests to which were added (1) Pleistocene migrants from Asia, (2) plants that returned during interglacials and in post-glacial time from unglaciated areas south of the continental icesheets, (3) in-situ survivors of Quaternary glaciation in northern refugia, and (4) the newly evolved taxa of Pleistocene and Holocene age (Murray 1981).

2.3 Evolution of Diversity

The evolution of new taxa must have occurred many times over the 1.5 million years of the Quaternary when climatic fluctuations forced so many significant geological and biological changes throughout the world. To some, however, the few endemic genera in the arctic vascular plant flora is generally explained by the youthfulness of the arctic flora, with too little time undisturbed to have acquired endemic elements. Nevertheless, there are numerous endemics at the rank of species throughout the Arctic (cf. McJannet et al. for a recent review of arctic Canada) as well as some genera among the arctic grasses. For some of these endemic taxa, we presume Tertiary and Pleistocene refugial origins; others we suppose to be no older than Holocene, in as much as they are confined to landscapes of post-glacial age. There seems to be no lack of capacity within certain groups for phyletic diversification, but that capacity has less to do with the Arctic than with the inherent genetic potential of the genera in question.

Possibilities for genetic diversification, evolutionary change, and an increase in total diversity vary from very little to very great. At one extreme are two monotypic genera of the Polygonaceae, *Oxyria* and *Koenigia*, which are quintessential examples of morphological conservatism. They have enormous geographic ranges, being essentially circumpolar and with extensions far south; *Koenigia islandica* is bipolar. Both are relatively restricted in terms of habitat. *Oxyria digyna* is physiologically differentiated and morphologically plastic (Mooney and Billings 1961) but taxonomically monotypic (although some of the larger, robust forms have been given names, ex. *O. elatior*). These two genera are very distinct from others in the family and have enormous ranges, consequently, we presume ancient origins and phyletic stasis for both and conclude, therefore, that time alone does not assure change. The genus *Poa* is perhaps one of the best examples of the opposite circumstance of numerous species with infraspecific taxa, phenotypic plasticity, broad ecological tolerance, and labile genomes with reticulate patterns of relationships among taxa of relatively recent origin.

2.4 Breeding Mechanisms

Asexual methods of producing seeds and other propagules have been noted as frequent and useful for plants in arctic environments where the length of the reproductive season is constrained by low temperatures and nasty weather. These observations, although true in some particular cases – there is generally low recruitment by seed among rhizomatous monocots, and some species-rich genera of dicots are agamospermous – we cannot generalize for the entire arctic flora (Murray 1987). Even the most persistently asexual taxa equipped with highly germinable bulbils, as in the cases of *Saxifraga cernua* and *S. foliolosa* (Molau 1992), are not restricted to this mode of reproduction.

Nevertheless, agamospermous genera such as *Antennaria*, *Potentilla*, and *Taraxacum* are very well represented in the arctic flora by numerous taxa. We are coming to understand just how complex the relationships can be among taxa in these genera, especially where the plants are facultatively apomictic and there are exchanges of genes between diploids and triploids and between diploids and tetraploids (Jennisens et al. 1985; Richards 1986; Bayer 1990, 1991). Therefore, wherever the apomictic populations are sympatric with the sexual ones, genetic variation can be greater in the apomicts than where they are allopatric and accrue genetic diversity solely from mutation and migration (Bayer 1991). Notwithstanding the low genetic diversity within populations, the numerous agamic microspecies in *Antennaria*, for example, collectively occupy ecological niches both similar to and intermediate with those of their sexual progenitors, so the agamic polyploid complex provide fill-in taxa well suited to colonize and exploit new and novel habitats of the post-glacial arctic environments (Bayer et al. 1991).

Taxonomic problems abound when an aggregate of apomicts is formed by multiple hybridizations among sexual relatives. Taken one at a time, the agamic microspecies have been described, named, and treated taxonomically as independent units. Since repeated hybridizations have produced many forms that blend together, when considered as a whole, the morphological discontinuities become relatively insignificant, and consistent taxonomic discrimination of the microspecies is virtually impossible (Bayer 1990). The species richness recorded will vary, therefore, according to one's species concept.

Among the sexually reproducing species the question is one of the dominant mode, selfing or outcrossing or mixed mating. Reproductive traits vary, but Molau (1993) found correlations of traits among sexual taxa with flowering phenology. He divided life histories into *pollen-risking* and *seed-risking* strategies: the former early-flowering, outbreeding plants with low relative reproductive success and low autodeposition (of pollen) efficiency and the latter with late-flowering, inbreeding plants with high relative reproductive success and high autodeposition. In *Draba*, the arctic diploids tend to be autogamous, which retains their fitness (but constrains change) for a relatively narrow range of extremely stressful habitats to which they are well adapted. Under changing conditions this conservatism could be detrimental, except that the types of sites to which they are adapted have not disappeared; under climate change they may, however, become less abundant (Brochmann and Elven 1992).

The genus *Primula* is well known for its syndrome of genetic and morphological features that enforce outcrossing, but some successful hybrid, polyploid derivatives acquired mutations conferring self-fertility and have subsequently become homostylous. These have become more widespread than their diploid, self-incompatible ancestors. Most likely, the principal selective force was the reproductive advantage obtained by a release of the plants from dependence upon insect pollinators, when these became critically scarce during periods of climatic stress (Kelso 1992).

2.5 Polyploidy

Polyploidy occurs in 30 to 40% of higher plants, hence, it is basic to diversity itself, especially of grasses, sedges, and willows, which are well represented in arctic environments (Stebbins 1993). The discussion of arctic polyploids derives from two different but complementary perspectives: one at the scale of the population at which the mechanisms of formation and the fitness accruing from polyploidy are important and the other at the scale of the regional flora for which the relative abundance of polyploids vis-a-vis diploids has been the focus. Understanding (1) the role of polyploidy in the evolution of diversity in the arctic flora and (2) the reasons for different proportions of polyploids becomes a problem of (a) understanding the mechanisms of ploidy increases in plants, (b) rationalizing the advantages of the polyploid genome, and (c) sorting out the empirical data with respect to the frequency of polyploids, all the while recognizing the role of phyletic and geographic history. Genetic, ecological, and geographic perspectives become intertwined.

Tischler and Hagarup (cited in Stebbins 1950) were among the first to observe that there seemed to be a larger proportion of polyploid taxa at higher latitudes, even altitudes (see Johnson et al. 1965 and Löve and Löve 1967 for reviews of these data). In seeking answers to the geographic question, it was assumed that polyploidy per se must confer special physiological and ecological abilities for survival in the harsh arctic environment not possessed by their diploid ancestors. A corollary was that polyploids were superior at moving from south of the ice sheet margins and successfully exploiting new, post-glacial landscapes in the far north (Stebbins 1950) or that polyploids were better able to endure the periglacial environment of arctic refugia, hence, they persisted in the far north and were nearest and, therefore, the first to exploit new post-glacial environments (Löve 1959).

Polyploid drabas, for example, are stress-tolerant competitors (sensu Grime 1979) of such environments as snow beds or they are ruderal in behavior and inhabit screes, riverbanks, and young moraines, thus a broader ecological amplitude than the diploids. Given the ability of these species to occupy harsh microsites and the complexity of spatial heterogeneity in tundra habitats, several species of *Draba* often grow sympatrically. The potential for crossing is high among the polyploid taxa practicing mixed mating, consequently, there is high infraspecific genetic variation, mutiple genotypes, and greater biochemical complexity, hence, the great ecological amplitude (Brochmann and Elven 1992).

Both laboratory experiments and electrophoretic analyses of naturally occurring hybrids provide evidence of gene flow and the recovery of fertility of crosses even between different ploidy levels (Brochmann et al. 1992c, 1993). This result is not entirely unsuspected, for we are dealing here with polyploid taxa with varying degrees of shared genomes, which account for a higher degree of compatibility among them than among diploids. Furthermore, polyploid *Draba* of hybrid origin, have, in at least one case, been shown to have formed recurrently; that is, they have multiple geographic origins. Truly surprising is the demonstration that some of the hybrids are polyphyletic; that is, they have multiple genomic contributions from different species (Brochmann et al. 1992a,b).

Stebbins (1984, 1985) has now demonstrated that the greatest numbers of polyploids are not in the northernmost reaches of the Arctic, but actually at mid-northern latitudes, not in the areas of refugia, but where there had been glacial advances and retreats and, consequently, also the greatest changes in species ranges. Dynamic climate and landscapes provided an opportunity for the fragmentation of species ranges during glacial advances and the greatest likelihood of secondary contact of these isolated populations, when the glaciers waned and species expanded their distributions and reoccupied former ranges.

He concluded that it is reproductive stabilization through polyploidization of the new hybrids subsequent to secondary contact that accounts for the larger numbers of polyploids. This explanation is strengthened by what we know (and assume) of the genotypic and reproductive differences among populations of species at the edges (as opposed to centers) of their ranges, where the genetic differences can be the greatest and the hybrid outcomes often novel and especially suited to exploit new environments (cf. Bayer 1991 and references cited therein).

Given that the plants can survive the arctic environment, the historical development and the age of a flora – Tertiary, Pleistocene, or entirely postglacial – are important factors determining the relative abundance of polyploids as Johnson and Packer (1965) showed with their analysis of the flora at Ogotoruk Creek in northwestern Alaska. Many polyploids can be explained simply by the preponderance of grasses and sedges in the arctic flora, families of plants that are rich in polyploid taxa regardless of their geography. After subtracting the contribution from grasses and sedges, the proportion of polyploids was less than one would predict on the basis of latitude alone. This dilution, they concluded, was due to the strong Tertiary element, which survived locally, even in situ, and included many ancient diploid taxa. Moreover, the diploids (and paleopolyploids) favored the warm, dry, stable Tertiary uplands, whereas the polyploids were clearly more numerous in the colder, wetter, disturbed lowlands, which had become increasingly important during the Quaternary.

It is not polyploidy per se but the recovery of fertility, increased self-compatibility so useful in extreme habitats (Molau 1993), heterozygosity buffered against change (Mosquin 1966), protection against inbreeding depression (Brochmann and Elven 1992; Brochman 1993) and increased ecological amplitude (Johnson and Packer 1965; Brochmann and Elven 1992) that determine successful genotypes. In the Arctic, geomorphic instability presents a variety of disturbed habitats "favorable for hybridization, hybrid establishment, and establishment of allopolyploid derivatives" (Pojar 1973), with greater stabilized genetic variability than is found in diploids.

2.6 Ecotypes

Ecological versatility of arctic plants has long been appreciated by students of tundra vegetation whose plant lists and relevés show well how many of the same widespread species appear again and again, but in different habitats on different landforms, hence, in different communities (Griggs 1934). Indicator species are few and, in many cases, it is on quantitative measures of the component taxa that one community is distinguished from another (Beschel 1970).

Whereas Savile (1972) and others have presumed genetic homogeneity among arctic plants, heterogeneity is the hallmark of polyploids, and the unseen heterogeneity of ecotypes is a reservoir of genotypic variation upon which selection acts, fitness is adjusted, and adaptations become fixed (cf. Crawford et al. 1993). Stebbins (1984, 1985) has always stressed that ecotypically divergent taxa are more likely to retain the capacity to hybridize and bring to the process of hybridization genomes that are reproductively compatible yet adaptively different; therefore, variation at the population level is an important element in the secondary contact-hybridization process and plant diversity (see also McGraw, this Vol.).

The morphological species is a composite or abstraction of its many populations and expressions, resulting from phenotypic plasticity or ecotypic variation. Many important ecotypic differences of adaptive significance are unaccompanied by parallel morphological features that would enable and require formal taxonomic recognition, and without experimental evidence they cannot be distinguished from phenotypic plasticity (Bazzaz and Sultan 1987; Sultan 1987). Ecotypes of this sort do not become part of our tally of diversity, therefore, measures of species richness greatly underestimate genetic diversity.

Phenotypic plasticity is often discussed in the context of heterophyllous aquatic plants and the adaptive significance of capillary and laminate leaves. There are other good examples, however. The widespread and polymorphic *Stellaria longipes* complex has the vigor of weeds, and the included taxa exploit numerous habitats, frequently disturbed ones, presumably because of the diverse gene pool of polyploid origin genetically enriched by outcrossing. Depending upon taxonomic philosophy, the complex consists of four, six, or nine taxa in North America. Examination of many specimens shows that there were only weak north-south tendencies among some of the characteristics generally considered taxonomically important. The differences among taxa are quantitative and continuous (Chinnappa and Morton 1976, 1984), making it very difficult to assign all specimens unequivocally to species. Crosses among species produce fertile offspring, thus, there is a remarkable absence of major incompatibility among taxa.

When clones of the various taxa were subjected to altered temperatures and photoperiods in greenhouse environments, habit, leaf shape, number of flowers, and bracts (scarious or green) were found to vary to such an extent that the greenhouse phenotypes crossed the line from one "species" to another. The high degree of morphological plasticity of several features gives individuals the capacity to change in ways that are adaptive without the need for major changes in gene frequencies of the populations. Thus, these taxa are particularly well equipped to deal with habitats highly unpredictable over spans of time less than the life of the individual (Chinnappa 1985) as are found in the Arctic.

2.7 Conclusions

1. Is there anything special or different in evolutionary terms among these plants or in these environments? I think the answer is no. Molau (1993) concluded that the only generalizations more or less universal for tundra plants are perennial habit and long life span of genets and the lack of genetically determined self-incompatibility in plants with bisexual flowers. Instances of preformed floral buds are not uncommon (Sørensen 1941). Arctic plants photosynthesize, respire, absorb nutrients and grow as rapidly as temperate zone plants (Chapin and Shaver 1985; Billings 1987), so in physiological terms, it is the sum of adaptations conferring the ability to do this at low temperatures that distinguish these plants from others. There is little that is qualitatively different from other regions of the globe, except for the historical and ecological contexts in which the biological and evolutionary processes have operated.

   Statements, such as reproductive effort is lower than in temperate counterparts (McGraw and Fetcher 1992), but that allocation to seed production can be higher (Chester and Shaver 1982), are based on few examples, most if not all monocots. Chapin and Shaver (1985) contend that, in the very long term, recruitment by seeds is as important to the establishment and maintenance of tundra plants as in other ecosystems. Annual variation of seed production and seeding survival can be enormous, therefore, it is the uncertainty of suitable environmental conditions from year to year that characterizes the Arctic (cf. Philipp et al. 1990).
2. Accounts of evolutionary diversification among arctic taxa, analyzed taxon by taxon, do share certain features: immigration of the basic stock from diverse

sources, cycles of environmental change and the consequent restriction and isolation of taxa followed by expansion, secondary contact, and the stabilization of hybrids by polyploidy. Breeding systems promote or constrain the possibilities of genetic recombination, which in turn determine phenotypic patterns and taxonomic diversity. Genetic differentiation may or may not be accompanied by parallel and proportionate morphological discontinuities of taxonomic significance (cf. Engell 1977).

The genus *Primula* in North America is an exemplar, consisting of Asian and American immigrants and autochthonous elements of Quaternary and Holocene derivation. The dynamic landscape history during the Quaternary gave rise to geographic isolation and genetic differentiation of populations, which upon subsequent secondary contact yielded hybrid forms. These hybrids survived in the many novel habitats created by climatic change and became stabilized by polyploidy. "Separation and isolation, chance mutations acted upon by selection, hybridization, and polyploidy have all played important roles." (Kelso 1992).

3. General circulation models predict that climate change will be experienced first at northern latitudes, hence, our concern to understand how warming (or cooling) will be reflected in natural ecosystems. It would appear that the vascular plant flora is to a significant extent the product of climate changes in the past, but this does not necessarily protect it from profound alterations. Nevertheless, it does seem that perennial habit, polyploidy, mixed mating, genetic variation (ecotypes), and plasticity among populations confer an ability to accommodate change. Certainly, some plants adapt less well than others; autogamous ecological specialists could be lost from the system if their habitats disappear. On the other hand, plants now relatively rare could actually become more common if climate change increases their habitat. Perhaps more dramatic will be the changes in the associations of plants, the communities, as species, adjusting to new conditions and change their positions on the landscapes.

Acknowledgments. My thanks to F.S. (Terry) Chapin III for his patient editing by which I was able to greatly improve the manuscript. Barbara Murray read an early draft, and I am grateful for her helpful comments.

References

Bayer RJ (1990) Investigations into the evolutionary history of the *Antennaria rosea* (Asteraceae: Inuleae) polyploid complex. Plant Syst Evol 169: 97–110

Bayer RJ (1991) Patterns of clonal diversity in geographically marginal populations of *Antennaria rosea* (Asteraceae: Inuleae) from subarctic Alaska and Yukon Territory. Bot Gaz 152: 486–493

Bayer RJ, Purdy BG, Lebedyk DG (1991) Niche differentiation among eight sexual species of *Antennaria* Gaertner (Asteraceae: Inuleae) and *A. rosea*, their allopolyploid derivative. Evol Trends Plant 5: 109–123

Bazzaz FA, Sultan SE (1987) Ecological variation and the maintenance of plant diversity. In: Urbanska KM (ed) Differentiation patterns in higher plants. Academic Press, London, pp 69–93

Bennike O, Bøcher J (1990) Forest-tundra neighbouring the North Pole: plant and insect remains from the Plio-Pleistocene Kap København Formation, North Greenland. Arctic 43: 331–338

Beschel RE (1970) The diversity of tundra vegetation. In: Fuller WA, Kevan PG (eds) Proc Conf on Productivity and conservation in northern circumpolar lands. Int Union Conserv Nat Publ 16: 85–92

Billings DW (1987) Constraints to plant growth, reproduction, and establishment in arctic environments. Arct Alp Res 19: 357–365

Brochmann C (1993) Reproductive strategies of diploid and polyploid populations of arctic *Draba* (Brassicaceae). Plant Syst Evol 185: 55–83

Brochmann C, Elven R (1992) Ecological and genetic consequences of polyploidy in arctic *Draba* (Brassicaceae). Evol Trends Plant 6: 111–124

Brochmann C, Soltis PS, Soltis DE (1992a) Recurrent formation and polyphyly of Nordic polyploids in *Draba* (Brassicaceae). Am J Bot 79: 673–688

Brochmann C, Soltis PS, Soltis DE (1992b) Multiple origins of the octoploid Scandinavian endemic *Draba cacuminum*: electrophoretic and morphological evidence. Nord J Bot 12: 257–272

Brochmann C, Stedje B, Borgen L (1992c) Gene flow across ploidal levels in *Draba* (Brassicaceae). Evol Trends Plant 6: 125–134

Brochmann C, Borgen L, Stedje B (1993) Crossing relationships and chromosome numbers of Nordic populations of *Draba* (Brassicaceae), with emphasis on the *D. alpina* complex. Nord J Bot 13: 121–147

Chapin FS III, Shaver GR (1985) Arctic. In: Chabot BF, Mooney HA (eds) Physiological ecology of North American plant communities. Chapman and Hall, London, pp 16–40

Chester Al, Shaver GR (1982) Reproductive effort in cotton grass tussock tundra. Holarctic Ecol 5: 200–206

Chinnappa CC (1985) Biosystematics of the *Stellaria longipes* complex (Caryophyllaceae). J Cytol Genet 20: 46–58

Chinnappa CC, Morton JK (1976) Studies on the *Stellaria longipes* Goldie complex — variation in wild populations. Rhodora 78: 488–502

Chinnappa CC, Morton JK (1984) Studies on the *Stellaria longipes* complex (Caryophyllaceae) — biosystematics. Syst Bot 9: 60–73

Crawford RMM, Chapman HM, Abbott RJ, Balfour J (1993) Potential impact of climatic warming on arctic vegetation. Flora 188: 367–381

Edlund SA, Alt BT (1989) Regional congruence of vegetation and summer climate patterns in the Queen Elizabeth Islands, Northwest Territories, Canada. Arctic 42: 3–23

Engell K (1977) Morphology and cytology of *Polygonum viviparum* in Europe. Bot Tidsskr 72: 113–118

Griggs RF (1934) The problem of arctic vegetation. J Wash Acad Sci 24: 153–175

Grime JP (1979) Plant strategies and vegetation processes. John Wiley, New York

Hedberg O (1992) Taxonomic differentiation in *Saxifraga hirculus* L. (Saxifragaceae) — a circumpolar arctic-boreal species of central Asiatic origin. Bot J Linn Soc 109: 377–393

Hultén E (1937) Outline of the history of arctic and boreal biota during the Quaternary period. Bokforlags Aktiebolaget Thule, Stockholm

Hultén E (1958) The Amphi-Atlantic Plants. Almqvist and Wiksell, Stockholm

Jenniskens M-JPJ, den Nijs JCM, Sterk AA (1985) Crossability and hybridization of taxa of *Taraxacum* section *Taraxacum* from central and western Europe. Proc K Ned Akad Wet C 88: 297–338

Johnson AW, Packer JG (1965) Polyploidy and environment in arctic Alaska. Science 148: 237–239

Johnson AW, Packer JG, Reese G (1965) Polyploidy, distribution, and environment. In: Wright HE, Frey DG (eds) The Quaternary of the United States. Princeton University Press, Princeton, pp 497–450

Kelso S (1992) The genus *Primula* as a model for evolution in the Alaskan flora. Arct Alp Res 24: 82–87
Löve A (1959) Problems of the Pleistocene and Arctic. Publ McGill Univ Mus 1: 82–95
Löve A, Löve D (1967) Polyploidy and altitude: Mt. Washington. Biol Zentralbl 86: 307–312
Matthews JV Jr, Ovenden LE (1990) Late Tertiary plant macrofossils from localities in Arctic/Subarctic North America: a review of the data. Arctic 43: 364–392
McGraw JB, Fetcher N (1992) Response of tundra plant populations to climatic change. In: Chapin FS III, Jefferies RL, Reynolds JF, Shaver GR, Svoboda J (eds) Arctic ecosystems in a changing climate. Academic Press, New York, pp 359–376
McJannet CL, Argus GW, Edlund S, Cayouette J (1993) The rare vascular plants in the Canadian Arctic. Canadian Museum of Nature. Sylloqeus 72: 1–79
Molau U (1992) On the occurrence of sexual reproduction in *Saxifraga cernua* and *S. foliolosa* (Saxifragaceae). Nord J Bot 12: 197–203
Molau U (1993) Relationships between flowering phenology and life history strategies in tundra plants. Arc Alp Res 25: 391–402
Mooney HA, Billings WD (1961) Comparative physiological ecology of arctic and alpine populations of *Oxyria digyna*. Ecol Monogr 31: 1–29
Mosquin T (1966) Reproductive specialization as a factor in the evolution of the Canadian flora. In: Taylor RL, Ludwig RA (eds) The evolution of Canada's flora, Toronto University Press, Toronto, pp 43–65
Murray DF (1981) The role of arctic refugia in the evolution of the arctic vascular flora – a Beringian perspective. In: Scudder GGE, Reveal JL (eds) Evolution today. Hunt Institute, Pittsburgh, pp 11–20
Murray DF (1987) Breeding systems in the vascular flora of arctic North America. In: Urbanska KM (eds) Differentiation patterns in higher plants. Academic Press, London, pp 239–262
Murray DF (1992) Vascular plant diversity in Alaskan arctic tundra. Northwest Environ J 8: 29–52
Murray DF, Murray BM (1982) Bryological field guide Dalton Highway, Alaska, Yukon River to Prudhoe Bay. Univ Alaska Mus Stud 1: 1–46
Packer JG (1969) Polyploidy in the Canadian Arctic Archipelago. Arc Alp Res 1: 15–28
Philipp M, Böcher J, Mattsson O, Woodell SRJ (1990) A quantitative approach to the sexual reproductive biology and population structure in some arctic flowering plants: *Dryas integrifolia*, *Silene acaulis* and *Ranunculus nivalis*. Medd Grønl Biosci 34: 1–60
Pojar J (1973) Levels of polyploidy in four vegetation types of southwestern British Columbia. Can J Bot 51: 621–628
Polunin N (1951) The real Arctic: suggestions for its delimitation, subdivision, and characterization. J Ecol 39: 308–315
Rannie WF (1986) Summer air temperature and number of vascular species in arctic Canada. Arctic 39: 133–137
Richards AJ (1986) Plant breeding systems. George Allen & Unwin, London
Savile DBO (1972) Arctic adaptations in plants. Can Dep Agric Monogr 6: 1–81
Smith AC (1969) Systematics and appreciation of reality. Taxon 18: 5–13
Sørensen T (1941) Temperature relations and phenology of the northeast Greenland flowering plants. Medd Grønl 125: 1–305
Stebbins GL (1950) Variation and evolution in plants. Columbia University Press, New York
Stebbins GL (1984) Polyploidy and distribution of arctic-alpine flora: new evidence and a new approach. Bot Helv 94: 1–13
Stebbins GL (1985) The origin and success of polyploids in the boreal circumpolar flora: a new analysis. Trans Bot Soc Edinburgh, 150th Anniv Suppl: 17–31
Stebbins GL (1993) Concepts of species and genera. In: Flora of North America, vol 1. Oxford University Press, New York, pp 229–246
Sultan SE (1987) Evolutionary implications of phenotypic plasticity in plants. Evol Biol 21: 127–176
Tolmachev AI (1960) Der Autochthone Grundstock der arktischen Flora und ihre Beziehungen zu den Hochgebirgsfloren Nordund Zentralasiens. Bot Tidsskr 55: 269–276

Weber WA (1965) Plant geography in the southern Rocky Mountains. In: Wright HE, Frey DG (eds) The Quaternary of the United States. Princeton University Press, Princeton, pp 453–468

Weber WA (1987) Colorado Flora: Western Slope. Colorado Associated University Press, Boulder

Weber WA, Wittmann RC (1992) Catalog of the Colorado Flora: a biodiversity baseline. University Press of Colorado, Niwot

Wilson EO (1988) Biodiversity. National Academy Press, Washington, DC

Young SB (1971) The vascular flora of St. Lawrence Island with special reference to floristic zonation in the arctic regions. Contrib Gray Herb 201: 11–115

Yurtsev BA, Tolmachev AI, Rebristaya OV (1978) The floristic delimitation and subdivision of the Arctic. In: Yurtsev BA (ed) The arctic floristic region. Nauka, Leningrad, pp 9–104 (in Russian)

3 Patterns and Causes of Genetic Diversity in Arctic Plants

J.B. McGraw

3.1 Introduction

Low species diversity in arctic plant communities is low genetic diversity at one level. However, in this chapter I address the question of levels of within-species genetic variation in tundra plants. As studies in biogeography and physiological, population, community and ecosystem ecology typically ignore this level of variation, much less is known about within-species diversity than diversity of species in arctic and alpine (as well as other) communities. However, I will argue that in species-poor ecosystems such as tundra, within-species diversity may be as important as among-species diversity in terms of maintaining the integrity or persistence of tundra ecosystems in the face of climate change. Thus, studies of genetic variation in tundra plants may give important clues about the response of the entire ecosystem to environmental shifts.

3.2 Present Patterns of Genetic Diversity

3.2.1 Among-Population Variation

Just as total species diversity of tundra may be due to both among- and within-community variation, total genetic diversity within species consists of among-population and within-population components. More research has focused on among-population variation since variation at this level is often more readily discernible.

Mooney and Billings (1961) performed common environment and simulated reciprocal transplant studies to examine latitudinal variation in physiology, morphology, phenology, and reproduction in 16 populations of *Oxyria digyna.* These populations ranged across a remarkable 23° of latitude, from Elephant's Back, California (38°42′) in the south to Point Barrow, Alaska (71° 17′) in the north. Although the sites were variable in soils and vegetation, consistent differences were found between northern (arctic) and southern (alpine) populations. Many potentially ecologically significant differences were shown for these population groups, including: (1) differences in morphology; stamen number,

Department of Biology, West Virginia University, Morgantown, WV 26506-6057,USA

Ecological Studies, Vol. 113
Chapin/Körner (eds) Arctic and Alpine Biodiversity

inflorescence branching rates, and presence of rhizomes, (2) differences in development, including flower and perennating bud response to photoperiod, and (3) differences in physiology, including chlorophyll content, photosynthetic and respiratory response to temperature and light. Mooney and Billings (1961) concluded that the great geographic range of *Oxyria digyna* was due in large part to the existence of genetic variation in physiological traits within the species. Thus, from an ecological standpoint, this species includes significant genetic diversity. Interestingly, from a taxonomic standpoint, this species exhibits an extremely high level of phylogenetic stasis (see Murray, this Vol.). Ecotypic differentiation therefore does not necessarily signify a taxon undergoing rapid speciation, although it undoubtedly occurs in such groups. Rather, in arctic plants, genetic diversity within a species is a common mechanism by which individual species occupy a wide range of environments.

Callaghan's (1974) study of *Phleum alpinum* covered a 123° latitudinal gradient from Disko Island, Greenland, to South Georgia in the Antarctic. *In situ* demographic comparisons and common garden studies showed that these populations were distinctive. The arctic population appeared to be genetically depauperate, and experienced huge fluctuations in population numbers from year to year. In contrast, the antarctic population appeared to contain more plasticity and genetic variation, such that the species occupied many more microhabitats there, and was not nearly so vulnerable to climatic variation.

In a series of studies, Chapin demonstrated ecotypic differentiation among arctic and alpine populations of *Carex aquatilis* ranging from Barrow, Alaska, to Niwot Ridge, Colorado (Chapin 1974, 1981; Chapin and Chapin 1981; Chapin and Oechel 1983). These studies showed clear growth and flowering advantages of local populations in field reciprocal transplant experiments (Chapin and Chapin 1981). Associated with differences in success were differing suites of belowground traits, depending on the site. Northern populations had larger root:shoot ratios, smaller tiller size, and greater phosphate absorption potentials than southern populations.

Shorter latitudinal gradients reveal significant ecological genetic variation as well. Observations of *Eriophorum vaginatum* were made in 34 populations over a ca. 600-km gradient from Smith Lake, near Fairbanks, Alaska (65 °N latitude) to Prudhoe Bay, Alaska (70 °N latitude) (Shaver et al. 1986). Some traits (tillering rates, inflorescence density, nutrient concentrations) did not vary across this gradient, although tiller size (and as a result, overall plant size) declined with increasing latitude. In a parallel reciprocal transplant study of five sites, genetic differences in most traits became obvious, including differences in leaf length and width, total leaf number, and two indices of tiller size. Often, genetic variation accounted for more of the observed variation than the environment (Shaver et al. 1986). Fetcher and Shaver (1990) examined the response of the tiller size index in the reciprocal transplant study, showing that in general, northern populations were less plastic in response to the 600-km climatic gradient than southern populations. A repeat of the measurements on these gardens in 1993 (10 years after the first measurements) confirmed that this pattern was still present and that the differences in plasticity were genetically based (N. Fetcher and J.B. McGraw, unpubl. data). Fetcher and Shaver proposed that this could imply a genetic limitation to ecosystem-level responses to climatic warming. As they pointed out, this is analogous to limitation of productivity by the species composition of a community (Chapin et al. 1986; Vitousek and Walker 1989). In the same way that dramatic shifts in ecosystem productivity may require invasion of new species in unproductive sites, in northern populations, large changes in ecosystem productivity *without* species replacement may require invasion of new genotypes (from the south or lower elevations), or selection of the most responsive genotypes from within a population. This limitation ignores an alternative – local population density responses to climate change with no shifts in species or genotypes (McGraw and Fetcher 1992) – however, it remains an intriguing idea that deserves further experimental testing.

Similar levels of genetic differentiation occur across elevational gradients, despite the increasing probability of gene flow due to reduced distance among populations. One example is illustrated by the distribution of distinct populations of the *Stellaria longipes* complex (MacDonald and Chinnappa 1988, 1989; also discussed by Murray, this Vol.). Polyploid groups and ecotypic differentiation with respect to patterns and amount of plasticity allow this species to traverse diversity environments including tundra, boreal forest, prairie, and sand dunes. Ecotypic variation in reproductive response to temperature is important in *Poa alpina* (Hermesh and Acharya 1987).

Galen et al. (1991) performed a particularly thorough study of genetic variation across a shorter elevational gradient (500 m) in the alpine perennial, *Polemonium viscosum*, using a combination of isozyme, common garden and reciprocal transplant studies to infer patterns of gene flow and selection. They found that summit populations were more compact, had higher leaf production rates, and showed less resistance to herbivory than populations at the tree line (krummholz). Results of all three kinds of experiment were consistent with the hypothesis that these differences were adaptive, and were brought about by a change in the selection regime across the gradient. Such conclusions are not unlike those of Turesson (1922), who performed common garden experiments with plants from elevational gradients, but the combined approaches of Galen et al. (1991) place them on a much firmer ground.

Ecological genetic variation over short "mesotopographic" gradients (Billings 1973) is common in tundra plants. For example, *Dryas octopetala* in Alaska has ecotypes that form steep clines over distances of 1–100 m across snowbank gradients (McGraw and Antonovics 1983a, b). Restricted gene flow among populations, combined with large differences in the selection regime across these gradients, has led to markedly different populations. Manipulations of environmental factors and reciprocal transplant experiments demonstrated that there are large genetic differences in morphology (leaf size, shape, longevity, plasticity and petiole length), photosynthetic capacity (and response of this to the environment), growth (leaf area change and branching), and fitness response to nutrients, light, and competition (McGraw and Antonovics 1983a, b; McGraw 1985a, b, 1987a, b).

The long-term behavior of different ecotypes in a reciprocal transplant experiment may serve as a model of how selection may affect genetic diversity and distribution of genotypes in a changing climate. In this type of study, experimental introductions of ecotypes into nearby, environmentally different sites simulates dispersal, albeit accelerated, in a manner which is realistic in that the distance of movement is easily traversed by wind- or animal-dispersed seeds. Survival and growth of *D. octopetala* ecotypes in the reciprocal transplant study have been followed over 15 growing seasons (Fig. 1). Alien genotypes (open circles or squares, Fig. 1), which have experienced a rapid environmental change, have been gradually dying in both sites, resulting in a gradual return to the original genetic constitution of the site. No flowering of alien genotypes has occurred at either site, adding to this strong selection against nonnative genotypes. In both ecotypes, distinct patterns of selection were observed over time; rapid mortality of snowbed plants on the fellfield was followed by 8 years of no mortality. The plastic morphological response of this ecotype may have allowed the remaining individuals to persist over this period, although the plasticity was insufficient to permit a viable population, since this group has recently declined as well. For fellfield plants, the snowbed environment did not immediately cause mortality, but after 3 years, a period of mortality ensued, followed by a period of 6 years in which the remaining plants survived at a high rate. If the climate were to change, different results of the transplant experiment could be expected. Perhaps the initial period of (selective?) mortality would result in a remaining set of genotypes that could successfully exploit the new habitat. In fact, at a site intermediate between the extreme fellfield and snowbed sites, survival of both ecotypes was 100% over 15 growing seasons and both were capable of producing flowers (data

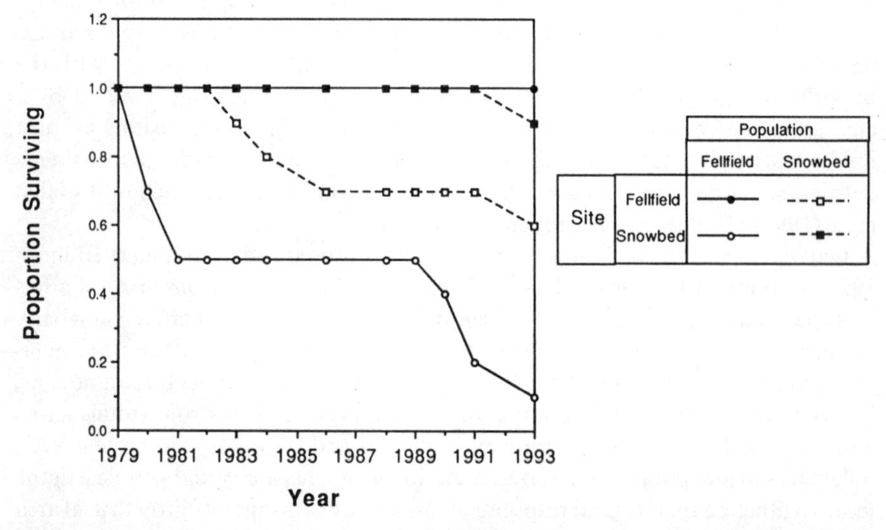

Fig. 1. Survival over 15 growing seasons of two ecotypically differentiated populations of *Dryas octopetala* planted at two sites along a snowbed environmental gradient

not shown). In the event of climatic warming, this intermediate zone could become more like the fellfield, with a longer growing season and less snow cover, and the selective regime would shift accordingly. However, if precipitation increases and the snowbed expands, the selective regime would favor the snowbed ecotype in this zone. Thus, with climatic change, relative distributions of ecotypes may change. In fact, with climate change, one of the native habitats could disappear entirely, resulting in extinction of the ecotype. However, it would seem less likely that both habitats occupied by the ecotypes of a species would disappear, thus, the existence of ecotypes adapted to different environments improves the probability that the species as a whole will survive.

Waser and Price (1985) showed similarly strong selection in a reciprocal transplant study with *Delphinium nelsonii* over 50 m gradients in an alpine meadow in Colorado. They started their plants from seed in their reciprocal transplant gardens and followed them through most of their lives, allowing them to determine λ, a measure of population growth rate for each population in home and away sites. Such short-distance, adaptive differentiation could lead to the curious phenomenon of "outbreeding depression" (Price and Waser 1979; Waser and Price 1983), whereby matings between plants over a distance may result in progeny with reduced fitness because the two parents are adapted to different environments. Peterson and Philipp (1986) demonstrated similar demographic shifts across gradients of a few hundred meters in response to variation in copper levels in Greenland soils.

There are probably few limits on the ability of plant species to achieve some degree of genetic differentiation along environmental gradients. Selection regimes can change radically over short distances, and many mechanisms prevent gene flow. Thus, the major limitation becomes the existence of appropriate genetic variation. Shaver et al. (1979) suggested that ecotypic differentiation occurred in *Carex aquatilis* from the center to the edge of an ice-wedge polygon. I have observed systematic gradients in *Dryas octopetala* phenotypes across distances of less than 50 cm. One should be cautious about interpreting all such differences as genetic, however, since environmental effects can directly cause differences in phenotype through plasticity, as well as genetic effects and these environmental effects can persist for a long time in perennial plants. In the cases of *Dryas octopetala* and *Carex aquatilis*, their ability to spread vegetatively over one to many meters suggests the possibility of environmental effects on phenotype within a single individual. Long-term experiments are required to tease apart the confounded effects of environment and genetics across such sharp environmental gradients.

3.2.2 Within-Population Variation

Between-population genetic differentiation has attracted the attention of researchers because different populations of the same species can show sharp differences in their phenotype and environmental associations. However, studies of within-population variation have been rare. This is unfortunate, for while the existence of ecotypes in different habitats may ensure the long-term persistence of a species, the ability of a given ecotype or species to adapt to environmental change will largely be a function of within-population genetic variation.

Studies of electrophoretic variation have shed light on the population histories of selected species. For example, Hermanutz et al. (1989) examined electrophoretic variation in *Betula glandulosa* near its northern limit on Baffin Island, Northwest Territories, demonstrating the existence of 15 three-locus genotypes. In most populations, within-population variation was low, but between population variation was high. Combined with data from pollen records and the lack of deviation of allele frequencies from Hardy-Weinberg expectations when population data were pooled, this was taken to mean that dwarf birch at these sites was once more widespread, and sexually reproducing. However, at present, the populations appear to be shrinking. They are maintained primarily by vegetative propagation, and the present genetic constitution of each population is a random, isolated remnant of the former widespread population.

Odasz et al. (1991) infer an even longer history from isozyme data in geographically isolated populations of *Pedicularis dasyantha*. The geographical pattern of frequencies at a single gene locus, combined with data on the breeding system of the species, suggested that this species may have survived a population bottleneck at several locations on Spitzbergen, with its attendant restriction of genetic variation in each isolated population during the last glacial period. The present pattern of variation suggests that this was followed by expansion from these refugia during the Holocene.

Although enzyme polymorphisms can act as markers to trace evolutionary histories and understand genetic structure, the most relevant genetic variation for understanding population response to expected environmental changes is genetic variation in traits that are ecologically relevant. Unfortunately, few studies have examined traits in this manner, despite the fact that many arctic and alpine plants are readily cloned, and therefore broad-sense heritabilities can be readily determined for quantitative traits such as photosynthetic capacity, nutrient uptake rates, tiller morphology, etc. Estimating narrow-sense heritabilities requires breeding designs requiring production of offspring with known parentage and growth of the offspring generation; this procedure is therefore unwieldy for many tundra plants, due to their longevity. However, for shorter-lived species this approach would be useful as well.

A complete accounting of genetic variation within tundra plant populations must also include variation found within seed banks (McGraw and Fetcher 1992). Approximately half of all arctic and alpine tundra species have dormant, viable populations of seeds in the soil. In a cottongrass tundra site in central Alaska which had been bulldozed, then naturally recolonized by germination of buried seed, I compared *Eriophorum vaginatum* adults derived from the seed bank with adults in the surrounding undisturbed vegetation (McGraw 1993). Using a reciprocal transplant study to control environmental differences, I found persistent differences in morphology (leaf length), plasticity (response of leaf length to site), growth response to site variation, and rate of flowering. The nature of these differences suggested that seed bank genotypes may in fact be early-successional genotypes that persisted as offspring of the last large disturbance at the site (probably fire). A survey of the extant genotypes would have underestimated the within-population genetic variation in *E. vaginatum* at this site.

Since the seed pool can store genotypes from the past, if it becomes stratified over time, it can serve as a record of genetic change in the extant, seed-producing population (McGraw and Vavrek 1989). Such a record would give an indication of whether within-population genetic variation is sufficient to allow adaptation to environmental change. Just such a historical record was reconstructed for two species of tundra plants near Eagle Summit, Alaska: *Luzula parviflora* and *Carex bigelowii*. By excavating under solifluction lobes, McGraw et al. (1991) showed that viable seeds of these two species could be retrieved from the former surface that had been buried. Radiocarbon dating of wood fragments suggested the lobes moved at a rate of approximately 1 cm y^{-1}. The seeds found in the oldest buried soils were germinated in the lab, then their seed coats were removed and dated directly by accelerator mass spectrometry. The oldest population of seeds dated were an average of 197 years old.

Old and young seeds from under the solifluction lobes were grown in the controlled environments and cloned for use in experiments under greenhouse, growth chamber and field conditions. In *Carex bigelowii*, genetic differences in leaf length and number of leaves per tiller were found (Vavrek et al. 1991). When grown in different temperature regimes, the older subpopulation performed better under low temperatures, while the recent subpopulation performed better under high temperatures (Vavrek et al. 1991). These results suggested that the population at this site had evolved over the course of 200 years to the point that they were ecologically distinct as well as morphologically different on average. Moreover, since the older seeds were products of the "Little Ice Age", when the climate was cooler, their superior performance under low temperatures, along with the reversal in this pattern under warmer conditions, suggests that the aboveground population may have evolved during two centuries in an adaptive manner. In *Luzula parviflora*, genetic differences were also found, primarily in size and growth characters and there was a trend toward differential response to nutrient variation (Bennington et al. 1991). In *L. parviflora*, performance in controlled environments revealed that plants from the older subpopulation were larger and grew more rapidly than plants from the young subpopulation. Plants of both species, from young and old subpopulations, were transplanted back to the field as an additional test to relative survival, growth and reproduction, with the null hypothesis that if extant populations evolve in an adaptive manner, plants from the younger seed subpopulations should outperform those from the older subpopulation. Preliminary results support this hypothesis for *Carex bigelowii*, but the differences are not as clear for *Luzula parviflora* (J.B. McGraw, unpub.).

3.3 Causes of Present Genetic Diversity Patterns

The studies described above show that genetic variation within arctic plant species is abundant. What is the origin of this variation? The answer to this question for many tundra species undoubtedly lies in their Pleistocene history. At present, we know few details of this history for any particular species, unlike the case for many species of trees (Davis and Zabinski 1992). However, based on the geological and paleobotanical evidence, we know that at glacial maxima, arctic tundra species persisted in severely contracted zones south of the ice sheets and in isolated refugia (see Murray, this Vol). For a given species, this means that populations could have become isolated from one another for long periods. If these populations experienced differing selection pressures during this time, but did not evolve reproductive isolating mechanisms, distinct ecotypes could be the result. Later, as the glaciers retreated, these populations would potentially come into contact, but remain distinct due to restricted gene flow caused by habitat separation.

Another mechanism of differentiation could operate during deglaciation and may be continuing today: ecological release followed by genetic differentiation. As reinvasion of the deglaciated landscape proceeded, plant diversity was far below the equilibrium number of species. Moreover, diversity undoubtedly remained far below the equilibrium due to extinctions during the previous glaciations and limited rates of migration of some species are reinvasion occurred. In the island biogeography model of species diversity (MacArthur and Wilson 1967), the condition that promotes speciation is long-term existence of a community below its equilibrium species number, such that the existing species experience ecological release, i.e., they invade habitats that under equilibrium conditions would be occupied by competitors. The populations that have occupied environments that are atypical for the rest of the species then evolve under a selection regime which is different in both abiotic and biotic factors. Ultimately, new species could be produced if isolating mechanisms were selectively favored (Antonovics 1968a, b). In the longterm, then, evolution may bring about significant increases in plant biodiversity. However, after 10000 years (since the last glaciation), the arctic tundra may still be in a transitional state, with many populations being genetically differentiated, but not yet reproductively isolated from their parent populations. Indeed, with repeated glaciation and deglaciation, this could be a long-standing condition.

3.4 Genetic Response to Future Climate Change

As climate change occurs, it seems likely that there will be strong selection pressures on populations of tundra species; due to the direct effects of climate and due to many possible indirect effects, including invasion of new species. However, since tundra is relatively low in diversity, the availability of invaders that can successfully take hold in an altered climate may be limited. Adaptation to climate change may therefore rely heavily on the existence of within-population genetic variation of tundra plants, about which we know little. Models of adaptation to climate change show that the probability of extinction decreases with increasing genetic variability for relevant traits (Pease et al. 1989). If the example of *Phleum alpinum* is typical (Callaghan 1974), demographic, and ultimately ecosystem stability, will depend greatly on the level of genetic variation within populations (McGraw and Fetcher 1992). Results of studies of time-stratified buried seed populations suggest that microevolutionary change on the century time scale can occur in long-lived perennial tundra plants.

Climate change can affect other factors important to microevolution such as covariances among traits, genetic drift, mutation and gene flow, about which we know even less for any given species (Holt 1990). For example, subtle changes in the expression of genetic variability under environmental stress can alter the heritability of traits in a given population (Hoffman and Parsons 1991).

The role of ecotypic differentiation in the response of species to climate change is also not straightforward. On the one hand, because ecotypic differentiation effectively extends the local, elevational, or latitudinal range of a species, the extinction probability for the species as a whole might be expected to decrease; climate change is unlikely to detrimentally affect every environment in which ecotypes of a species exist. On the other hand, a species that is genetically uniform on a local scale, but is ecotypically differentiated on a long environmental gradient, may be unable to adapt to climate change, particularly if the change is rapid relative to the life span of the species and gene flow among populations is low. Davis and Zabinski (1992) describe such a possible scenario for beech in the eastern USA, noting that extinction of northern ecotypes could occur even though climate change might not exceed the tolerance limits of southern ecotypes. The same phenomenon cold occur for arctic and alpine species that range over great distances. This possibility emphasizes the urgent need for studies of within-population variation in ecologically important traits, as well as studies of potential rates of gene flow.

3.5 Conclusions

1. A broad selection of studies of tundra plants suggests a high level of genetic variation among populations, which effectively expands the range of tundra species, often across great physical distances or sharp environmental gradients. No published studies clearly demonstrate a *lack* of adaptive genetic variation in response to the environment. This could be due to the fact that negative results are rarely published, but more likely in this instance it is due to the near-ubiquitous occurrence of ecotypes. Ecotypic differentiation is likely to be important for species survival in the face of climate change.
2. Within-population variation in tundra plants has been studied much less than among-population variation, yet this level of variation will most likely be critical in determining the fate of many species in a changing climate. Identification of characters that are relevant to survival and reproduction under expected future climates and quantification of heritable variation in these characters will help greatly in assessing the potential role of microevolution in the response of plants to climate change.

References

Antonovics J (1968a) Evolution in closely adjacent plant populations. 5. The evolution of self-fertility. Heredity 23: 219–238

Antonovics J (1968b) Evolution in closely adjacent plant populations. 6. Manifold effects of gene flow. Heredity 23: 507–524

Bennington CC, McGraw JB, Vavrek MC (1991) Ecological genetic variation in seed banks. II. Phenotypic and genetic differences between young and old subpopulations of *Luzula parviflora*. 79: 627–643

Billings WD (1973) Arctic and alphine vegetation: similarities, differences and susceptibility to disturbance. Bioscience 23: 697–704

Callaghan TV (1974) Intraspecific variation in *Phleum alpinum* L. with specific reference to polar populations. Arc Alp Re 6: 361–401

Chapin FS III (1974) Morphological and physiological mechanisms of temperature compensation in phosphate absorption along a latitudinal gradient. Ecology 55: 1180–1198

Chapin FS III (1981) Field measurements of growth and phosphate absorption in *Carex aquatilis* along a latitudinal gradient Arc Alp Re 13: 83–94

Chapin FS III, Chapin MC (1981) Ecotypic differentiation of growth processes in *Carex aquatilis* along latitudinal and local gradients Ecology 62: 1000–1009

Chapin FS III, Oechel WC (1983) Photosynthesis, respiration, and phosphate absorption by *Carex aquatilis* ecotypes along latitudinal and local environmental gradients. Ecology 64: 743–751

Chapin FS III, Vitousek PM, Van Cleve K. (1986) The nature of nutrient limitation in plant communities. Am Nat 127: 48–58

Davis MB, Zabinski C (1992) Changes in geographical range resulting from greenhouse warming: effects on biodiversity in forests In: Peters RL, Lovejoy TE (eds) Global warming and biological diversity. Yale University Press, New Haven, pp 297–308

Fetcher N, Shaver GR (1990) Environmental sensitivity of ecotypes as a potential influence on primary productivity. Am Nat 136: 126–131

Galen C, Shore JS, Deyoe H (1991) Ecotypic divergence in alpine *Polemonium viscosum*: genetic structure, quantitative variation, and local adaptation. Evolution 45: 1218–1228

Hermanutz LA, Jones DJ, Weis IM (1989) Clonal structure of Arctic dwarf birch (*Betula glandulosa*) near its northern limit. Am J Bo. 76: 755–761

Hermesh R, Acharya SN (1987) Reproductive response to three temperature regimes of four *Poa alpina* populations from the Rocky Mountains of Alberta, Canada. Arc Alp Res 19: 321–326

Hoffman AA, Parsons PA (1991) Evolutionary genetics and environmental stress. Oxford University Press, New York

Holt RD (1990) The microevolutionary consequences of climate change. Trends Eco Evo 5: 311–315

MacArthur RH, Wilson EO (1967) The theory of island biogeography. Princeton University Press, Princeton

MacDonald SE, Chinnappa CC (1988) Patterns of variation in the *Stellaria longipes* complex: effects of polyploidy and natural selection. Am J Bot 75: 1191–1200

MacDonald SE, Chinnappa CC (1989) Population differentiation for phenotypic plasticity in the *Stellaria longipes* complex. Am J Bot 76: 1627–1637

McGraw JB, Antonovics J (1983a) Experimental ecology of *Dryas octopetala* ecotypes. I. Ecotypic differentiation and life cycle stages of selection. J Ecol 71: 879–897

McGraw JB, Antonovics J (1983b) Experimental ecology of *Dryas octopetala* ecotypes. II. A demographic model of growth, branching, and fecundity. J Ecol 71: 899–912

McGraw JB (1987) Experimental ecology of *Dryas octopetala* ecotypes: relative response to competitors. New Phytol 100: 233–241

McGraw JB (1985b) Experimental ecology of *Dryas octopetala* ecotypes. III. Environmental factors and plant growth. Arc Alp Res 17: 229–239

McGraw JB (1987a) Experimental ecology of *Dryas octopetala* ecotypes. IV. Fitness response to transplanting in ecotypes with differing plasticity. Oecologia 73: 465–468

McGraw JB (1987b) Experimental ecology of *Dryas octopetala* ecotypes. V. Field photosynthesis of reciprocal transplants. Holarc Ecol 10: 308–311

McGraw JB, Vavrek MC (1989) The role of buried viable seeds in arctic and alpine plant communities In: Leck MA, Parker VT, Simpson RL (eds) Ecology of soil seeds banks. Academic Press, San Diego

McGraw JB, Fetcher N (1992) Response of tundra plant populations to climatic change. In: Chapin FS III, Jefferies RL, Reynolds JF, Shaver GR, Svoboda J (eds) Arctic ecosystems in a changing climatic Academic Press, San Diego pp 359–376

McGraw JB, Vavrek MC, Bennington CC (1991) Ecological genetic variation in seed banks. I. Establishment of a time transect. J Eco 79: 617–625

McGraw JB (1993) Ecological genetic variation in seed banks. IV. Differentiation of extant and seed bank-derived populations of *Eriophorum vaginatum.* Arct Alp Re 25: 45–49

Mooney HA, Billings WD (1961) Comparative physiological ecology of arctic and alpine populations of *Oxyria digyna* Ecol Monogr 31(1): 1–29

Odasz AM, Karkkainen K, Muona O, Wein G (1991) Genetic distances between geographically isolated *Pedicularis dasyantha* populations in Spitzbergen, Svalbard Archipelago, Norway: evidence of glacial survival? Nor Geol Tidsskr 71: 219–222

Pease CM, Lande R, Bull JJ (1989) A model of population growth dispersal and evolution in a changing environment. Ecology 70: 1657–1664

Petersen PM, Philipp M (1986) Growth and reproduction of *Viscaria alpina* on Greenland soils with high and low copper concentrations. Arc Alp Re 18: 73–82

Price MV, Waser NM (1979) Pollen dispersal and optimal outcrossing in *Delphinium nelsonii* Nature 277: 294–296

Shaver GR, Chapin FS III, Billings WD (1979) Ecotypic differentiation in *Carex aquatilis* on ice-wedge polygons in the Alaskan coastal tundra. J Ecos 67: 1025–1046

Shaver GR, Fetcher N, Chapin FS III (1986) Growth and flowering in *Eriphorum vaginatum*: annual and latitudinal variation. Ecology 67: 1524–1535

Turesson G (1922) The species and variety as ecological units. Hereditas 3: 100–113

Vavrek MC, McGraw JB, Bennington CC (1991) Ecological genetic variation in seed banks. III. Phenotypic and genetic differences between young and old seed populations of *Carex bigelowii*. J Eco 79: 645–662

Vitousek PM, Walker LR (1989) Biological invasion by *Myrica faya* in Hawaii: plant demography, nitrogen fixation, and ecosystem effects. Ecol Monogr 59: 247–265

Waser NM, Price MV (1983) Optimal and actual outcrossing in plants, and the nature of plant-pollinator interaction. In: Jones CE, Little RJ (eds) Handbook of experimental pollination biology. Van Nostrand Reinhold, New York, pp 341–359

Waser NM, Price MV (1985) Reciprocal tranplant experiments wtih *Delphinium nelsonii* (Ranunculaceae): evidence for local adaptation. Am J Bot 72: 1726–1732

4 Alpine Plant Diversity: A Global Survey and Functional Interpretations

CH. KÖRNER

4.1 Introduction

High-altitude vegetation is often treated as a special type of tundra, but this is an inadequate simplification. The major common features of real tundra and "alpine vegetation" are the absence of trees, the short stature of plants, and the low annual mean temperature. Most other components of the alpine environment may differ substantially from arctic tundra environments (Table 1; Billings 1973, 1979a). The term "alpine" is used here exclusively for the vegetation above the natural subalpine tree line. Often this boundary is unsharp and is fragmented over several hundred meters of altitude. Where an upper tree line is missing, as in many arid mountain regions, the approximate level of the tree line in the nearest more humid mountains is taken as a rough guideline. At the polar end of the alpine vegetation, there is no clear distinction between the arctic-alpine and the arctic-lowland flora. Depending on region, most arctic-alpine vegetation north of 65° to 70 °N is possibly better included in the term arctic (similar climate and species composition).

For simplicity, I will subsume here the more open and fragmented plant cover above the closed alpine vegetation belt under the same term, although this highest ranging vegetation is often termed subnival or nival vegetation (Ozenda 1988). Furthermore, "alpine" is used here for high-altitude vegetation above the tree line from all parts of the world and is thus not restricted to the mountain ranges of the European Alps only. With this wide definition alpine is synonymous to "oreophytic", a neutral term preferred by some authors (cf. Agakhanjanz and Breckle, this Vol.).

The global land area covered by alpine vegetation is fragmented into many mountain regions and extends from the Arctic to the Antarctic (Fig. 1). In some subtropical mountains, higher plants have been found at altitudes above 6000 m (references in Miehe 1989; Halloy 1991; Grabherr, this Vol.). The complementary latitudinal extremes are polar plant species growing north of 80° northern latitude (Rikli 1917). The alpine floras of the world are nested within a great variety of regional floras, partly explaining the great overall biodiversity among alpine vegetation. However, even within a single mountain region (e.g. the Great Caucasus, the Venezuelan Paramos), plant species diversity of the alpine zone alone may outrange species diversity of the total arctic tundra. Geographic

Institute of Botany, University of Basel, Schönbeinstr. 6, 4056 Basel, Switzerland

Ecological Studies, Vol. 113
Chapin/Körner (eds) Arctic and Alpine Biodiversity

Table 1. Similarities and differences in life conditions in the arctic and alpine life zone during the growing season. (Note: "alpine" includes mountains of all latitudes)

	Environment	
	Arctic	Alpine
Length of growing season	Shorter	Longer
Maximum radiation	Low	High
Radiation sum per day	Similar	
Daily mean temperature	Similar	
Difference between max. and min. temperature	Small	Large
Maximum temperature	Low	High
Minimum temperature	Similar or lower in alpine	
Diurnal variation in temperature	Small	Large
Atmospheric vapour pressure (2 m)	Similar	
Vapour pressure difference leaf to air	Low	High
Mechanical soil stability	Higher	Lower
Soil Carbon pool	Greater	Lower
Cryogenic soil processes in "summer"	Lower	Greater
Cryogenic soil processes in "winter"	Greater	Lower
Soil permafrost under closed vegetation	Present	Absent
Soil moisture	High	Moderate
Soil pH	Lower	Higher
Regional isolation of floras	Lower	Higher
Habitat fragmentation	Low	High

isolation, glaciation and a varied history of species migration and/or evolution, all together lead to high degrees of taxonomic richness of genetic complexity and speciation (including endemism). Even on a micro-scale alpine plant diversity is high, and 200 to 300 species can be found in areas of less than 10 km^2, suggesting that additional characteristics of the alpine environment favour species richness.

4.2 How Much Land Is Covered by Alpine Ecosystems?

At all scales biodiversity can only be defined within an area or space. Thus, before discussing patterns of alpine biodiversity on a global, regional, and habitat scale, it is necessary to consider the land area covered by alpine ecosystems. No geographical data base exists that accounts for alpine vegetation. Most geographical statistics treat alpine vegetation either as mountain pasture (where cattle grazing is still possible), or non-vegetated bare land. In fact, national statistics rank large parts of alpine vegetation in the same categories as rocks and glaciers, sensu useless, i.e. unproductive. This is even true for countries whose economy strongly depends on alpine tourism and hydroelectric use of these ecosystems, such as Austria and Switzerland. An assessment of this discrepancy for the Central Alps (Körner 1989) revealed that 50% of the land rated as bare and useless in official statistics bears some of the biologically richest alpine ecosystems.

Using published records (Hermes 1955; Troll 1973; Wardle 1974) and personal observations from many regions of the world, a potential (idealistic) altitudinal

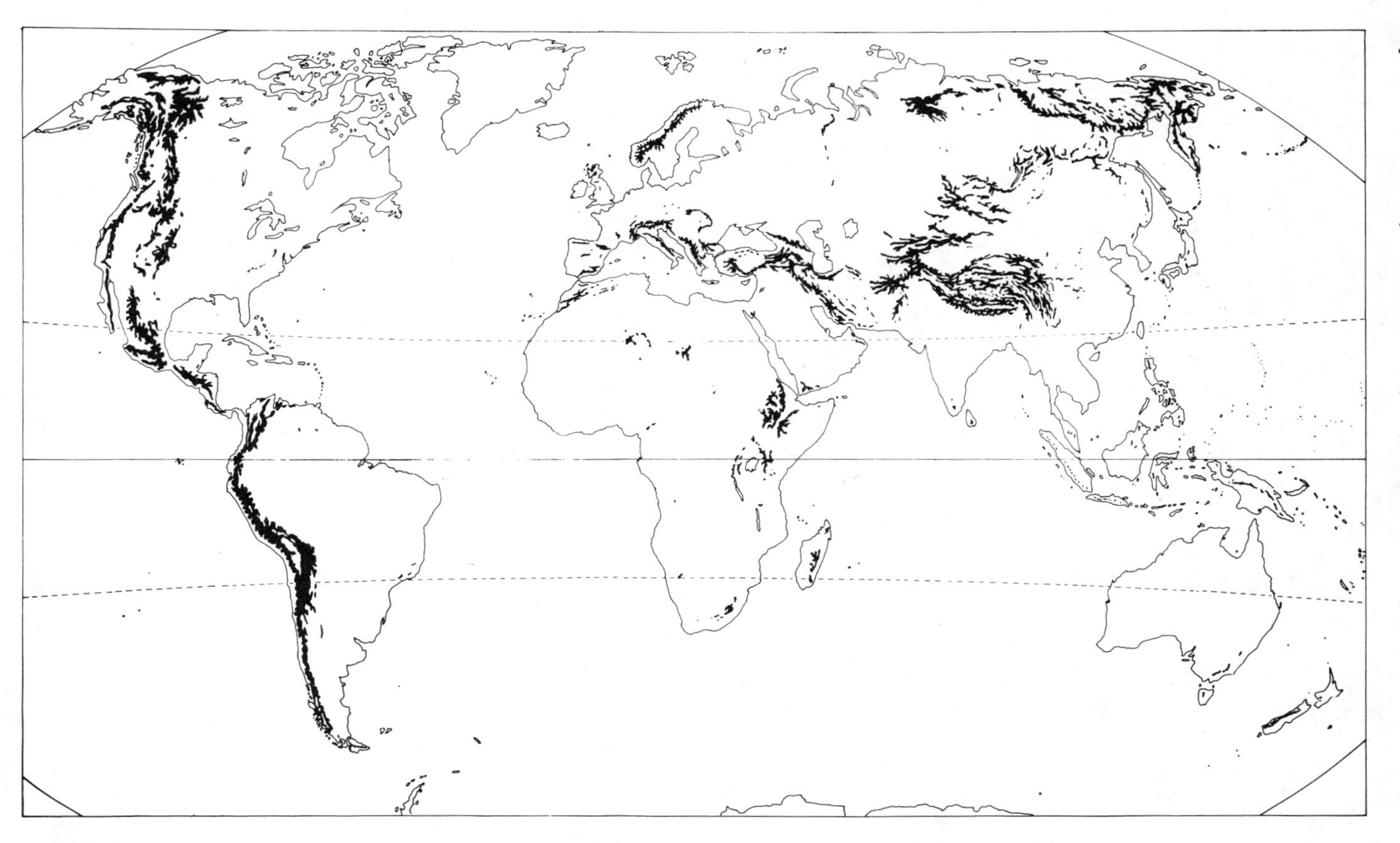

Fig. 1. Geographical distribution of the alpine life zone

range of the alpine life zone at various latitudes of the globe can be approximated (Fig. 2). This rather simple approach neglects that parts of these areas may be arid, rocky or glaciated; it also neglects that these altitudinal ranges of the alpine belt change with the degree of oceanity, with lower tree lines in coastal areas and on islands compared to more continental mountains. Furthermore, the range shown in Figure 2 does not account for the somewhat lower upper limits of plant growth in the tropics compared to the subtropics (Halloy 1989). With these uncertainties, the resultant land area that globally falls into the alpine life zone between 60 °N and 60 °S approaches approx. 4.5×10^6 km^2 (derived from global GIS). It is rather uncertain which part of the tundra-type vegetation in mountains north of 60 °N should be rated as alpine. Walter and Breckle (1986) introduced the term "mountain tundra" for these areas. For the present purpose, I assumed that 50% of the land area that occurs worldwide between 60° and 70 °N at altitudes between 600 and 1300 m (2.4×10^6 km^2, the largest part of which is in eastern Siberia) contributes to the arctic-alpine zone (i.e. 1.2×10^6 km^2), bringing the total area to 5.7×10^6 km^2. According to a detailed quantitative analysis for the Central Alps, approx. 30% of the land falling within the "alpine life zone", as defined above, is truly bare of any vegetation (Körner 1989). Applying this areal reduction to the global scale, one may assume a total area of 4×10^6 km^2 of

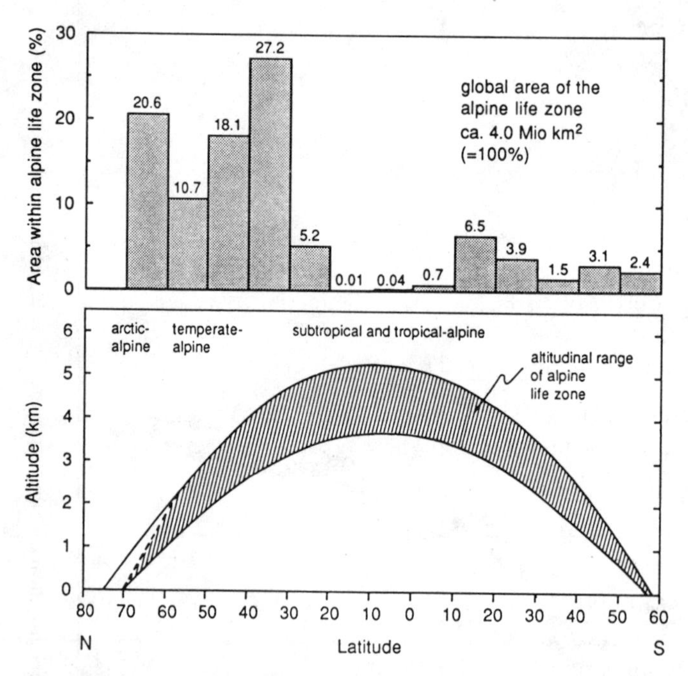

Fig. 2. An estimation of the latitudinal distribution of the alpine life zone. The *lower diagram* illustrates the mean lower and upper altitudes of the alpine life zone, the *dashed line* on the *left* accounts for the uncertainity between arctic-alpine vegetation vs. so-called mountain tundra (see text). The *upper diagram* illustrates the relative contribution of each 10° latitudinal range to the total global area covered by alpine vegetation (as defined by the lower diagram)

vegetation-covered land to fall within the alpine life zone. Hence, after subtracting glaciers, mountain deserts and otherwise truly bare terrain, roughly 3% of the global land area is possibly covered by some sort of alpine vegetation. Assuming biomass and soil carbon pools of 0.5 and 5 kg C m^{-2} respectively (various sources), the global alpine carbon pool amounts to about 2 and 20 Gt C for biomass and soil carbon, or ca. 1% of the global terrestrial biosphere carbon pool of about 1800 Gt C.

Estimates for the land area covered by treeless arctic tundra (depending on definition and including barren parts) vary around 7×10^6 km^2 (ca. 5% of the land area). Thus, arctic and alpine ecosystems together cover about 8% of the global land area, which is similar to the area covered by boreal forests (ca. 8.1%) and the total agricultural land area of the world (9.4%).

The global latitudinal distribution of the alpine life zone is strongly asymmetric. Using the potential altitudinal and latitudinal ranges as shown in Figure 2, 82% of the total alpine area is in the Northern Hemisphere. Arctic-subantarctic alpine (23%), cool-temperate alpine (32%) and warm-temperate alpine systems (29%) are the three largest categories. The subtropical and tropical alpine life zone together represent 16% of the total alpine land area.

4.3 Plant Species Diversity in the Alpine Life Zone

It is difficult to estimate the number of plant species found in the alpine life zone due to the following reasons:

1. Published floras are often restricted to political areas and not to orographically or phytogeographically defined regions.
2. Floras of adjacent regions are largely redundant and therefore the number of taxa must not be summed (cf. Agakhanjanz and Breckle, this Vol.).
3. Taxonomic inventories often lack a clear distinction between alpine and below-alpine floras.
4. Floras of important mountain areas are not available.

With these constraints in mind, I will briefly describe the alpine species richness on a global, regional, and micro-scale, whereby the accuracy increases, the smaller the considered scale.

4.3.1 The Global Scale

Similar to observations of other vegetation types, the plant diversity within the alpine life zone decreases from the equator to the pole (Table 2). Over much shorter distances, comparable altitudinal reductions in species richness are observed (Grabherr, this Vol.). Latitudinal gradients have also been illustrated in the tundra belt for plants (Matveyeva 1988) and animals (Chernov, this Vol.).

Hot spots of alpine plant diversity are the subtropical and the tropical Andes, the Caucasus region and the central Asian mountains. For instance, ca. 830 species out

Table 2. The latitudinal variation in species diversity in the alpine life zone of Eurasia for regions of similar size

Region	Latitude	No. of species
Great Caucasus	41–43 °N	>1000
Alps	46–47 °N	650
Scandinavian mountains	59–67 °N	250

of 1500 listed by Polunin and Stainton (1984) for the western part of the Himalaya (Nepal) grow in the alpine zone (> 3900 m; estimated from Polunin's and Stainton's numbers for elevational ranges of species). The total alpine flora of the Asian high mountains may be two to three times as large. The Great Caucasus is inhabited by substantially more than 1000 alpine taxa (Agakhanjanz and Breckle, this Vol.). Vareschi (1970) assumes a species number of ca. 1000 for the Venezuelan Paramos. The most clearly defined alpine floras are those of relatively isolated mountains such as the New Zealand Alps (650 species, with 90 species of the genera *Celmisia* and *Aciphylla* alone) and the Snowy Mountains of southeastern Australia (200 species). The European Alps are inhabited by approximately 650 truly alpine taxa (Ozenda 1993). No compendium of the North American alpine flora exists (possibly somewhere between 500 and 1000 species within the USA), but numerous regional floras for certain parts of the Rocky Mountains, the Great Basin Mountains, and the Sierra Nevada have been published (see below).

After checking several floras and trying various ways of compilation, it appears plausible to assume a total alpine flora of the world in the order of 8000 to 10 000 species. These belong to approx. 100($\pm$ 10) families and about 2000 genera of higher plants. Hence, one-fourth to one-third of all plant families of higher plants include alpine representatives, and the mean number of alpine species per plant family is approx. 80 to 100. Obviously, there are large differences in species diversity among families in the alpine zone. The majority of alpine taxa are found within Asteraceae and Poaceae, with Brassicaceae, Cyperaceae, Caryophyllaceae, Gentianaceae, Rosaceae and Ranunculaceae forming further prominent families. Among shrubby species, Ericaceae and Asteraceae (> 1000 species) dominate in the lower alpine belts of many mountain areas of both hemispheres. Compared to their otherwise worldwide great abundance, Fabaceae and Orchidaceae are strongly underrepresented. Although typical tropical families such as Arecaceae, Araceae, Moraceae and Piperaceae are absent, alpine taxa do exist in several families, otherwise known for their preference for warm climates, such as Rubiaceae (e.g., *Coprosma* sp.) and Cactaceae (e.g. *Tephrocactus* sp.).

Comparable numbers for the arctic flora (tundra sensu latu) are in the order of 1500 species, including possibly 500 arctic-alpine species. Murray (1992) estimates approx. 500 species in the arctic part of Alaska. Of course, these numbers vary when subspecies are included. As a rule of thumb, it may be estimated that the subarctic and arctic lowland flora is composed of approx. one-tenth of the total number of plant species of the alpine floras of the world. The taxonomic link between arctic and alpine floras is most obvious within the families Ericaceae and

Cyperaceae. Typically, Asteraceae represent a smaller fraction of all species in arctic compared to alpine vegetation.

4.3.2 The Regional Scale

As a rule of thumb, it appears that regional floras of the alpine zone of mountains extending over a few km^2 up to about 20 km^2 are in the order of 200 to 300 species (Table 3). Often, numbers do not increase appreciably until several hundred km^2 are exceeded. Wohlgemuth (1993) analyzed 215 different areas in the Alps, each with an average areal extent of 49 km^2 (Fig. 3). The mean number of species per

Table 3. The number of phanerogam plant species for some distinct mountain regions

	No. of species (phanerogam families)	
North America		
Ruby Range (Colorado)	220 (35)	Hartman and Rottman (1987)
Teton Range (Wyoming)	260 (36)	Spence and Shaw (1981)
Beartooth Plateau (Wyoming/Montana)	210 (–)	Johnson and Billings (1962)
Upper Walker River (Sierra Nevada)	280 (–)	Lavin (1983)
South America		
Cumbres Calchaquies	200 (49)	Halloy (1983)
Australia		
Snowy Mountains	250 (40)	Costin et al. (1979)
Europe		
Scandinavian Mountains	250 (29)	Nilsson (1986)

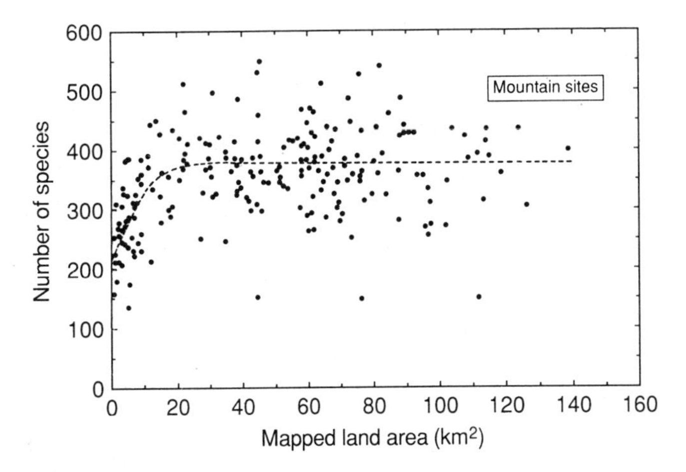

Fig. 3. The correlation between the number of plant species and the size of mountainous land area censused in the Swiss Alps (Wohlgemuth 1993). Note: Each *point* represents data for different, non-overlapping areas. For areas less than 20 km^2, the correlation coefficient for the linear regression is 0.51. Data points cover an altitudinal range of 1000 m, and the mean number of species per mapping area was 352 (minimum 135, maximum 548). (Reproduced with permission from Birkhäuser Verlag, Basel)

plot (including some subalpine elements) was 352 species, indicating that on average in each of these plots half of the total Swiss alpine flora could be found. Island peaks with only a few km^2 of alpine terrain such as the San Francisco Mountains in Arizona may bear much fewer species (50 to 80 alpine taxa, depending on definition; Moore 1965; Rominger and Paulik 1983). Surprisingly, regional floras in the Arctic also include approx. 200 to 300 species.

4.3.3 The Micro Scale

Alpine plant species diversity on a global scale is rather moderate compared to many other life zones, in particular the lowland humid tropical zone. At the regional scale, alpine ecosystems match species diversity of most areas in the temperat zone and exceed that of arctic lowland species diversities, but tropical and subtropical lowland ecosystems will still outrange alpine species diversity in many cases. However, at the micro-scale, species diversities within the alpine life zone are unbeaten.

For example, Table 4 lists the species and life forms found in the arctic-alpine life zone of northern Sweden (Kärkevagge valley, 900 m) on one single m^2 of land.

Table 4. Plant species diversity within 1 m^2 of arctic-alpine terrain (Sweden, 68 °N, Kärkevagge valley, 900 m)

Deciduous dwarf shrubs	**Hemiparasites**
Salix hastata	*Bartsia alpina*
Salix reticulata	*Euphrasia frigida*
Salix polaris	
Dryas octopetala	**Insectivore plants**
Vaccinium uliginosum	*Pinguicula alpina*
Evergreen dwarf shrubs	**CAM plants**
Vaccinium vitis-idea	*Rhodiola rosea*
Cushion plants	**Tall cryptogames**
Silene acaulis	*Equisetum variegatum*
Saxifraga oppositifolia	*Equisetum scirpoides*
	Selaginella selaginoides
Graminoids	
Carex rupestris	**Bryophytes** ca. 10 species
Carex vaginata	
Carex bigelowii	**Lichens**
Carex atrata	Fruticose 4 sp
Luzula spicata	Foliose 2 sp.
Juncus trifidus	Crustose Ca.5 sp.
Festuca ovina aggr.	
Perennial dicots	**Algae**
Viola biflora	
Parnassia palustris	**Fungi**
Saussurea alpina	
Polygonum viviparum	
Thalictrum alpinum	
Potentilla crantzii	Total number of plant species
Saxifraga cernua	(without Algae and Fungi) ca. 50

This incredibly rich vegetation is only a few 100 m apart from flat land tundra with only 15 species, almost exclusively Cyperaceae and Juncaceae species, resembling a single life form. This hillside vegetation certainly exemplifies one of the richest species mixes ever recorded per unit land area. Of course, one must admit that the number of species would not increase dramatically, if the considered area were to be increased. The total set of all plant species in this small valley may be 150. Such high species densities on the smallest scale may be found in the lower alpine belt of many mountains.

4.4 Plant Functional Groups in Alpine Ecosystems

Although the species composition of alpine floras may vary regionally and globally, the structural inventories of alpine floras are rather similar in all parts of the world. Among higher plants the three most important growth forms found in the alpine life zone are (1) graminoid tussocks, (2) narrow dwarf shrubs, and (3) herbaceous, mostly rosette-forming dicotyledonous species (cf. Billings 1974). More than 90% of the species of the alpine life zone show some sort of vegetative, i.e. clonal, propagation. Annuals are very rare, in most regions less than 3 out of 100 species are annual (Billings 1988). Cushion plants frequently found in some (the more windy) regions are rare or even absent in other regions. Their greatest abundance is in the temperate and subpolar regions, in particular in the subantarctic mountains, while they are almost absent in some tropical alpine floras (Hedberg 1964, and pers. observ.) where another peculiar growth form, the giant rosettes are found (afro-alpine and tropical Andes).

The more humid a region, the greater the biomass fraction of cryptogams (lichens, mosses) will be. Locally, cryptogams can represent 90 to 100% of the alpine vegetation cover, with moss genera like *Racomitrium* and *Polytrichum* or fruticose lichens like *Cetraria* or *Cladonia* often dominating.

Ranking the alpine growth forms with respect to their overall contribution to alpine biomass, tussock grasses would represent the largest, and herbaceous dicotyledonous plants the smallest fraction. Ranking these growth forms with respect to their contribution to species diversity, herbaceous dicots would by far outrange the two other growth forms. These functional groups still bear a large variability in morphologies (Halloy 1990), allocation patterns (Körner and Renhardt 1987), and physiological characteristics (references in Körner and Larcher 1988). For instance, within alpine forbs, the ratio between aboveground and belowground biomass varies by one order of magnitude, and species exhibiting the most extreme differences may grow jointly at the most exposed sites high above the glacier region (Körner and Renhardt 1987). Flowering rhythms vary enormously even within a small area (Fig. 4). Leaf nitrogen content and photosynthetic capacity (Körner 1989; Körner and Pelaez Menendez-Riedl 1989) vary by a factor of 3 to 5, with forbs exhibiting the highest and evergreen shrubs the lowest values. The photosynthetic responses to short-term doubling of ambient CO_2 also vary substantially, ranging from almost zero to more than 50% enhancement (Körner and Pelaez Menendez-Riedl 1989). However, recent in situ CO_2 enrichment

Flowering spectra of alpine plants

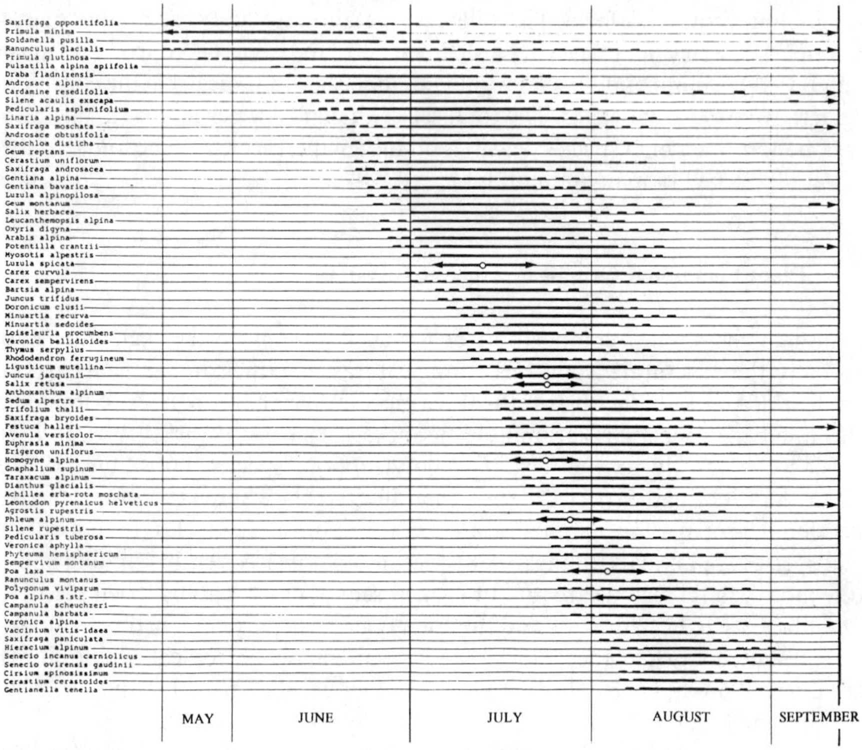

Fig. 4. The diversity of flowering periods within 1 ha of summit terrain on Mt. Glungezer (2600 m), Tyrolian Alps of Austria. (Bahn and Körner 1987)

experiments in the Alps revealed no stimulation of biomass production at the community level (Körner et al. 1995). With respect to rates of photosynthesis and growth, graminoids do not fit into any category. Some species of Poaceae and Cyperaceae meet the peak values of forbs (e.g. *Poa*, some *Carex*), while some others (e.g. *Nardus*, *Chionochloa*) have similarly low values as evergreen shrubs. As mentioned above, legumes do not represent a prominent component in truly alpine flora and are absent in the uppermost region of higher plant life. A number of alpine C4 (some Poaceae; e.g. Pyankov and Mokronosov 1993) and CAM (Cactaceae, Crassulaceae; Larcher and Wagner 1983) species do exist, but again their contribution to overall diversity and to biomass is rather small. Although most alpine plants will fit into one of the functional groups mentioned above, 'the' alpine plant type does not exist. However, morphotype spectra, as the ones developed by Halloy (1990), permit a comparison of alpine floras independently of their taxonomic composition. In fact, Halloy could demonstrate similar spectra (i.e. morphological diversities) in floristically contrasting alpine communities of the Andes, New Zealand and the Alps. Similarly, virtually identical spectra of leaf weight ratio (about 22% of total plant mass) were found in herbaceous alpine communities across the globe (Körner 1994a).

4.5 Causes of Alpine Plant Diversity

4.5.1 The Global Scale – The Historical and Geographical "Sieve"

Here, I will only refer to the large body of literature on the possible historical origins of today's alpine taxa (e.g. Braun-Blanquet 1923; Merxmüller 1954; Billings 1974, 1978, 1979b; Ozenda 1988, chapters by Ammann as well as Agakhanjanz and Breckle, this Vol.). According to these sources, the current floras in most cases represent a mix of ancestral elements (mostly Tertiary), "immigrants" (of various ages) and new evolutionary lines. Glaciation, fragmentation and the speed of tectonic uplift represent the major selective events at the continental scale. Many mountain areas are famous for their large number of endemics, but these are largely confined to the lower, climatically more favourable alpine belts, while the species found at the highest elevations often also have the widest geographical distribution (e.g. Breckle 1974). The east-west orientation, the more pronounced fragmentation, and the greater separation from the arctic circle of the Eurasian mountain ranges compared to the north-south oriented Cordilleras cause them to show greater regional variation and speciation. However, there are also significant longitudinal barriers that affected species migration in the alpine life zone of North America. Loope (1969), for instance, reported significant differences between Great Basin mountain floras and the Sierra Nevada of California (see also Billings 1974, 1978).

4.5.2 The Regional Scale – The Frost and Grazing "Sieve"

As mentioned earlier, at the regional scale (within a certain mountain or smaller parts of a mountain range), approx. 200 to 300 species are found throughout the world. Why not 50 or 1000? Why are the numbers so similar for the Australian Snowy Mountains, the Cumbres Calchaquies in Argentina, the Great Basin Mountains, mountains in Switzerland, Alaska, or Scandinavia (Table 3)? Neither the causes for this surprising consistency nor the functional implications for these ecosystems are understood.

What all these species and life forms have in common (independent of latitude) are (1) the accumulation of photosynthetic active tissue within a few centimetres above the ground, (2) the position of buds or meristems close to, at, or below the ground, and (3) the ability of their fully active aboveground tissues to survive minimum temperatures down to −6 °C, in some cases down to −15 °C (rarely below; for references, see Körner and Larcher 1988). Besides the ability of tissues to survive certain low temperatures, frost survival has a second, at least as important component, namely the ability of plants to replace lost tissue. This depends (1) on the existence of meristems or regeneration buds and (2) on the fraction of tissue lost compared to that preserved (the carbon balance of the plant). Both these aspects of frost resistance are strongly selective for the prostrate life forms found in the alpine zone. In contrast, alpine species differ widely in almost all other morphological and physiological features, including their absolute frost resistance in the dormant state, temperature and nutrient requirements for their basic life processes, water relations etc. (see below). Frost resistance is the primary environmental filter through which species had to pass in the course of evolution and migration in order to contribute to alpine floras and, as mentioned above, this has a physiological and a morphological component.

An important additional point, not commonly mentioned, is that all alpine ecosystems of the world are naturally grazed, even though larger grazers may have become recently extinct in some regions such as in New Zealand. The severe grazing pressure reaches the highest peaks, and alpine grazers are known to be quite selective (Järvinen 1984; Diemer 1992).

4.5.3 The Micro-Scale – Effects of Vectors and Relief

As soon as slope forces come into play and shape the landscape, exposition, soils, mineral nutrients, water availability and microclimate change over short distances which translates into small-scale mosaics of varying life conditions. This is also true for the arctic life zone, where similar "explosions" of diversity are observed on steep slopes (see Walker, this Vol.). The most important of these influences, emerging from the environmental micro-fragmentation, are variable radiation and – in temperate and subpolar mountain ranges – changing patterns of snow distribution and, hence, spatial variations of seasonality. These, in turn, feed back on moisture and nutrient availability. Despite the fact that the shape of the alpine land surface can be considered constant within the time frame of interest here, its influence on snow distribution is co-determined by wind direction and thus varies from year to year, causing spatial and annual variations in seasonality over relatively small areas. Small-scale "change" is thus one of the most important factors of alpine life conditions. Because of these strong exposure and radiation-controlled micro-environments, the true climate experienced by alpine plants cannot be predicted from standard meteorological data (for references, see Körner and Larcher 1988). Beyond the tree line, the macro-climate deviates from the micro-climate to an extent that elevation per se also becomes a very poor predictor of life conditions or the occurrence of certain plants or plant life forms.

Such steep spatial climatic gradients are clearly reflected in plant species distribution which, in turn, can be related to plant physiological or developmental features. Another example is frost resistance. Snowbed plants do not survive even moderately low temperatures when snow is removed in mid-winter; some of these plant species would be killed even at low altitudes if they were not protected by snow. In contrast, plant species growing on exposed crests may be dipped in liquid nitrogen without damage when fully hardened in winter (Larcher 1980). Since these and the following aspects of alpine plant functioning have been treated in various earlier publications, I refer the reader to the review by Körner and Larcher (1988) and will not include this extensive literature here.

Wind and water distribution, both controlled by relief and canopy density, are the main factors causing horizontal allocation of nutrients in sloping terrains.

Small-scale allocation of litter (from local sources to local sinks) and substrate leaching create mosaics of rather different nutrient supply and physical soil quality. Dense growth forms (most pronounced in cushion plants) must be seen as litter traps, supporting microbiological activity orders of magnitude greater than in adjacent, uncovered soil.

The micro-scale processes that generate these mosaics of plant communities prevent the dominance of single types of communities over larger areas as long as the slope processes continue. Thus, hot "micro-deserts" with CAM plants may be found in close proximity to wet, snowbed communities. Small-scale disturbances that create new surface patterns will create opportunities for seedling establishment and thus maintain diversity. Any levelling or smoothing of the landscape will immediately reduce species diversity.

4.6 Effects of Biodiversity on Alpine Ecosystem Functioning

Approximately 10% of the world's human population lives in mountain regions, but more than 40% depend in some way on mountain resources (e.g. water; cf. Messerli 1983). Changes in the functioning of "up-slope" ecosystems may thus have substantial impact on welfare "down the slope". Biodiversity is most likely to influence a number of functions of high-altitude ecosystems, but the available evidence is scarce. I will mention three examples, in which alpine species or community richness appear to have an impact on ecosystem function; one is from the European Alps, one from the equatorial Andes, and one from the Australian Snowy Mountains.

4.6.1 Biodiversity Feedbacks on Soils

Among the many functions of an ecosystem, the most obvious one (from an anthropogenic point of view), the biomass production, is not of particular relevance at high altitudes, where the greatest risk for life is the loss of adequate substrates for plant growth. Soil development and soil preservation depend completely on the presence and persistence of a plant cover, the quality of which must be such that erosion is prevented. Thus, it is crucial that certain taxa persist in space and time in sufficient abundance, which may be rather independent of the accumulation of matter. In some cases, fast biomass production may even lead to species extinction and subsequent damage of soils. A high diversity of plant species, in this case also a high diversity of rooting patterns, is important to fulfill the various mechanical functions and consolidate the ground during the plant succession process, finally leading to the establishment of late successional mosaics of plant communities. These, in turn, require a certain species redundancy within functional types in order to assure slope stability in the case of elimination of a certain species by the environment, pathogens or herbivores. A good example for the potential risk in highly specialized ecosystems of low diversity with only one or few species covering the slope is the case of espaliers of the holarctic-alpine *Loiseleuria procumbens* (Ericaceae). This prostrate, extremely slow growing evergreen dwarf shrub produces extremely acid soils and massive layers of raw humus. In addition, its dense, cushion-like growth habit creates a very favourable micro-climate (Cernusca 1976), enabling the species to dominate on windswept slopes. However, *Loiseleuria* is extremely sensitive to mechanical disturbance (Körner 1980) and the addition of mineral nutrients (in particular nitrogen) (Körner 1984), both leading to the rapid extinction of the species and thus removal of the soil protection. There are only a few species that are able to inhabit bare *Loiseleuria* soils and provide some transitory soil protection (e.g. *Juncus trifidus*). The presence of these species in the community is thus of key significance for the prevention of erosion. This is one example where the presence of low abundance species in a community suddenly acquires great functional significance.

4.6.2 Diversity of Plant Morphologies Warrant Frost Survival of the Plant Cover

A minimum plant soil coverage is not only required for erosion protection, it also assures nutrient cycling, carbon uptake, water vapour discharge and food for consumers. In the tropics, where plants at high altitudes may experience summer and winter every day, frost episodes may severely affect plants. Squeo et al. (1991) showed for the Andean Paramos that there is a large variation in species-specific frost resistance related to different morphologies. The taller species are less resistant than the smaller ones, which are also climatically better protected from frost. This diversity of species assures that some, perhaps the slower growing smaller ones, survive and protect the regrowth of the taller ones, thus assuring ecosystem integrity over time.

4.6.3 The Control of Tree Migration by Alpine Community Diversity

It is well established that the presence of trees at high altitudes has a significant influence on surface albedo, soil development, water retension, avalanche protection etc. The actual position of the timberline and the uppermost outposts of forest trees strongly depends on the community diversity in their target area. Biotic interactions affect the establishment of seedlings and these interactions may be more important than a few degrees of change in temperature. Egerton and Wilson (1993) demonstrated that mosaics of diverse shrublands support tree establishment above the tree line in the Australian Snowy Mountains, whereas continuous carpets of tussock grasses are prohibitive. Similar observations with respect to tussock grass competition (mainly via root competition) were made by Wardle (1971) and A.F. Mark (pers. comm.) at the Nothofagus tree line in New Zealand. Hence, high structural diversity of lower alpine plant communities facilitates tree migration. More uniform, grass-dominated vegetation will constrain it.

4.7 Alpine Biodiversity and Climate Change

Because "change" i.e. temporal and spatial variation of life conditions, is the essential driving force of biodiversity in alpine vegetation at the micro-scale, a change in global climate would affect these patterns, but not necessarily the overall diversity within a certain region. An increase in the mean air temperature by 2 to 3 K may not be of particular significance for a plant growing at 4000-m altitude in a rocky, south-facing mountain slope. Here, the climate is always too cold for life according to atmospheric data, but may be too hot on sunny days even today because of extreme insolation. Direct influences of changes in temperature on photosynthesis and respiration of alpine plants are negligible because of broad response curves and acclimated behaviour (Körner and Larcher 1988; Körner, 1992, 1994b). If direct effects of changes in ambient temperature come into play, this would be most effective during the night by influencing growth and developmental processes (Körner and Woodward 1987; Körner and Pelaez Menedez-Riedl 1989). In general, indirect effects of temperature, particularly the length of the growing season and snowfall patterns, will have much greater influence.

Given the large variations in micro-climate, climate change is unlikely to affect overall biodiversity at high altitudes. However, the abundance of certain communities will decrease, while that of others will increase (see Grabherr, this Vol.). Ecoclines are not likely to move proportionally to climate change, since species interactions buffer such systems over a substantial thermal amplitude (Armand 1992; Slatyer and Noble 1992; Woodward 1993; Körner and Hättenschwiler 1995). Accordingly, the few available palaeorecords for alpine biodiversity indicate surprisingly little variation (Bortenschlager 1993; Ammann, this Vol.). Bortenschlager's pollen diagrams from the highest situated mire in the Alps (2750 m, i.e. 700 m above the tree line) suggest that the maximum amplitude of the position change of the alpine tree line in the Alps was 150 to 200 m within the last 10 000 years, which included periods during which people were growing grapes in Scotland and found coastal Greenland very "green". The rather invariable position of the tree line over the past ca. 5000 years in the Rocky Mountains (Ives and Hansen-Bristow 1983) also supports the view that pronounced altitudinal variations of this prominent ecological boundary are rather unlikely, unless temperatures will exceed the 1–3 K currently predicted for a "doubled CO_2" world.

4.8 Conclusions

High species diversity of alpine vegetation can be explained by:

1. The great diversity of climatic and floristic zones in which alpine vegetation occurs.
2. The present and past isolation of many mountain ranges that causes a high degree of endemism as well as an abundance of relict species.
3. The strong fragmentation of the land surface by slope forces which leads to

   (a) slope x radiation interaction with drastic effects on the thermal regime,
   (b) slope x wind interactions which (in extratropical mountains) controls snow distribution and, hence, micro-scale seasonality,
   (c) micro-scale partitioning of mineral nutrients and water.

4. Although clonal growth dominates in alpine vegetation as in tundra, alpine vegetation, when fragmented, seems to have a greater potential for establishment of sexual offsprings, particularly after small-scale disturb- ances.
5. In alpine areas of temperate and higher latitudes, the most significant influences on vegetation will not be mediated by changing mean temperature itself, but via a combination of amounts and patterns of snowfall, wind, and cloudiness, all affecting the length of the snow-free period.
6. Since today's micro-scale differences in alpine climate largely exceed those differences currently predicted for a future atmosphere, a substantial resilence of alpine biodiversity may be expected, although local variations in abundance of certain community types are likely.
7. In many mountain regions, in particular the Alps, direct anthropogenic influences on alpine diversity via land use certainly outrange those of possible climatic effects.

Acknowledgments. I am grateful to D. Billings (Duke University) for references and copies of floristic studies by US authors, to W. Cramer (Potsdam Institute for Climate Research) for providing GIS data on the alpine life zone, and to B. Ammann, S. Breckle and G. Grabherr for valuable comments to the manuscript.

References

Armand AD (1992) Sharp and gradual mountain timberlines as a result of species interaction. In: Hansen AJ, di Castri F (eds) Landscape boundaries. Springer, Berlin Heidelberg New York, pp 360–378

Bahn M, Körner Ch (1987) Vegetation und Phänologie der hochalpinen Gipfelflur des Glungezer in Tirol. Ber Naturwiss Med Ver Innsbruck 74:61–80

Billings WD (1973) Arctic and alpine vegetations: similarities, differences, and susceptibility to disturbance. BioScience 23:697–704

Billings WD (1974) Adaptations and origins of alpine plants. Arct Alp Res 6:129–142

Billings WD (1978) Alpine phytogeography across the Great Basin. In: Intermountain biogeography, a symposium. Great Basin Nat Mem 2:105–117

Billings WD (1979a) Alpine ecosystems of western North America. In: Johnson DA (ed) Special management needs of alpine ecosystems. Society for Range Management, Denver pp 6–21

Billings WD (1979b) High mountain ecosystems. In: Webber P J (ed) High altitude geoecology. American Association for the Advancement of Science, Westview Press Boulder, pp 97–125

Billings WD (1988) Alpine vegetation. In: Barbour MG, Billings WD (eds) North American terrestrial vegetation. Cambridge University Press, New York, pp 391–420

Bortenschlager S (1993) Das höchst gelegene Moor der Ostalpen "Moor am Rofenberg" 2760 m. Festschrift Zoller, Diss Bot 196:329–334

Braun-Blanquet J (1923) Über die Genesis der Alpenflora. Verh Naturforsch Ges Basel 35: 243–261

Breckle SW (1974) Notes on alpine and nival flora of the Hindu Kush, east Afghanistan. Bot Not 127: 278–284

Cernusca A (1976) Energie- und Wasserhaushalt eines alpinen Zwergstrauchbestandes während einer Föhnperiode. Arch Meteorol Geophys Bioklimatol Ser B 24: 219–241

Costin AB, Gray M, Totterdell CJ, Wimbush D J (1979) Kosciusko alpine flora. CSIRO and Collins, Melbourne

Diemer M (1992) Population dynamics and spatial arrangement of *Ranunculus glacialis* L., an alpine perennial herb, in permanent plots. Vegetatio 103: 159–166

Egerton JJG, Wilson SD (1993) Plant competition over winter in alpine shrubland and grassland, Snowy Mountains, Australia. Arct Alp Res 25: 124–129

Halloy SRP (1983) High mountain climatology and edaphology in relation to the composition and adaptations of biotic communities (with special reference to the Cumbres Calchaquies, Tucuman, Argentina). Diss Univ Nacional de Tucuman, Argentina. University Microfilms Int. (Ann Arbor) Cat no 8502967

Halloy S (1989) Altitudinal limits of life in subtropical mountains: what do we know? Pac Sci 43: 170–184

Halloy S (1990) A morphological classification of plants, with special reference to the New Zealand alpine flora. J Veg Sci 1: 191–304

Halloy S (1991) Islands of life at 6000 m altitude: the environment of the highest autotrophic communities on earth (Socompa Volcano, Andes). Arct Alp Res 23: 247–262

Hartman EL, Rottman ML (1987) Alpine vascular flora of the Ruby Range, West Elk Mountains, Colorado. Great Basin Nat 47: 152–160

Hedberg O (1964) Features of afroalpine plant ecology. Acta Phytogeogr Suec 49: 8–89

Hermes K (1955) Die Lage der oberen Waldgrenze in den Gebirgen der Erde und ihr Abstand zur Schneegrenze. Kölner Geographische Arbeiten 5, Köln

Ives JD, Hansen-Bristow KJ (1983) Stability and instability of natural and modified upper timberline landscapes in the Colorado Rocky Mountains, USA. Mountain Res Dev 3: 149–155

Järvinen A (1984) Patterns and performance in a *Ranunculus glacialis* population in a mountain area in Finnish Lapland. Ann Bot Fenn 21: 179–187

Johnson PL, Billings WD (1962) The alpine vegetation of the Beartooth Plateau in relation to cryopedogenic processes and patterns. Ecol Monogr 32: 105–135

Körner Ch (1980) Zur anthropogenen Belastbarkeit der alpinen Vegetation. Verh Ges Ökol (Göttingen) 8: 451–461

Körner Ch (1984) Auswirkungen von Mineraldünger auf alpine Zwergsträucher. Verh Ges Ökol (Göttingen) 12: 123–136

Körner Ch (1989) Der Flächenanteil unterschiedlicher Vegetationseinheiten in den Hohen Tauern: eine quantitative Analyse großmaßstäblicher Vegetationskartierungen in den Ostalpen. In: Cernusca A (ed) Struktur und Funktion von Graslandökosystemen im Nationalpark Hohe Tauern. Veröffentl Österr MaB-Programm 13:33–47, Wagner, Innsbruck

Köner Ch (1992) Response of alpine vegetation to global climate change. CATENA Suppl 22:85–96

Körner Ch (1994a) Biomass fractionation in plants–a reconsideration of definitions based on plant functions. In: Roy J, Garnier E (eds) A whole plant perspective on carbon-nitrogen interactions SPB Acad Publishing, The Hague, pp 213–225

Körner Ch (1994b) Impact of atmospheric changes on high mountain vegetation. In: Benniston M (ed) Mountain environments in changing climates. Routledge, London-New York, pp 155–166

Körner Ch, Larcher W (1988) Plant life in cold climates. In: Long SF, Woodward FI (eds) Plants and temperature. Symp Soc Exp Biol 42:25–57. The Company of Biol, Cambridge

Körner Ch, Pelaez Menendez-Riedl S (1989) The significance of developmental aspects in plant growth analysis. In: Lambers H, Cambridge ML, Konings H, Pons TL (eds) Causes and consequences of variation in growth rate and productivity of higher plants. SPB Acad Publ, The Hague, pp 141–157

Körner Ch, Renhardt U (1987) Dry matter partitioning and root length/leaf area ratios in herbaceous perennial plants with diverse altitudinal distribution. Oecologia 74:411–418

Körner Ch, Woodward FI (1987) The dynamics of leaf extension in plants with diverse altitudinal ranges. II. Field studies in *Poa* species between 600 and 3200 m altitude. Oecologia 72:279–283

Körner Ch, Diemer M, Schäppi B, Zimmermann L (1995) The response of alpine vegetation to elevated CO_2. In: Koch G, Mooney HA (eds) "Terrestrial ecosystem response to elevated CO_2." Physiological Ecology Series. Academic Press (in press)

Larcher W (1980) Klimastress im Gebirge – Adaptationstraining und Selektionsfilter für Pflanzen. Rheinisch-Westf Akad Wiss Vortr N 291:49–88
Larcher W, Wagner J (1983) Ökologischer Zeigerwert und physiologische Konstitution von *Sempervivum montanum*. Verh Ges Ökol (Göttingen) 11:253–264
Lavin M (1983) Floristic of the upper Walker River, California and Nevada. Great Basin Nat 43:93–130
Loope LL (1969) Subalpine and alpine vegetation of northeastern Nevada. PhD thesis, Duke University, Durham, NC
Matveyeva NV (1988) The horizontal structure of tundra communities. In: Werger MJA, van der Aart PJM, During HJ (eds) Plant form and vegetation structure. SPB Academic Publishing, The Hague, pp 59–65
Merxmüller H (1954) Untersuchungen zur Sippengliederung und Arealbildung in den Alpen. Verein z Schutz d Alpenpflanzen und-tiere 19, München
Messerli B (1983) Stability and instability of mountain ecosystems: introduction to a workshop sponsored by the United Nations University. Mountain Res Dev 3:81–94
Miehe G (1989) Vegetation patterns on Mount Everest as influenced by monsoon and föhn. Vegetatio 79:21–32
Moore TC (1965) Origin and disjunction of the alpine tundra flora on San Francisco Mountain, Arizona. Ecology 46:860–864
Murray DF (1992) Vascular plant diversity in Alaskan arctic tundra. North West Environ J 8:29–52
Nilsson O (1986) Nordisk fjällflora, Bonniers, Göteborg
Ozenda P (1988) Die Vegetation der Alpen im europäischen Gebirgsraum. Gustav Fischer, Stuttgart
Ozenda P (1993) Etage alpin et Toundra de montagne: parenté ou convergence? Fragm Florist Geobot Suppl 2:457–471
Pisek A, Larcher W, Vegis A, Napp-Zinn K (1973) The normal temperature range. In: Precht H, Christophersen J, Hensel H, Larcher W (eds) Temperature and life. Springer, Berlin Heidelberg New York, pp 102–194
Polunin O, Stainton A (1984) Flowers of the Himalaya. Oxford University Press, Oxford
Pyankov VI, Mokronosov AT (1993) General trends in changes of the earth's vegetation related to global warming. Russian J Plant Physiol 40:515–531
Rikli M (1917) Die den 80° n erreichenden oder überschreitenden Gefässpflanzen. Vierteljahresschr Naturforsch Ges Zürich 62:169–193
Rominger JM, Paulik LA (1983) A floristic inventory of the plant communities of the San Francisco Peaks Research Natural Area. USDA Forest Service, General Technical Report RM-96, Fort Collins
Slatyer RO, Noble IR (1992) Dynamics of montane treelines. In: Hansen AJ, di Castri F (eds) Landscape boundaries. Springer, Berlin Heidelberg New York, pp 346–359
Spence JR, Shaw RJ (1981) A checklist of the alpine vascular flora of the Teton Range, Wyoming, with notes on biology and habitat preferences. Great Basin Nat 41:232–242
Squeo A, Rada F, Azocar A, Goldstein G (1991) Freezing tolerance and avoidance in high tropical Andean plants: is it equally represented in species with different plant height? Oecologia 86:378–382
Troll C (1973) The upper timberlines in different climatic zones. Arct Alp Res 5:A3–A18
Vareschi V (1970) Flora de los Paramos de Venezuela. Universidad de los Andes, Merida, Venezuela
Walter H, Breckle SW (1986) (eds) Ökologie der Erde. Band 3: Spezielle Ökologie der gemässigten und arktischen Zonen Euro-Nordasiens. UTB Gustav Fischer, Stuttgart
Wardle P (1971) An explanation of alpine timberline. NZ J Bot 9:371–402
Wardle P (1974) Alpine timberlines. In: Ives JD, Barry RG (eds) Arctic and alpine environments. Methuen, London, pp 372–402
Wohlgemuth T (1993) Der Verbreitungsatlas der Farn-und Blütenpflanzen der Schweiz (Welten und Sutter 1982) auf EDV: die Artenzahlen und ihre Abhängigkeit von verschiedenen Faktoren. Bot Helv 103:55–71
Woodward FI (1993) The lowland-to-upland transition-modelling plant responses to environmental change. Ecol Appl 3:404–408

5 Origin and Evolution of the Mountain Flora in Middle Asia and Neighbouring Mountain Regions

O. Agakhanjanz[1] and S.-W. Breckle[2]

5.1 Introduction

The flora of Middle Asia is very rich and diverse (Walter and Breckle 1986; Breckle and Agakhanjanz 1994). This is also true for the mountain floras of this orographically very complex region (Fig. 1). We will discuss the relationships of the various mountain floras and their diversity, how these floras developed and to which extent migration of plant species contributed to diversity. We will also investigate the tectonic history and more recent glacial history as decisive factors for the development of species richness in the mountains.

5.2 Number of Species in the Mountains

In most cases, the flora of the mountains is richer than in the neighbouring plains. The comparison of data is complicated, because some data are available according to administrative rather than natural regions and because the differentiation between Middle Asia, Central Asia and Inner Asia is not perfectly defined. Russian literature mentions Middle Asia as the Turan area where a Mediterranean precipitation regime occurs (Fig. 2, with cool, humid winters and hot arid summers), in comparison to the more eastern areas (Central and Inner Asia) of Dzhungaria, Kashgaria, Tibet, Takla-Makan, and Mongolia with maximum precipitation in summer and cold, dry winters. European literature does not always differentiate between these regions. In this text, we assign the western Tien Shan and the western Pamir and Alai ridges to Middle Asia. The inner Tien Shan and most parts of the central and eastern Pamir belong to Central Asia. In Kamelin's statistics (1973), the eastern Pamir and the inner Tien Shan are missing (Fig. 3). The Himalaya and Kashmir will only be treated marginally.

Including the plains, the flora in all of Middle Asia (south of 43°N, where the line between subtropical and temperate climate is drawn) includes 7500 species of vascular plants (Kamelin 1973), 2000 of which inhabit the plains. Of these, 800 also occur in the foothills of the mountains (Fig. 3). The flora of Middle Asian

[1] Department of Physical Geography, Pedagogic University of Belarus, Sovetskaja 18, 220809 Minsk, Belarus/CIS; and DFG-Guest Professor at the Department of Ecology in Bielefeld, Germany
[2] Department of Ecology, P.O. Box 100310, Bielefeld University, 33501 Bielefeld, Germany

Ecological Studies, Vol. 113
Chapin/Körner (eds) Arctic and Alpine Biodiversity

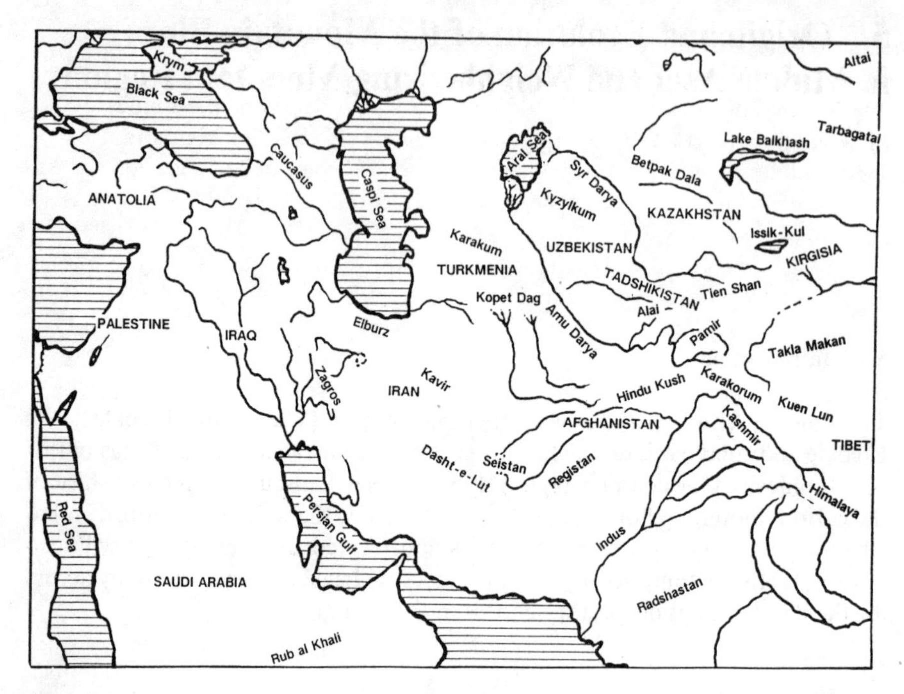

Fig. 1. Geographic outline of Middle and Central Asia

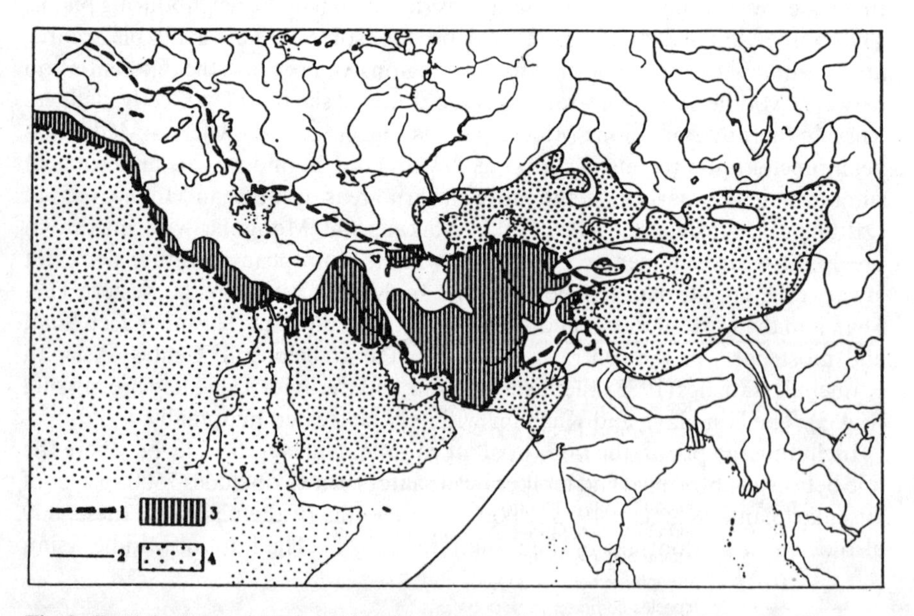

Fig. 2. Precipitation regimes in arid regions of Eurasia (Agakhanjanz 1981). *1* Border of the area with Mediterranean precipitation regime; *2* isohyet with 300 mm/year; *3* arid areas with Mediterranean precipitation; *4* areas with another precipitation regime

mountain regions alone (without Tien Shan and eastern Pamir) consists of about 5500 species. Subtracting the 800 species that occur in both the plains and the mountains, 4700 mountain species remain. Thus, 670 appear at the alpine, 1300 at the subalpine levels. We use the term *oreophytic* to describe plants that occur above the timberline, generally the alpine and subalpine zone. Since 325 species are common at both levels, the total number of oreophytic species is 1645 (Fig. 3D, E).

The various regions with their largely overlapping floras consist of 3276 species in Kirgisia, 5239 in Kasakhstan, 2659 in Usbekistan, more than 4200 in Tadshikistan, 2800 in Turkmenistan, 1800 in the Turkmenian part of the Kopet Dag Mountains, 2588 in the basin of Seravshan and 2000 species in western Pamir. Further examples of the number of species for some of these mountain regions are 800 for eastern Pamir, 1870 for central Tien Shan, 2812 for western Tien Shan, 2230 for northern Tien Shan, and 2128 species for the Dzungar Mountains. The

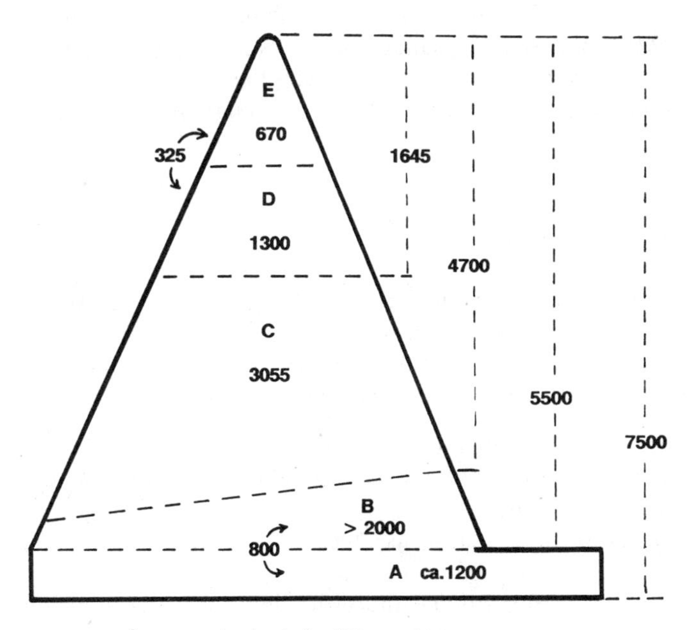

A : lowlands (< 500m a.s.l.)
B : lower altitudinal belt with lowland species at the foothills (about 500 - 1000m)
C : altitudinal belt of the mountain deserts, ephemeral associations and woods, or forests (about 1000 - 3000m)
D : subalpine belt (about 3000-3500m)
E : alpine belt (> 3500m)
D+E : oreophytic belt (> 3000m)
← → : species common to both belts
figures : number of species

Fig. 3. Number of species of the flora in Middle Asia south of 43°N lat. (Agakhanjanz 1980)

endemism of Middle Asian mountain flora is approx. 25–30%, with about 62 endemic genera and more than 1600 endemic species.

The flora of the Caucasian area is especially rich with 6350 species. Of this, about 25.2% are endemic (12 endemic genera and about 1500 endemic species). Even in relatively small areas, the flora is very rich: Armenia 3000 species (in the Aras basin around Jerevan about 1452 species), Georgia 4028, Azerbeidshan 4200, Dagestan 2641, northern Caucasus 3800 species (including the 2750 species from the Stawropol region). The southern slopes of the Greater Caucasus exhibit less than 200 species that occur only above the timberline, whereas from the south-eastern part about 1150 and from central Caucasus about 800 species are known above the timberline.

The flora in other mountain regions is also rich: the Krim peninsula exhibits 2356 species (including the 1373 species from the Yalta Nature Reserve). Especially rich in species are the various mountains of Iran (about 70% out of 6000 species in the mountains), Afghanistan (about 80% out of 4000 species in the mountains) and Kashmir (about 85% out of 5000 species in the mountains). The mountains in Mongolia have only 2272 species, the high mountains in West Sayan 601 species, the larger Altai mountains 1900 species. Even in relatively low mountains and the plateau plains of Middle Siberia (between Yenissey and Lena), there were 2306 species recorded.

5.3 Effective Factors for Evolution

Using correlations between the present flora and environmental data of various regions, one can estimate the historical ranges of species and vegetation distribution. For arid mountains, such correlative estimates were tested using the following input variables (Agakhanjanz 1981; Agakhanjanz 1992):

- the rate and extent of mountain uplift (orogenesis);
- mixing of the ancestral plant formations (first in the plains and foothills, later in the lower and middle mountains);
- the degree of fragmentation and isolation as a result of orogenesis (speciation, endemism);
- the rate and extent of floristic migration within altitudinal belts (depending on the size of connected mountain systems).

Mountain uplift caused (1) species extinctions due largely to the limits set by frost resistance (Larcher 1980, 1981; Larcher and Bauer 1981; Sakai and Larcher 1987), (2) evolutionary adjustment to the new life conditions, or (3) it provided access from older mountain ranges. The latter requires migration corridors. When migration corridors are missing, as in isolated ranges like the Greater Balkhan, floristic depauperation can be observed, with biodiversity decreasing abruptly with altitude (Model I: isolated mountains).

The florogenesis in larger, more extensive mountains is different (Model II: mountain systems). During orogenetic uplift, the various altitudinal belts of neighbouring mountains are not strictly isolated from each other (so-called open altitudinal belts). Within the same belt with a similar thermal climate regime, the migration of species continues during further uplift. Between the adjacent plains, along the foothills of the mountains and along the lower mountains, the migration of species was and is still far-reaching. For example, out of the 2000 species occurring in the plains of Middle Asia about 40% of the species had a chance to penetrate the lower mountain regions within the lower altitudinal belt (Fig. 3).

These processes (Table 1; Fig. 3) are the reason that regions with tectonic uplift are rapidly enriched by species migration. It explains the floristic richness of the middle altitudinal belts, where most of the migrational connections are still active. A specific floristic pattern for each mountain area enhances this richness.

For Model I (isolated mountains), in the Greater Balkhan or the Kopet Dag Mountains, the relative number of species in the middle belt is lower than in mountain systems (Model II, Table 1). The floristic richness of the Middle Asian mountains is highest in the middle altitudinal belt. This is normally the belt below the tree line. This tree line, however, is not always a strict line bordering the forests or indicating where forests were formerly growing and where today open thickets or scattered trees are found, but it makes the various altitudinal belts of different mountains comparable.

The middle altitudinal belts are rich in species due to the complicated ecological and genetic exchange processes between the immigrating and the local species. Some of the major processes are introgressive hybridization, competition between species, oppression of the original species and their repression to refuge territories and on other stands, as well as the adaptation to new conditions.

Table 1. Percentages of floristic richness within the three main elevational belts of some Asian mountains (Agahanjanz 1981)

Mountain region	No. of species	Floristic richness		
		Plains, foothills	Montane belt	Subalpine, alpine belts
Model I				
Greater Balkhan	475	49	32	19
Kopet Dag	1800	41	39	20
Model II				
Greater Caucasus, SE slope	4110	26	46	28
Greater Caucasus (southern slopes)	3703	27	68	5
Central-Caucasus, NW part	2300	27	38	35
Warzob-Bassin (Hissar Mts.)	1680	28	53	19
Shachristan valley (Turkest. Mts.)	838	30	48	22
Western Pamirs	1480	33	48	19
Western Tien Shan	2812	26	40	34
Middle Asian Mts.	5500	26	48	26

5.4 Models for Mountain Florogenesis

The floristic conditions in the Greater Balkhan and in the Kopet Dag are the only examples of isolated mountains in Middle Asia (Model I) that are high enough to develop a distinct zonation of altitudinal belts (Table 1). Even the separation of the Kopet Dag from the Elburz Mountains is not strict, and migration of species took place actively in the past and may still be occurring today. All other mountains are part of connected mountain systems, whose florogenesis fits Model II, and various factors are responsible for the specific pattern of the flora.

5.4.1 Tectonics

During the ongoing tectonic uplift, the formation of species in the upper altitudinal belts originated from the rich ancestral flora. In some mountains, the oreophytic belt is very high, and the elimination of many species by cold stress cannot be compensated by the local formation of new species. This is obvious from the low number of species of that belt (Table 1), e.g., in the Hindu Kush mountains (Breckle 1974, 1975), where the number of species decreases from 377 (4000 m) to 162, 36 and finally to 2 species (5500 m), along 500-m altitudinal intervals. A similar sequence from the eastern Pamirs occurs between 3500 and 5000 m; i.e. 522, 106, 12, and 7 species.

The block orogenesis (without vulcanism), especially in Middle and Central Asia (north of the Himalayas), exhibits a pronounced acceleration (Fig. 4). Recent seismic and geophysical measurements indicate an orogenetic uplift of the Pamir of about 5 to 10, in some places 20 mm/year; in the Caucasus, about 7 mm/year. Such movements are comparable to the speed of plant adaptation and the formation of a distinct pattern in high mountain plant species.

The florogenetic models are consistent with the orogenetic processes: Block orogenesis is typical of all mountains in Middle Asia in contrast to the vulcanic mountain regions in the Caucasus or in Kamchatka. Block orogenesis enables the ancestral plant formations to adapt themselves to changing ecological conditions during the uplift (where the whole mountain blocks are lifted tectonically: lift model).

5.4.2 Influences During the Pleistocene

Compared to tectonic uplift, which is normally a slow process that is often in equilibrium with erosion processes, glaciation of high mountains can be rather fast. The Pleistocene glaciations covered some regions almost totally by ice, eliminating most of the ancestral flora. Normally, three types of mountain glaciation can be distinguished:

1. *Shield Glaciation.* The ice masses are thick enough to cover the whole relief. The relief of the ice surface does not reflect the relief of the covered mountain below (today, in the Antarctic or Greenland).

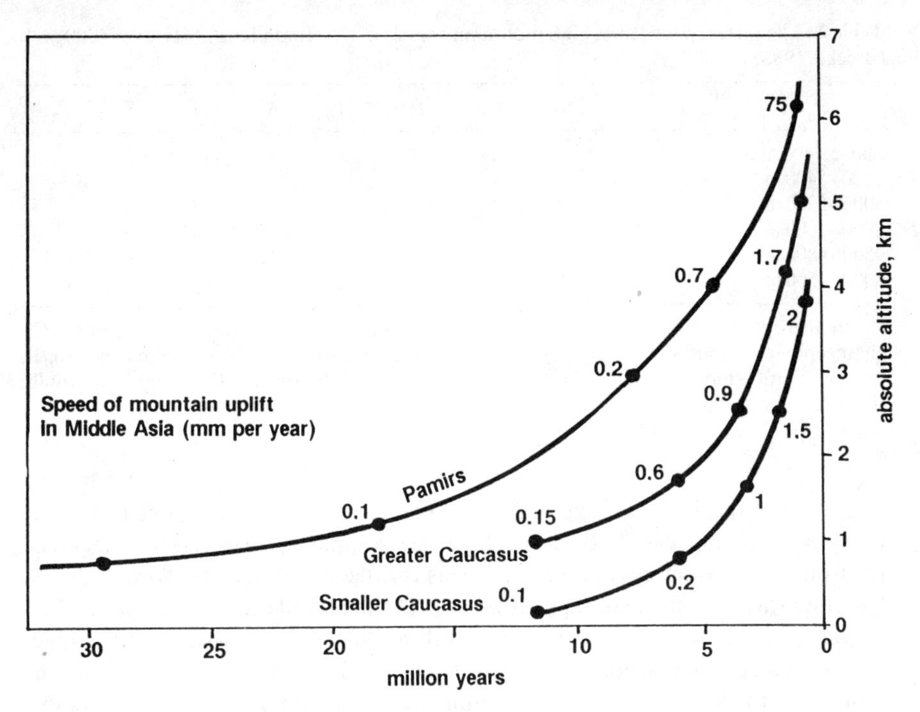

Fig. 4. Uplift speed of mountains in the last 30 million years (mm/year)

2. *Blanket Glaciation.* The ice masses of the continental glaciers cover most of the area. They are, however, not so thick, thus the mountain relief is traced by the ice and the highest peaks are free of ice (nunataks). This was the case in higher plains (today, in Svalbard).
3. *Valley Glaciation.* The glaciers from the upper parts of the mountains are slowly flowing from the nival area to the lower valleys (today, in many higher mountain systems, Alps, Hindu Kush, the Pamirs, Himalayas).

Shield glaciation leads to the almost total extinction of the flora. Nunataks can function as refugia for single species, but normally not for large-scale vegetation. Valley glaciation, however, is ideal for further evolution of plant species since the ancestral flora remains partly intact. In the highlands of the eastern Pamir, the maximum glaciation was only a blanket glaciation.

After the retreat of the glaciers during the late Glacial and early Postglacial times, a new allochthonous flora with low percentages in endemism developed (in E Pamir: 4.6% endemics). In the Middle Asian mountains, only valley glaciation during the Pleistocene took place, thus enabling many species to withstand and to develop a higher degree of endemism, mainly on relict stands.

During the glaciation periods,the migration of species was primarily vertical, so that vegetational belts migrated rather distinctly up and down or were partly eliminated. Refugial stands and other isolated areas were "opened", and

Table 2. Life forms spectrum of high mountain species in the Hindu Kush Mts. over 4000 a.s.l. (Breckle 1988)

	P	NP	Ch	H	B	A	G	S	Y
above 5400 m a.s.l	–	–	10	70	10	–	10	–	–
5200–5400 m	–	–	16	74	5	–	5	–	–
5000–5200 m	–	–	12	71	8	6	4	–	–
4800–5000 m	–	–	12	72	6	5	4	–	–
4500–4800 m	–	1,6	11,7	69,2	4,0	6,5	6,5	–	0,4
4000–4500 m	0,1	2,8	17,1	61,2	2,3	9,1	7,2	0,1	0,1

P, Phanerophytes (trees, larger woody plants); NP nano-phanerophytes (shrubs); Ch chamaephytes (dwarf shrubs, thorny cushions); H, hemikcryptophytes (perennial herbs); B, biennial plants (mostly rosette plants); A, annual species (therophytes); G, geophytes (bulb, rhizome, tuber geophytes); S, parasites (incl. semiparasites); Y, water plants

migration of species was enhanced. However, the upper vegetation belts became impoverished in species. In the foothills, a very complex flora developed, originating from various sources and regions. This is called the "*syncretic flora.*" Almost free migration occurred along the foothills of the mountains.

During the interglacial periods (at least three different very long periods, longer than the glaciation periods), the syncretic flora had a chance to migrate from the foothills to the lower parts of the mountains, as the glacier retreated and the snow limit rose. Further movement of the flora to higher mountain areas was possible by adaptation and by a temperature regime (mainly the summer temperatures), which was not too different from that of the original region. This mechanism explains the similarity of high mountain floras in upper altitudinal belts of distant mountain systems (e.g. Middle Asian and Caucasian mountains, as well as the alpine and nival flora of the Hindu Kush and some arctic circum polar plant species; Breckle 1973, 1983; Breckle and Frey 1974).

After another glacial period, the next autochthonous cycle of floristic evolution started, and again some endemics developed. Today, the recent flora of the mountains consists mainly of widely distributed and widely adapted plant species. The natural selection of plant species in the high mountains was very specific. In the highest vegetation belts, hemicryptophytes are the dominant life form with >70% of the species in the Hindu Kush (Table 2) and in other mountain areas.

5.4.3 Endemism

The degree of isolation of the various altitudinal belts differs among mountain systems, explaining the different degrees of endemism of the respective mountain flora. In some mountains (especially the smaller, less massive mountains), the upper belts are more isolated and, therefore, the degree of endemism is rather high (Model I). In larger, higher and more extensive massive mountain systems, the degree of endemism is lower (Model II).

Table 3. Percentage of endemic species within the total flora of the respective elevational belt of various mountain systems

Mountain	Percentage of endemics		Reference
	Subalpine	Alpine/nival	
Hindu Kush	20	10	Breckle (1974)
Hindu Kush	> 22	< 2	Breckle (1988)
West-Tien Shan	22	17	Pavlov (1980)
Central Caucasus	31,5	18,5	Galuschko (1976)
Kopet Dag	18,0	7,7	Kamelin (1973)
East Pamirs	9,8	4,6	Agachanjanz (1981)
West Pamirs	3,4	1,7	Agachanjanz (1981)
Dzungar. Alatau	1,4	0,7	Goloskokov (1984)

During interglacial periods, extensive migrations of species (species from the alpine and subnival belts) along the mountain ridges were possible. Lower in the subalpine belt, barriers to migration of plant species were more severe, so the ecological isolation is more rigorous. Thus, the percentage of endemics in the lower subalpine belt is distinctly higher than in the alpine belt (Table 3). This phenomenon also explains why the subalpine belt is regionally much more different than the alpine.

5.4.4 Geographic Distribution of Species

It is important to know the geobotanical connections of the various areas and mountain belts in order to verify the florogenetic models. In the Hindu Kush mountains, the Central Asian and the Himalayan floristic elements dominate (Table 4). The endemics, however, are missing in the highest belts. Only in the lower alpine belts are there large percentages of endemics. This again is a strong indication of an open alpine belt and a more closed subalpine belt with respect to plant migration.

Table 4. Percentage of the various phytogeographic groups within the flora of the upper vegetation belts in the Hindu Kush Mts. (Breckle 1988)

	GL	NH	ZA + HY	ZA	HY	END
over 5400 m a.s.l	–	10	40	20	30	–
5200–5400	5	15	45	15	20	–
5000–5200	6	16	38	14	24	2
4800–5000	4	14	36	14	28	4
4500–4800	3	10	32	14	28	13
4000–4500	2	10	28	11	27	22

GL, cosmopolitan mountain species; NH Northern Hemisphere mountains and arctic region; ZA, Central and Middle Asian mountains; HY, Himalayas; END, endemics of the Hindu Kush

The typical geographic belts in some more isolated mountains of Afghanistan were mentioned by Breckle (1975). In the Safed Koh Mountains in eastern Afghanistan as well as in the Nuristan ridges, the lower altitudinal belts are dominated by Himalayan vegetation types and plant species, the middle altitudinal belts by Iranian species and the alpine belt by widespread plant species of the Northern Hemisphere.

A similar "geobotanical layering" is known from the Himalayas. In the western Himalayas (Meusel 1972; Meusel and Schubert 1971; Schickhoff 1993), many different phytogeographic elements are present. Some are related to tropical-humid or tropical-semiarid, others to extratropical-humid (monsoonal regions) or to extratropical-semiarid areas. The latter group includes the pseudo-Mediterranean and the Irano-Turanian elements. This phytogeographic pattern is typical for an intermediate region. Though older "Tethys" elements, like *Nerium, Punica, Cercis, Cotinus coggygria,* are present in the western Himalayas, other species from younger evolutionary centers of the mediteranean vegetation as well as from Turkestan are lacking (Freitag 1971, 1982).

The development of endemism in the various altitudinal belts depends on the character of the forest belts. Dense forest belts, as in the climatically wetter mountains (e.g. Alps, Balkan, northern Caucasus, Siberian Mountains etc.), effectively separate alpine and lowland vegetation. The lowland vegetation may contain species which are also able to live in the upper belt. In this case, a high percentage of endemics is observed (2.7–14.5%) in the montane (forest) belt, a very low percentage, however, in the lower belts (0.1–2.0%).

If the forest belt is restricted only to slopes with favourable ecological conditions (in semiarid regions e.g. Armenia, Daghestan, in parts of the Middle Asian mountains, in the southern Karakorum etc.), the percentage of endemics in this open, montane forest belt is higher (1.1–15.0%), but very low in the foothills (0–0.2%)or in the alpine belt (0–0.7%).

If forests are absent (in arid regions, e.g. in most parts of the Pamirs, in northern Karakorum, Tsinchay-Tibet etc.), then the maximum of endemics is found in the subalpine belt (5–7%) and is lower in the alpine belt (2–4%), as is the case with the flora of the high Pamir. The endemism is low and, thus, the flora must be regarded mainly as allochthonous (Agakhanjanz 1992).

The schematic model of florogenesis must take into account all relevant ecological factors, especially the temperature changes in the Pleistocene (with a temperature minimum before the maximal glaciation), the metachronic character of the glaciations (phase shift of the glaciations after climatic changes), and the continuing orogenetic uplifting of the mountains, caused by ice isostasy. Other factors are: the river system, along which a rapid migration of species can occur; the system and direction of mountain ridges; the formation of soils according to parent material; the history and fate of the forest belt (as the main barrier to vertical species migrations) and the fate of stenochorous species (i.e. species having a narrow or small distributional area) (Fig. 5).

As a result of florogenesis in the mountains, species can show a very broad altitudinal amplitude (Fig. 6; Tables 5–7). The model of florogenesis attempts to

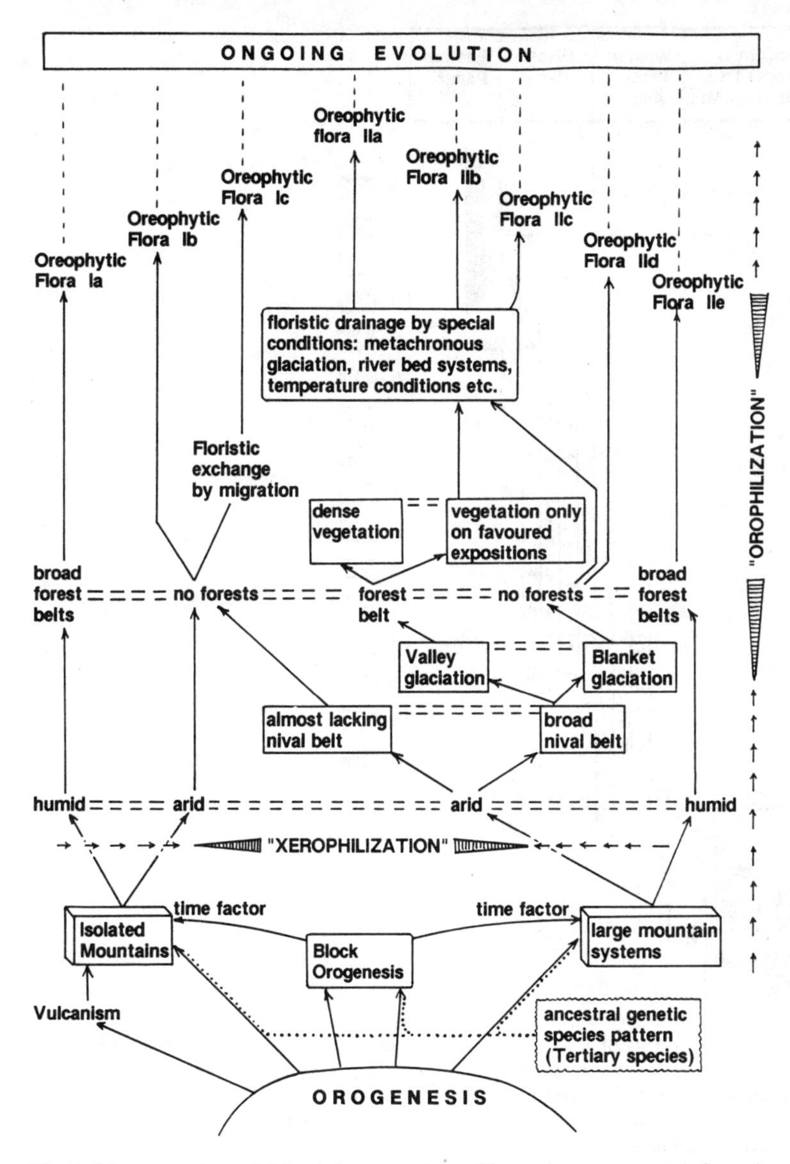

Fig. 5. Florogenesis model for Asian mountains. The various processes influencing adaptation of the vegetation to high mountain conditions (orophilization) are shown schematically: aridity (causing xerophilization); formation of nival belt; glaciation type, forest belt formation and other factors. Thus, various types of oreophytic flora evolve, belonging to types I or II

explain the formation of floras in different altitudinal belts, the floral structure, the floristic pattern in high mountain belts of remote mountains, the role of migration processes, or even the migration trajectories of single species or floristic elements (see Fig. 7, as an example of how evolutionary processes were affected in the course of long-distance migrations).

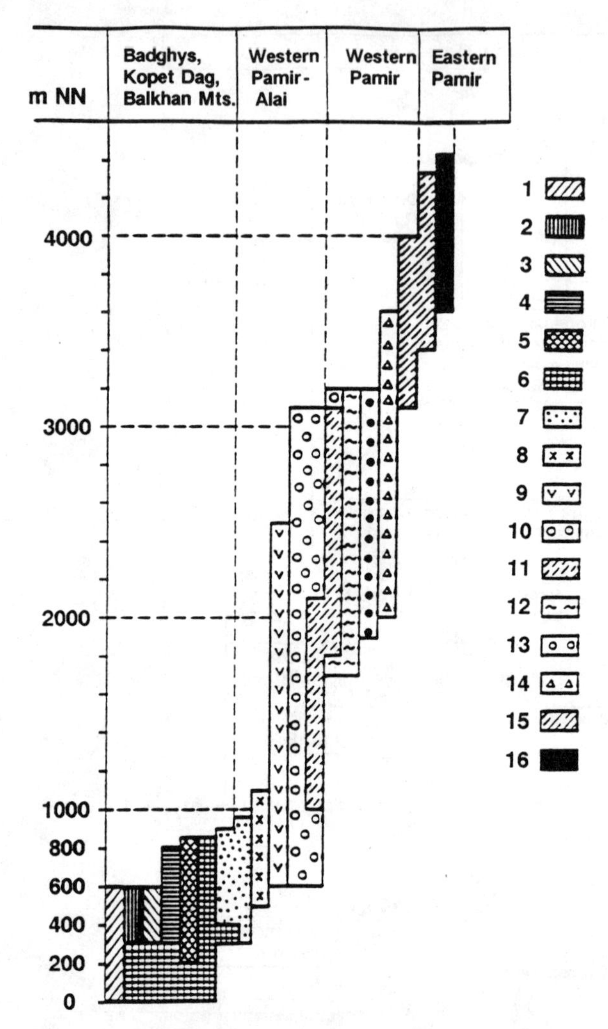

Fig. 6. The height vicariants of *Artemisia* species (subgen. *Seriphidium*) in different mountain regions of Middle Asia (Agakhanjanz 1981). **1** *Artemisia ciniformis*; **2** *A. kopetdagensis*; **3** *A. dumosa*; **4** *A. herba-alba*; **5** *A. badghysi*; **6** *A. maritima*; **7** *A. turanica*; **8** *A. cina*; **9** *A. tenuisecta*; **10** *A. tianschanica*; **11** *A. ferganensis*; **12** *A. vachanica*; **13** *A. knorringiana*; **14** *A. korshinskyi*; **15** *A. lehmanniana*; **16** *A. rhodantha*

5.5 The History of Vegetation

During the late Neogene (= Pliocene, about 6 million years age) and the Pleistocene, there were critical events for vegetation:

I. During the late Pliocene today's high mountain regions of the Himalayas, the Hindu Kush, the Pamirs and the Alai mountains, and the Tien Shan were hardly 3 km high. Therefore, the monsoon affected wide areas further north. The

Table 5. The altitudinal amplitude of a few plants species of meadowlands and river valleys in the Pamir-Alai (m a.s.l.)

Carex diluta	400	–	3600
Carex soongorica	400	–	4000
Calamagrostis pseudophragmites	600	–	4100
Blysmus compressus	1000	–	4100
Kobresia stenocarpa	2200	–	4200
Carex melanantha	2700	–	4600
Kobresia pamiroalaica	2900	–	4500
Calamagrostis tianschanica	3200	–	4500

Table 6. The altitudinal amplitude of a few plants species in more arid sites in the Pamir-Alai (m a.s.l.); in parentheses: the entire amplitude; the numbers in the middle: range of dominant occurrence

Salsola arbuscula		400	–	2500	(2900)
Carex pachystylis		400	–	2500	(3400)
Halogeton glomeratus		400	–	3600	(4100)
Kochia prostrata		2000	–	4300	
Ceratoides papposa	(400)	1000	–	4700	
Carex stenophylloides	(500)	1300	–	4400	(4700)
Piptatherum vicarium	(800)	1000	–	2600	(2900)
Stipa szowitsiana	(800)	1100	–	2700	(3100)
Stipa caucasica	(1100)	1500	–	4000	(4400)
Carex stenocarpa	(1400)	1900	–	4600	
Piptatherum laterale	(1800)	2200	–	4000	(4700)

Table 7. Examples of a few species ofthe Afghan Hindu Kush and the Afghan Wakhan area with a particularly large or small altitudinal amplitude

Species	Height amplitude	(m a.s.l.)
Stipa szowitsiana (=S. barbata)	T	500–3300
Poa bulbosa	T	500–3800
Salix pycnostachya	A	1500–3800
Cerastium cerastioides	T/A	2000–5100
Psathyrostachys caduca	T	2200–3900
Androsace villosa	T	2900–5200
Primula macrophylla	T/A	3400–5500
Phaeonychium surculosum	T	3700–5100
Ermania flabellata	T	3800–5200
Holosteum kobresioides	A	4500–4700

T, arid sites; A, meadowlands and humid sites

precipitation during the cooler season (the west wind depressions) caused winter rains. Today, this special climatic regime is restricted to a small area in eastern Afghanistan and western Pakistan.

Consequently, the Himalayan forests ranged to the Pamir and the Tien Shan in the north (some sources even say to southern Kazakhstan, partly also to

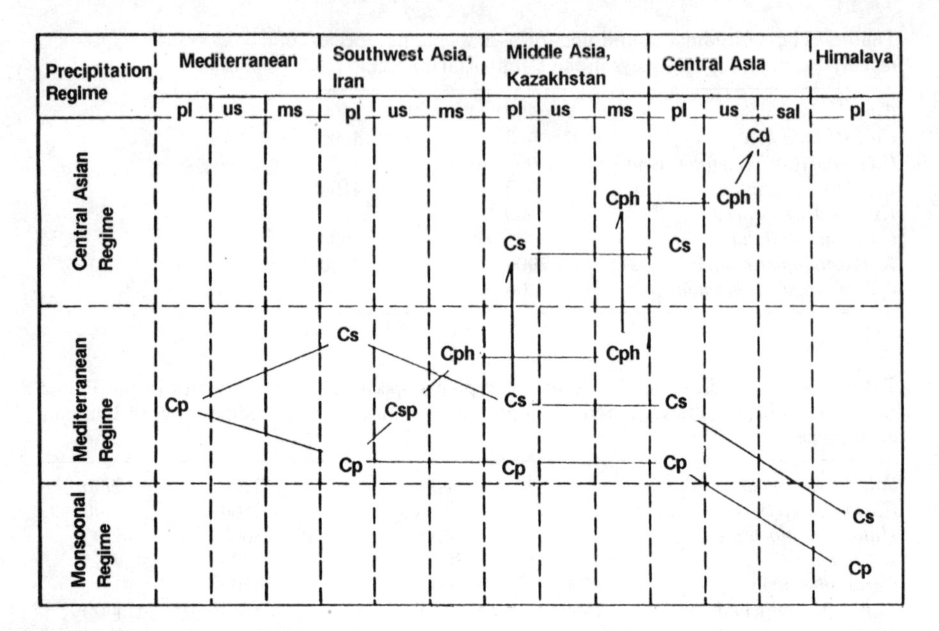

Fig. 7. The different vicariants of sect. *Boernera* V. Krecz. ex Egor. of the genus *Carex* along the edaphic gradients and development of species from the Mediterranean areas to the east (Agakhanjanz 1981). **Cd** *Carex duriuscula*; **Cp** *C. pachystylis*; **Cph** *C. physodes*; **Cs** *C. stenophylloides*; **Csp** *C. subphysodes*
pl *zonal soils* (*plakor sites*); **ms** mobile sands, dunes; **us** immobile sands; **sal** saline soils

Tsinchay-Tibet). The borders between the Himalayan and the Mediterranean forests were open (Meusel and Schubert 1971; Meusel 1972).

Northwards, with decreasing effects of the monsoon, a zone developed called the palaeoprairie, or palaeosavannah (Kamelin 1973; Agachanjanz 1981; Pakhomov 1982, 1991), and further south, semi-deserts and deserts developed with halophytes (Tethys area) on the plains of Iran and southern Afghanistan.

At the end of the Pliocene the mountains reached a height of more than3 km. In the mountains of the Pamirs and in the related mountains, a restricted glaciation of the mountains took place (Pakhomov 1982).

II. During the late Pleistocene about 20000 B.P., the mountains reached approximately their present height. The influence of the monsoonal climate to the north stopped and only the southern slopes of the Hindu Kush (Breckle 1971, 1973) and the Himalayas were directly influenced by summer monsoon. The overall conditions, therefore, were already similar to those today (Fig. 2). This is also indicated by the critical investigation of the pollen spectra from Lake Zeribar (Iran by Freitag 1977; van Zeist 1967). Some authors believe that this time exhibited the temperature minimum of the Pleistocene (Velichko 1982, 1993; Velichko et al. 1982; Pakhomov 1982). However, this would mean that the maximal glaciation does not coincide with the thermal minimum (upper Pleistocene). The maximal glaciation coincides with the maximum of precipitations, which occurred during the middle Pleistocene, when mountain glaciation seemed to have been at a maximum.

The syncretic flora of the foothills became mixed with the floristic elements of the plains, boreal species migrated far to the south and steppe elements became members of periglacial tundra vegetation. These far-reaching changes also influenced the arcto-montane and arcto-alpine species (Table 4: NH). Their evolution in many high mountains of Eurasia was favoured by the thermal uniformity of the very large periglacial areas. Their expansion to the various mountains is explained by florogenesis Model II. Forests had already retreated during the lower Pleistocene. In the course of the Pleistocene, the flora of Middle Asia lost most representatives of the genera *Quercus, Fagus, Tilia, Corylus, Pinus, Syringa* etc. due to extinction. This natural process of deforestation peaked during the dry and extremely cold upper Pleistocene phase (Cryoxeric phase), when pine trees completely disappeared (Agachanjanz 1980).

The depression of the altitudinal belts in the mountains was extreme during the upper Pleistocene. For example, in the Alai range, very low temperatures during the Pleistocene caused each altitudinal belt to be much lower than at present. The open *Betula* and *Juniperus* woods disappeared from the middle altitudes and remained on the about 2-km-lower foothills. This process, combined with the extensive expansion of boreal and periglacial floristic elements to the south, may explain the existence today of some coniferous woods (probably of different origin) in the Ustjürt highlands and in Iran, as well as the periglacial steppe relicts in the near Caucasus and in the Ukraine.

If we compare the changes in vegetation cover from the Pliocene, Pleistocene and today, it becomes obvious that in the last 5–6 million years, there were tremendous changes in vegetation zones in Asia (Frenzel 1967, 1987, 1992; Kahlke 1981).

During the Pleistocene, the evolution of species in the mountains was more strongly enhanced. The depression of the altitudinal belts during glacial times and the sometimes rapid recovery of the original ecological conditions during the interglacial periods accounted for a distinct "floristic drainage" of the respective mountains. Another process was the immigration of the boreal forests (with varying floristic compositions), which led to a strong repression or extinction of other vegetation units. During the interglacial periods, the new altitudinal belts were restored, and new vegetation units were formed: the mountain xerophytes, the open *Juniperus* woods, the mountains steppes and mountain deserts, the giant herb meadows with large umbellifers, and the alpine vegetation units.

At the end of the Pleistocene (about 18 000 years ago), the monsoon was totally blocked and restricted to the Indus and Ganges plains. The Mediterranean influence was reduced and the precipitation regime became irregular with a slight summer maximum. The Holocene, (the last 10 000 to 12 000 years) resembles the beginning of a typical interglacial period.

5.6 Conclusions

1. Orogenesis during the Cenozoic played a major role in the evolution of alpine species and the alpine flora. This orogenesis caused changes in the relief and

precipitation regime of different mountain areas. At the same time, aridity increased in the plains in Middle Asia and in parts of the developing mountains, with a decrease in forest species of the mountains, thus leading to a different system of altitudinal belts. Without considering orogenesis, it is impossible to understand the floristics of the mountains and the adjacent plains.

2. The migration of flora and vegetation during the glacial periods was a complex three-dimensional process that played a major role in the florogenesis of mountains. The ancestral flora was forced into isolation in the mountains. Long-distance migrations occurred, but were restricted to regions with similar precipitation regimes. The recent distribution of the precipitation regimes (Fig. 2) limits the migration of many species to the eastern border of the Mediterranean regime. The mountain junction in the Pamirs is an important factor for floristic migrations. The Pamir-Knot, with the eastern Pamir, the Tien Shan and the Karakorum, is linked with a northern mountain system to the Siberian mountains, which develop a new orogenesis, and with a southern-mountain system to the Himalayas continuing to southern China.
3. The ecological adaptability of plant species depends on their floristic history. Species formation and the speed of this process are determined by the adaptability of the ancestral species. Ecologically conservative taxa became extinct or could only survive in small relict stands.
4. The time scale of a glaciation period to interglacial periods has a ratio of 1:3. The speed of species formation in isolated altitudinal belts is one of the key factors explaining the diverse floras of the various mountain areas today. Isolation was strong enough for the evolution of endemics only in some areas. In most regions, however, species migration was prominent and strongly governed by the glaciation regime.

The specific florogenetic processes explain the very diverse floristic situation in the Asian mountains observed today.

Acknowledgments. Financial assistance from the German Research Council (DFG) for the guest professorship for O. Agakhanjanz at the Dept. of Ecology, and the support of Prof. Grotemeyer, are gratefully acknowledged. We thank especially Rolf and Margit Breckle who translated parts of the German manuscript, Mrs. Lydia Gebel, Irmingard Meier, and all the other helpful members of the Dept. of Ecology.

References

Agakhanjanz OE (1980) Die geographischen Ursachen für die Lückenhaftigkeit der Flora in den Gebirgen Mittelasiens. Petermanns Geogr Mitt H1:47–52

Agakhanjanz OE (1981) The arid mountains of the USSR (nature and geographical models of florogenesis). Mysl. Moskau. 272 pp (in Russian)

Agakhanjanz OE (1992) Vegetation belts at different stages of growth and degradation of mountain glaciations. Data of glaciological studies. 73, Mysl. Moscow, pp 18–23 (in Russian)

Breckle S-W (1971) Vegetation in alpine regions of Afghanistan. In: Davis P H, Harper P C, Hedge I C (eds) Plant life of south-west Asia. Proc Symp 1970, Edinburgh, pp 107–116
Breckle S-W (1973) Mikroklimatische Messungen und ökologische Beobachtungen in der alpinen Stufe des afghanischen Hindukusch. Bot Jahrb Syst 93:25–55
Breckle S-W (1974) Notes on alpine and nival flora of the Hindu Kush, East Afghanistan. Bot Not (Lund) 127:278–284
Breckle S-W (1975) Ökologische Beobachtungen oberhalb der Waldgrenze des Safed Koh (Ost-Afghanistan). Vegetatio (Acta Geobot) 30:89–97
Breckle S-W (1983) Temperate deserts and semideserts of Afghanistan and Iran. In: West NE (ed) Temperate deserts and semideserts. In: Goodall DW (ed) Ecosystems of the world, vol 5. Elsevier, Amsterdam pp 271–319
Breckle S-W (1988) Vegetation und Flora der nivalen Stufe im Hindukusch. In: Grötzbach E, (Hrsg) Neue Beiträge zur Afghanistanforschung. Schriftenreihe der Stiftung Biblotheca Afghanica Bd 6 (Liestal), pp 157–174
Breckle S-W, Agakhanjanz OE (1994) Ökologie der Erde, Band 3. Spezielle Ökologie der Gemäßigten und Arktischen Zonen Eurasiens. UTB Große Reihe, Fischer, Stuttgart, 720 pp, 2te Aufl
Breckle S-W, Frey W (1974) Die Vegetationsstufen im Zentralen Hindukusch. Afghanistan J (Graz) 1:75–80
Freitag H (1971) Die natürliche Vegetation Afghanistans. Beiträge zur Flora und Vegetation Afghanistans I. Vegetatio 22:285–344
Freitag H (1977) The pleniglacial, late-glacial and early postglacial vegetations of zeribar and their present-day counterparts. Palaeohistory 19:87–95
Freitag H (1982) Mediterranean characters of the vegetation in the Hindukush Mts., and the relationship between Sclerophyllous and Lauriphyllous forests. Ecol Mediterr 8:381–388
Frenzel B (1967) Die Klimaschwankungen des Eiszeitalters. Vieweg/Braunschweig, 141 pp
Frenzel B (1987) The history of flora and vegetation during the Quaternary. Prog Bot 49:354–380
Frenzel B (ed) (1992) Atlas of palaeoclimates and palaeoenvironments of the Northern Hemisphere. Fischer Stuttgart, 153 pp
Galuschko AI (1976) Analysis of the flora of the western parts of the central Caucasus. In: Flora of northern Caucasus and its history. Pedagog Univ Stavropol, pp 5—130 (in Russian)
Goloskokov VP, (1984) Flora of the Dsungarian Alatau. Ilm-Publ, Alma-Ata, 272 pp (in Russian)
Kahlke HD (1981) Das Eiszeitalter. Urania, Leipzig, 192 pp
Kamelin RK, (1973) Florogenetic analysis of the natural flora of middle Asian mountains. Nauka, Leningrad, 242 pp (in Russian)
Larcher W (1980) Klimastress im Gebirge – Adaptationstraining und Selektionsfilter für Pflanzen. Rheinisch-Westfael Akad Wiss Vortr 291:49–88
Larcher W (1981) Resistenzphysiologische Grundlagen der evolutiven Kälteakklimatisation von Sproßpflanzen. Plant Syst Evol 137:145–180
Larcher W, Bauer H (1981) Ecological significance of resistance to low temperature. In: Lange OL, Nobel PS, Osmond CB, Ziegler H (eds) Encyclopedia of Plant Physiology, New Series, vol 12A: Springer; Berlin Heidelberg New York, pp 403–437
Meusel H (1972) Semiarid elements in the flora and vegetation of western Himalayas. In: Rodin L (ed) Ecophysiological foundation of ecosystems productivity in arid zone, Nauka, Moscow, 48 pp (226–331)
Meusel H, Schubert R (1971) Beiträge zur Pflanzengeographie des Westhimalaya. 1. Die Arealtypen. Flora 160:137–194; 2. Die Waldgesellschaften. Flora 160:373–432; 3. Die pflanzengeographische Stellung und Gliederung des Himalaya. Flora 160:573–606
Pakhomov MM (1982) Paleogeography of the mountains in the eastern part of middle Asia during late cenozoicum and open questions of florogenesis. Additional report on the thesis. Nauka, Moscow, 48 pp (in Russian)
Pakhomov MM (1991) The correlation of pleistocene processes in middle Asia and the dynamics of the mountainous vegetation belts. Izvest Akad Nauk Ser Geogr N6, pp 94–103 (in Russian)

Pavlov VN (1980) The vegetation of the western Tien Shan. Moscow Univ, 248 pp (in Russian)
Sakai A; Larcher W (1987) Frost survival of plants – responses and adaptation to freezing stress. Ecological Studies 62: Springer, Berlin Heidelberg New York, 321 pp
Schickhoff U (1993) Das Kaghan-Tal im Westhimalaya (Pakistan). Bonn Geogr Abh 87:268
Serebrijanny LR, Pschenin GN, Khalmuchammedova RA (1976) Evolution of arid landscapes of Fergana during the holocene. In: The history of the biogeocenoses of the USSR during the Holocene. Nauka, Moskau, pp 221–229 (in Russian)
Van Zeist W (1967) Late quaternary vegetation history of western Iran. Rev Palaeobot Palynol 2, N1–4: 301–311
Velichko A (ed) (1982) Paleogeography of Europe during the last 10 000 years. Atlas-Monography, Nauka, Moskau, 156 pp + 14 maps (in Russian)
Velichko A (ed) (1993) Evolution of landscapes and climates of the northern Eurasia. Late pleistocene to holocene: elements of prognosis, vol 1. Regional palaeogeography. Nauka, Moscow, 102 pp, 8 maps (in Russian)
Velichko A, Spasskaya J, Khotinsky N (1982) Evolution of the environment at the USSR territory during Late Pleistocene and Holocene. Nauka, Moskau, 271 pp (in Russian)
Walter H, Breckle S-W (1986) Ökologie der Erde. Band 3, Spezielle Ökologie der Gemäßigten und Arktischen Zonen Eurasiens. UTB Große Reihe, Fischer, Stuttgart 587 pp (2 Aufl 1994: in press)

6 Diversity of the Arctic Terrestrial Fauna

YU.I. CHERNOV

6.1 Number of Species in the Arctic Fauna

It is presently impossible to say exactly how many animal species dwell in the Arctic. The reasons for this are (1) the lack of taxonomic and faunistic knowledge about many groups of arctic animals, especially the Protozoa, Nematoda, terrestrial Oligochaeta, a number of Acari taxa, and insect families of the orders Diptera, Hymenoptera, and Lepidoptera.; (2) incomplete collections and data on regional faunas; (3) a large number of incompatible synonyms used by taxonomists in Russia, North America, and Western Europe.

Exact numbers of species for the Arctic are available only for certain well-studied taxa, such as mammals, birds, and beetles. In these groups, the percentage of species dwelling in the Arctic compared to global numbers illustrate the patterns of species richness of the arctic fauna.

For the Arctic s. 1. (i.e. including forest tundra) Schwarts (1963) recorded 61 terrestrial mammal species. Including arctic marine species, there is a total of about 75 species or 1.8%of the world's mammal fauna, or 12–13%of the number of species found in the temperate zone (Chernov 1978). The avifauna of the tundra zone consists of approximately 200 species (2.3% of all species), but including the forest tundra and the marine polar basin 300 species that nest in the Arctic, including the forest tundra and the marine polar basin (Danilov 1966), i.e. 3.4%of the world's bird species, or 13–15% of the bird fauna of the temperate zone.

Reptiles are generally absent from the Arctic, with only a few species extending into the subarctic (Borkin et al. 1984). Amphibians are also poorly adapted for life in polar environments, but several species are abundant in the forest tundra and the southern tundras. Four *Rana*-frogs and one Uradela species are found in the Arctic, or more precisely, in the hypo-arctic fauna (Schwarts and Ishchenko 1971; Borkin et al. 1984).

The evaluation of species richness of the arctic insect fauna is also problematic. Danks (1981) listed about 1362 insect species (excluding collembolans) in northern America and, in a later paper (1990), the list grew to 1650 species. According to his estimate, 50% of the entire arctic insect fauna consists of Diptera, 780 of which were recorded from North America. We estimate that about 1500 Diptera species and 3000 species of insects inhabit the circumpolar

Institute of Animal Evolutionary Morphology and Ecology, Leninski Prospect, 33, Moscow 117071, Russia

Ecological Studies, Vol. 113
Chapin/Körner (eds) Arctic and Alpine Biodiversity

Arctic, including both the tundra zone and polar desert. This is about 0.3–0.4% of the world's insect fauna, if the global number is assumed to be 800 000 to 1 000 000 species. In the Arctic, most of the insect orders account for only 0.3% of the global species richness of a given taxon. For example, arctic beetles, the largest insect order, total only 0.1% of the total number of species in this order. By contrast, the order Diptera is represented by approximately 1% of the world's species number.

Springtails (Collembola) are an important part of the tundra fauna. Among invertebrates, they show perhaps the greatest potential for adaptation to arctic conditions. According to A. Babenko in my laboratory, 184 species of springtails are known from the northern tundra zone alone. Considering the well-known increase in biodiversity with decreasing latitude, the entire arctic fauna of this group may include 400–500 species or 7–8% of the world fauna of about 6000 species

The taxonomy of arctic Arachnida is rather incomplete. Danks (1981) mentions 112 species of spiders (Aranei) for arctic America, and Es'kov (1985) lists about 100 species for the Eurasian tundra zone. In contrast, Iu. Marusik (pers. comm.) found 170 spider species in the Chukchi Peninsula alone. The world's Aranei fauna includes at least 33 000 species. With a panarctic spider fauna of about 300 species, about 0.9% of all Aranei live in the Arctic. Data on mites (Acarina) are even less definite. According to Danks (1981), there are 349 mite species in arctic America. This number must at least be doubled for the Whole Arctic. In this case, the arctic mite fauna would comprise about 2% of the world's fauna of the order Acarina (at present time, about 30 000 species are known, but the true number is probably much greater).

The class Myriapoda, which is of great importance in forest ecosystems, is almost absent in the tundra and is represented by only a few Chilopoda species. There are not more than ten species in the arctic fauna; most are found only in the very southern parts of the subarctic.

Very common in tundra soils are tardigrades (Tardigrada) that are well adapted to arctic environments. Already in the 1930s about 40 species were recorded in Russian tundras (Bozhko 1936). However, their taxonomy is so vague that so far it is impossible to estimate even approximately the number of species of these very peculiar animals in the Arctic.

There are only a few molluscs (Mollusca) in the terrestrial tundra fauna, and these are slugs (Limacoidea). Together with freshwater forms, no more than one to two dozens species of molluscs are in the tundra zone, depending on the position of its southern boundary (Clarke 1973).

A very important component of the animal community in the Arctic is the class Oligochaeta. Among these the earthworms (Lumbricidae) are represented in the Eurasian tundra zone by no more than five species, including forms with uncertain status. The absence of earthworms in American tundras is still unexplained. Enchytraeidae have been studied in the Arctic since the beginning of this century, but even now their taxonomy is rather vague and no data on exact species number arc available. In the range of habitats at one tundra site, about 20 species were recorded (Piper and McLean 1982), and probably at least 70 species inhabit the tundra zone.

Even less definite information is available about the species richness of another group of soil worms, the Nematoda. Samples from a range of terrestrial communities in one tundra site yielded more than 160 species (Kuz'min 1973), and even in the polar desert about 50 species were found (Chernov et al. 1979). There is poor documentation of the species richness of these tiny animals. Perhaps at least 500 species of free-living (non-parasitic) species of this class inhabit the tundra zone. It is impossible at this stage to estimate the number of Protozoa living in soils and aquatic systems of the tundra region.

In summary, many classes of animals that form an essential part of the fauna of the boreal forest belt are completely or largely reduced in the arctic fauna. Among vertebrates, these include amphibians; among invertebrates, univalve molluscs (Gastropoda) and myriapods. The great majority of the highest taxa of the terrestrial fauna are represented in the Arctic by less than 1% of the world fauna of each group. Classes that demonstrate the highest potential of adaptation to the arctic environments constitute about 1–3% or more of the corresponding group in the world. These are, for example, birds and springtails. Naturally, if we consider the lower taxonomic levels – orders and families – the figures may be rather different. For instance, all species of the order of the diving birds Gaviiformes are in the Arctic.

As a whole, the arctic fauna perhaps constitutes no more (perhaps even less) than 1% of the global fauna. By rough approximation, the terrestrial arctic fauna includes 6000–7000 species.

6.2 Patterns and Causes of Biodiversity Changes in the Arctic

The main feature of arctic biological diversity is not so much its great poverty relative to other biomes, but its progressive decrease within the Arctic (Tables 1,2). For example, the number of breeding bird species decreases from more than 50 in the southern Arctic to less than 10 in northernmost regions, and leaf beetles decline from six to one species (Chernov et al. 1993). In general, when we discuss the level of species richness of any group of animals in the Arctic, we have in mind only its southern part, i.e. the subzones of southern and "typical" (i.e. low arctic) tundras (see Chernov and Matveeva 1979, 1986; Chernov 1985). In the high arctic tundra subzone and particularly in the polar desert, many groups are represented by very few, often by a single species, or are absent from the fauna. Thus, in the southern shrub tundras of Taimyr and Yamal, the concrete fauna (i.e. the gamma diversity of Whittaker 1977) of ground beetles (Carabidae) includes about 60 species, in typical tundras – 15 to 20, in high arctic tundras – 2 to 6 species, while this group is absent in the polar desert. Danks (1981) reported about 651 genera and 1362 species in the American Arctic, but in the "high Arctic" (arctic tundra subzone and polar desert according to Russian authors) there are respectively only 194 and 235, i.e. only one-sixth of all arctic species.

Table 1. Patterns of ecological-taxonomic diversity of orders of free-living insects in arctic landscapes. (After Chernov 1978)

Order	Forest tundra	Southern tundra	Typical tundra	Arctic tundra	Polar desert
Ephemeroptera	++	++	++		
Odonata	+	+			
Orthoptera	++				
Plecoptera	+++	+++	++	++	
Dermaptera	+				
Homoptera	+++	+++	++	+	
Heteroptera	++	++	++	+	
Thysanoptera	++	++	+		
Coleoptera	+++	+++	+++	++	+
Megaloptera	+	+	+		
Neuroptera	+	+			
Trichoptera	+++	+++	++	+	
Lepidoptera	+++	+++	+++	++	+
Hymenoptera	+++	+++	+++	++	++
Diptera	+++	+++	+++	+++	+++

+++, Relatively large number of species and adaptive types with essential ecological role; ++, group is represented by several species with important ecological role, but group is less dominant than at southern latitudes; +, few representatives with insignificant ecological roles.

Table 2. Patterns of species number in the main orders of insects in the American Arctic (After Danks 1990)

Order	Arctic America as a whole	Canadian arctic islands	Islands of the high Arctic
Hemiptera	66	18	6
Coleoptera	203	30	13
Lepidoptera	162	39	23
Hymenoptera	272	106	80
Diptera	787	307	200

The great changes in biodiversity within the Arctic reflect sharp thermal gradients that have no analogue in other biomes. In the Siberian sector of the tundra zone over a distance of 600 km, the mean July temperature decreases from 12 to 2 °C, whereas in the boreal forest belt a comparable 10 °C change in mean July temperature occurs over almost 2000 km, a range across which we find three natural life zones (Chernov 1975, 1985). It is the decreasing energy input within the Arctic that accounts for the strong dependence of biological diversity on temperature. The number of species in concrete faunas of diverse groups of animals is especially correlated with the mean air temperature of the warmest month (usually July, Figs. 1, 2). South of the tundra zone, at less extreme temperatures (mean July temperature >12 °C), the relation between the number of species and July mean temperature is weaker (Fig. 1). For many groups, the correlation coefficient reaches 0.9.

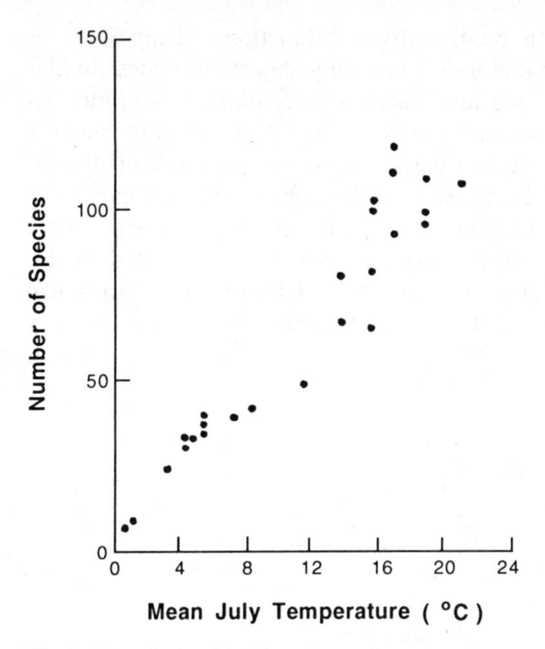

Fig. 1. The relationship between number of nesting bird species and July mean temperature, western and middle Siberia. (After Chernov 1989)

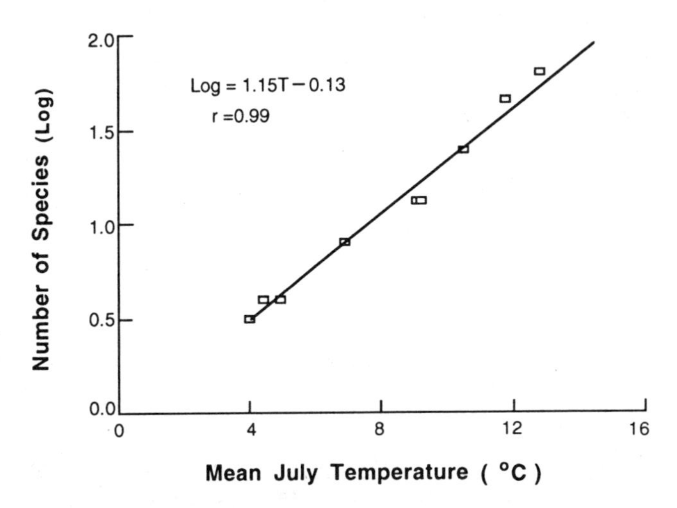

Fig. 2. Correlation between July mean temperature and number of ground beetle species in concrete faunas of Taimyr (original data)

The analysis of trends in arctic biodiversity as a function of temperature is complicated by the fact that zonal limits do not coincide with latitudes. Similar landscapes and types of ecosystems may occur at very different latitudes. At similar July temperatures, tundra areas may be quite different in other climatic parameters, and these differences will co-determine the level of biodiversity. For example, in northeastern Asia and in the western American Arctic, the number of species of many groups of animals at the same mean July temperature is greater than in central Siberia (Figs. 3,4). Several explanations can be suggested for this phenomenon (see Chernov 1989), but undoubtedly both the present environment and the historical development of the fauna determine the present level of biodiversity.

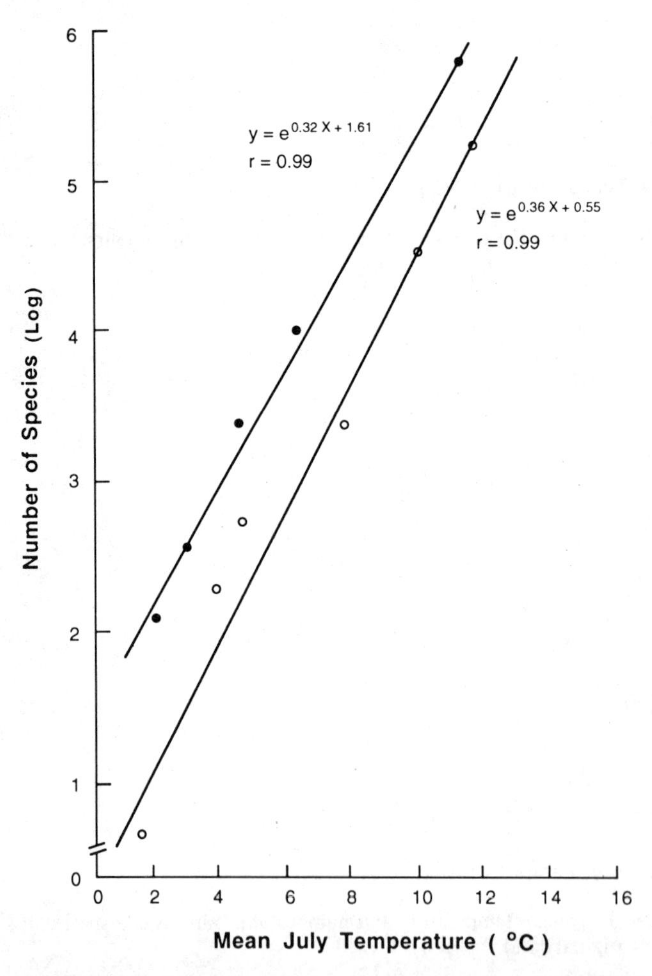

Fig. 3. Correlation between July mean temperature and number of spider species in concrete faunas in middle Siberian (*open circles*) and Beringian (northeastern Asia and western Arctic America) (*filled circles*) sectors of the Arctic. (After Chernov 1989)

It is noteworthy that the thermal dependence of the number of species in concrete faunas is best described by a logistic curve (Fig. 5), suggesting that the "concrete fauna" (gamma diversity) represents a rather rigidly organized system (Chernov 1989). The use of this level of species grouping may represent the most useful way to monitor arctic biodiversity.

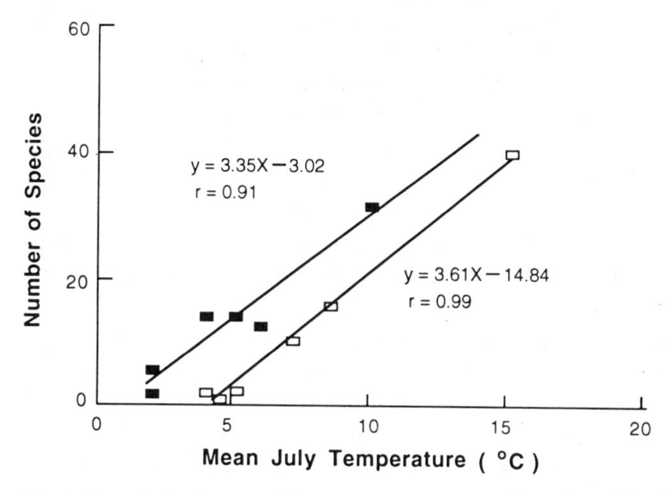

Fig. 4. Correlation between July mean temperature and number of day butterflies (Rhopalocera) in middle Siberian *(open squares)* and Beringian *(filled squares)* sectors of the Arctic. (After data from different literature sources and collections)

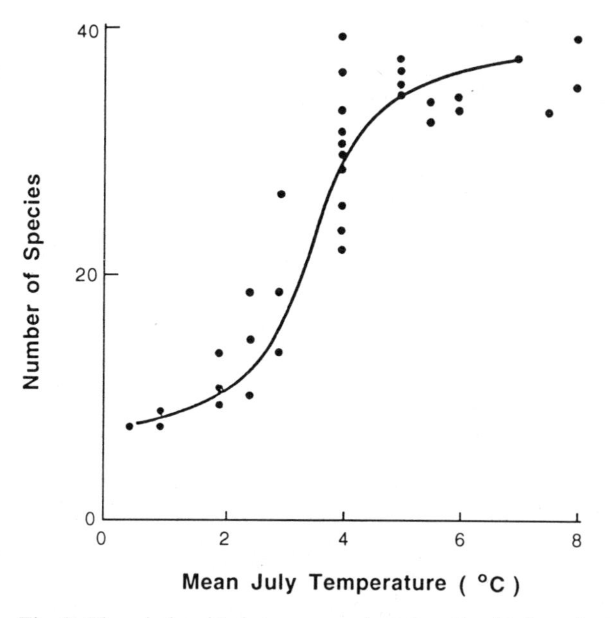

Fig. 5. The relationship between number of nesting bird species in concrete avifaunas and July mean temperature. (After Stishov et al. 1989; data for the whole Arctic are from different sources)

6.3 Taxonomic Composition and Functional Groups

Biological diversity does not decrease latitudinally in the same way in all taxa. Some taxa decrease in importance or even disappear at high latitudes (e.g. Coleoptera; Table 3). In other taxa, species richness changes in proportion to the overall decrease in animal diversity (e.g. Lepidoptera and Hymenoptera). A third group of taxa (e.g. Diptera) maintains a relatively high level of biodiversity, so that their relative contribution to the whole fauna increases at high latitudes. This peculiar feature of the arctic fauna reflects the presence of true arctic species with very specific adaptations and is expressed to differing degrees at different taxonomic levels: classes, orders, and families. These patterns do not exist in other biomes. For instance, the arctic avifauna shows a characteristic structure with a high percentage of plovers and related species (Charadriiformes) and anseriform-like birds (Anseriformes) along with a relatively low proportion of Passeriformes, which is the predominant order in all other zonal faunas. Another typical arctic feature is the low percentage of insect groups with incomplete metamorphosis (Hemimetabola), which are almost absent in the far north of the tundra zone, compared with a high proportion of Diptera. Specific changes in taxa composition were also recorded in leaf beetles (Chrysomelidae) (Chernov et al. 1993), tipulid Diptera (Lantsov and Chernov 1987), and springtails (Anan'eva et al. 1987). The trends for springtails, the most abundant invertebrate group of arctic fauna, are shown in Table 4.

Table 3. Percentage of four dominant orders of insects in the insect faunas of the world and the Arctic. (After Danks 1990)

Order	Whole world	Whole Arctic	High Arctic
Coleoptera	39	13	3
Lepidoptera	15	11	10
Hymenoptera	14	13	10
Diptera	16	50	61

Table 4. Collembolan taxa in natural zones of Eurasia, % of total number of collembolan species. (After Anan'eva et al. 1987)

Taxon	Polar desert	Arctic tundra	Mixed forests
Isotomidae	50	53	23
Hypogastruridae	40	12	15
Onychiuridae	10	12	6
Neanuridae	—	7	9
Lepidocertidae	—	5	10
Entomobryidae	—	5	12
Symphypleona	—	5	20
Others	—	1	5

The specific taxonomic compositions are to a large extent related to particular arctic environments, the structure of ecosystems, and types of biotic interactions. But, again, they cannot be completely explained by present conditions. The reasons are closely related to the global processes of evolution and landscape development (Chernov 1984, 1988).

Many of the peculiarities of arctic biomes result from their marginal position along the global climatic gradient. In fact, many of these peculiarities of biological diversity in the Arctic are manifestations of global trends that are expressed most clearly under conditions of polar climate and landscape. For example, the proportion of herbivorous and carnivorous animals changes from the tropics to poles in a predictable way. The diversity of herbivorous animals is highest in the tropics, where the bulk of specialized forms of this group is concentrated. In the temperate zone, the diversity of herbivorous animals is lower, and their abundance within the overall fauna appears to diminish. In the tundra zone, in all groups that contain both phytophages and carnivorous species, the percentage of carnivorous species is higher than in any other biome (Chernov 1992). This is particularly apparent in birds and Arachnida, as well as in some orders of insects, for instance, Coleoptera and Hemiptera. In birds, the greatest number of specialized phytophagous groups like Columbiformes, Psittaciformes, Bucerotes, Ramphastidae, Trochili, and many plant-eating passerine birds occur in the tropics. There are still several groups of typical phytophagous birds like pigeons, game birds (Galliformes), and anseriform birds (Anseriformes) in the boreal belt. In the Arctic, the only typically herbivorous birds are grouse (Tetraonidae) and geese (Anserinae), which have a rather diverse ecology, while carnivorous groups are represented by loons (Gaviiformes), diving ducks (Aythinae), seagulls (Lari), snipes (Limicolae), razorbills (Alcae), and some families of passerine birds. Also rather characteristic are some owl species (Strigiformes) and diurnal birds of prey (Falconiformes). Presently, it is difficult to provide evolutionary phylogenetic reasons for these correlations and their ecological consequences, but some possibilities were discussed by Chernov (1992).

6.4 Biotagenesis

The structural diversity and total species number of most advanced (apomorphic) taxa reach a maximum in the tropics and decrease toward the poles, whereas the proportion of groups with a relatively low phylogenetic level increases toward the poles. This phenomenon is discussed in detail in the Russian literature as part of the concept of global biotagenesis (Chernov 1975, 1978, 1984, 1988; Zherikhin 1978; Meyen 1986, 1987; Kafanov 1987). The relatively primitive, plesiomorphic or secondarily simplified groups with otherwise wide adaptive radiation and high ecological importance have their largest proportion in arctic biota. That is clearly seen in all the major taxa of both arctic fauna and flora (Chernov 1978, 1984, 1988; Chernov and Matveeva 1983). Birds are the best example (Fig. 6). The most apomorphous order Passeriformes is predominantly found in the south, and

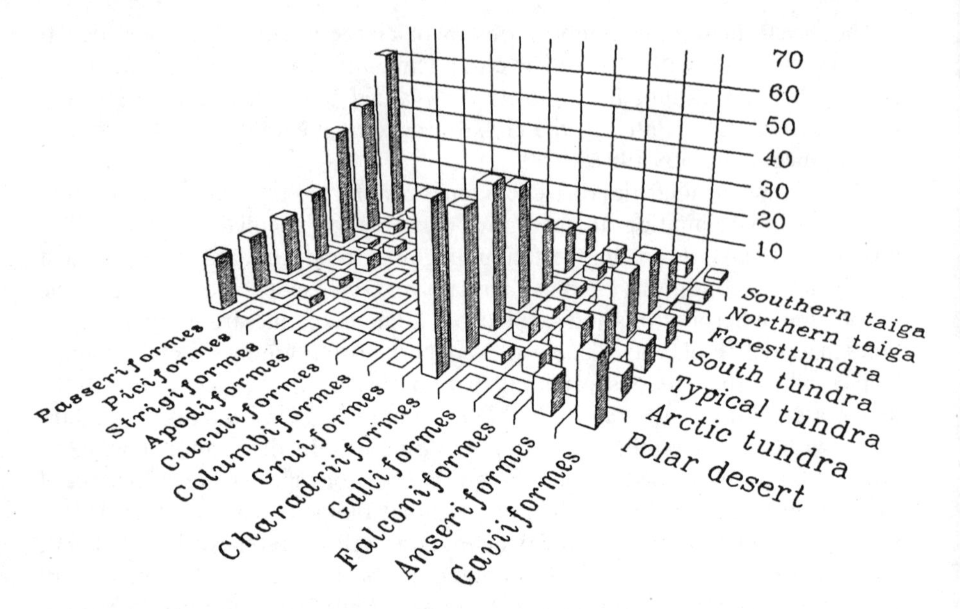

Fig. 6. Changes in proportions of number of species in different bird orders in concrete avifaunas south from the Taimyr along the Yenisei River

Charadriiformes, a group having a significantly lower level in the phylogenetic hierarchy, predominates in the north. Another good example is the pronounced polar abundance of Gaviiformes, the most primitive bird order. We find an analogous picture within Diptera (Fig. 7). The infraorder Tipulomorpha, which has the most plesiomorphic features, increases significantly in its importance at higher latitudes in general, particularly in the high Arctic.

All these examples demonstrate that the peculiar structure of biological diversity in the Arctic can only be understood if one takes into consideration both the global processes of evolution and current dynamics and differentiation.

6.5 Species Structure of Arctic Communities

The decrease in the number of species in communities is often compensated for by an increase in population density. The best arctic example of the inverse relationship between diversity and density is the lemming, where one species has an unprecedented large influence on the plant cover of the tundra zone. Similar patterns can also be observed above the species level. For example, the taxonomic diversity of soil microarthropods as a whole is greatly reduced in the Arctic, and several groups are absent, but springtails reach their highest density in the arctic tundra subzone and particularly in the polar desert. (Anan'eva et al. 1987).

Another interesting phenomenon in the arctic fauna and flora is the even distribution of certain species among diverse communities in the landscape. The

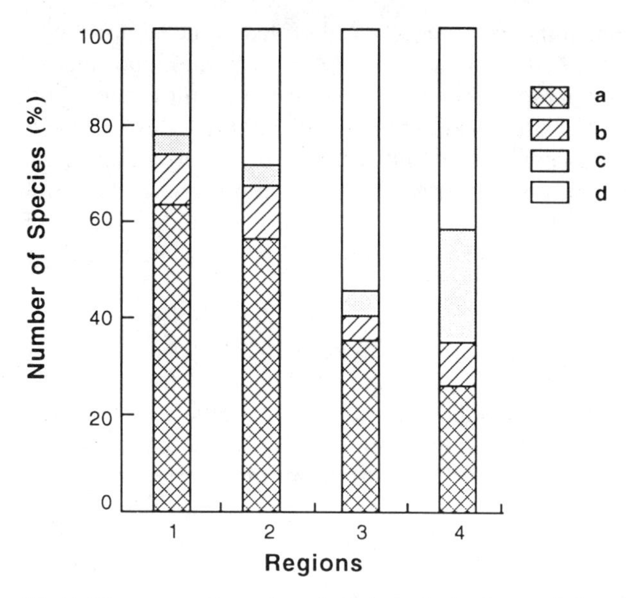

Fig. 7. The proportions of number of species in *Diptera* infraorders (after Danks 1981; with additions). *1* Polar desert; *2* High Arctic; *3* all Arctic; *4* world fauna. *a Tipulomorpha*; *b Bibionomorpha*; *c Asilomorpha*, *d Myiomorpha*

same set of species may occur in different habitats and, by changing their numbers and abundance, they form different combinations and community types (Matveeva and Chernov 1977; Chernov 1985). For example, there are about ten springtail species in the Taimyr polar desert on Cape Chelyuskin, and most of these occur in each community type or even in small fragments of a community (Chernov et al. 1979).

One consequence of low taxonomic diversity is superdominance, or the dominance by a few species in many different communities. This phenomenon is widespread on small oceanic islands (Chernov 1982) as well as in the Arctic (Chernov 1985). In the tundra zone, the most striking examples of superdominance are the cranefly *Tipula carinifrons* and the lemming *Lemmus sibiricus*. Thus, under extreme conditions with low species diversity, the width of the ecological niche of the remaining species is wider. Some aspects of this problem have been recently studied in birds (Chernov and Khlebosolov 1989; Ryabitsev 1993). Unfortunately, the mechanisms of segregation of ecological niches have received little attention in the Arctic, but this is a promising field of future polar ecology.

6.6 Intraspecies Diversity

Compensation for low taxonomic and ecological diversity can occur through an increase in morphological and ecological heterogeneity within species. This ecological evolutionary phenomenon is regularly observed in the fauna at high latitudes. A large percentage of circumpolar or almost circumpolar species show high variability. It is a paradox that arctic species, inhabiting extreme and rather monotonous environments, do not exhibit wide adaptive radiation and taxonomic and ecological differentiation, but instead are rather polymorphic and ecologically variable (see also Murray, this vol.). Many arctic plant and animal species are highly differentiated complexes that include distinctive forms, which deserve to be separated as species or at least superspecies (Mayr 1942; Stepanyan 1983).

Good examples of such agglomerative forms in the arctic fauna are: for birds, the herring gull *Larus argentatus*; for fish, the arctic char *Salvelinus alpinus*; for invertebrates the earthworm *Eisenia nordenskioldi*; and many arctic insects, e.g. the leaf beetle *Chrysolina septentrionalis* and the ground beetle *Carabus truncaticollis*. Unfortunately, entomologists have not given proper attention to this problem, although such study may yield much important information about species formation.

6.7 Correlation Between Forms of Diversity in Arctic Biomes

Whittaker (1977) mentions seven categories of biodiversity. Along with the decrease in taxonomic diversity of zonal faunas (epsilon diversity) at high latitudes, there is a relatively high richness of concrete, or local faunas (gamma diversity). This means that in a given locality within the Arctic one may find a higher proportion of fauna (or flora) of the overall biome than, for example, in a boreal or tropical site. In other words, at high latitudes each list of species of a concrete fauna (or flora) is more representative of the whole arctic biota than in more species-rich biomes. For instance, in concrete faunas of springtails in the typical tundra subzone, we recorded up to 100 species, which is about 25% of the entire arctic fauna of this group. For comparison, the same number of species recorded in concrete faunas of the forest zone comprise only about 5% of the forest collembolan fauna. Another example: 84 bird species nest in the arctic tundra subzone, of which at least half can be found in each local fauna (Stishov et al. 1989).

There are also correlations between alpha and beta diversities, i.e. between the internal diversity of a community and changes in diversity along environmental gradients within the landscape. Even though there are very few species, or even a single species, in each functional group in an arctic community, most of these species can be found within a given community, resulting in a relatively high alpha diversity. By contrast, beta diversity, i.e. the differentiation of communities along an environmental gradient, is relatively low in the Arctic, because there are relatively few community types. Many arctic communities, such as polygonal or frost-boil tundras, exhibit a complicated mosaic pattern (see Chernov 1978), due to the poor ability of the vegetation to withstand severe environments and to reduce the heterogeneity of microhabitats.

The comparative analysis of different forms of biodiversity in the Arctic is only beginning. The analysis of different types of biodiversity in the Arctic and the study of their dependence on climate and their role in the functioning of ecosystems have great potential. Such studies are very important for the development of theories explaining the global geographical pattern of biological diversity.

6.8 Conclusions

1. The Arctic contains about 1–3% of the world's fauna, depending on the taxonomic group. Each group decreases in abundance toward high latitudes, reflecting direct or indirect effects of temperature. Dipterans are less sensitive to temperature than are other groups and become relatively more abundant at high latitudes.
2. Within a given taxonomic group, evolutionarily primitive lineages tend to predominate in the arctic fauna.
3. The low number of species in the Arctic is compensated for by high densities of individuals (e.g. cranefiles and lemmings).
4. Decreasing heat input at high latitudes explains the strong correlation between summer temperature and biological diversity.
5. In the Arctic the relationships between levels of diversity are unusual. Although the diversity of local concrete faunas (gamma diversity) and landscape diversity (beta diversity) are low, a large proportion of these species is found within a given community (high alpha diversity).

References

Anan'eva SI, Babenko AB, Chernov IuI (1987) Nogokhvostki (Collembola) v arkticheskikh tundrakh Taimyra. [Springtails (Collembola) in arctic tundras of Taimyr]. Zool Zh 67:1032–1044 (English Summary)

Borkin LIa Belimov GT, Sedalishchev VT (1984) Novye dannye o rasprostranenii amfibii i reptilii v Yakutii. (New data on distribution of amphibians and reptiles in Yakutia). In: Borkin LIa (ed) Ekologiia i faunistika amphibii i reptilii SSSR i sopredel'nykh stran. Nauka, Leningrad, pp 89–101

Bozhko MP (1936) Tardigrada evropeiskoi chasti SSSR. (Tardigrada of the European part of the USSR). Trudy Kharkovskogo Universiteta, Jubilee sbornik. Kharkov, pp 5–36

Chernov IuI (1975) Prirodnaia zonal'nost'i zhivotnyi mir sushi. (Natural zonation and the terrestrial animal world). Nauka, Moscow, 222 pp

Chernov IuI (1978) Struktura zhivotnogo naseleniia Subarktiki. (Structure of the animal population in the subarctic). Nauka, Moscow, 167 pp. (English Summary)

Chernov IuI (1982) O putiakh i istochnikakh formirovaniia fauny malykh ostrovov Okeanii. (On the pathways and sources of fauna formation on small islands of Oceania). Zh Obshch Biol 13:35–47 (English Summary)

Chernov IuI (1984) Biologicheskie predposylki osvoeniia arkticheskoi sredy organizmami razlichnykh taksonov. (Biological preconditions of settling the arctic environments by different taxa). In: Chernov IuI (ed), Faunogenez i filotsenogenez. Nauka, Moscow, pp 154–174

Chernov IuI (1985) Sreda i soobshchestva tundrovoi zony. (Environments and communities in the tundra zone). In: Chernov IuI (ed) Soobshchestva Krainego Severa i chelovek. Nauka, Moscow, pp 8–22 (English Summary)

Chernov IuI (1988) Filogeneticheskyi uroven'i geograficheskoe raspredelenie taksonov. (Phylogenetic level and geographical distribution of taxa). Zool Zh 67: 1445–1458 (English Summary)

Chernov IuI (1989) Teplovye usloviia i biota Arktiki. (Heat conditions and Arctic biota). Ekologiia 2:49–57

Chernov IuI (1992) Kogo bolshe v tundre – khishchnikov ili fitofagov? (What are more numerous in the tundra – predators or phytophages?) In: Chernov IuI (ed) Tsenoticheskie vzaimodeistviia v tundrovykh ekosistemakh. Nauka, Moscow, pp 100–125

Chernov IuI, Khlebosolov EI (1989) Troficheskie svyazi i vidovaia struktura naseleniia tundrovykh nasekomoiadnykh ptits (Trophic relations and species structure of a population of tundra insectivorous birds). In: Chernov IuI (ed) Ptitsy v soobshchestvakh tundrovoi zony. Nauka, Moscow, pp 39–51

Chernov IuI, Matveeva NV (1979) Zakonomernosti zonal'nogo raspredeleniia soobshchestv na Taimyre. (The regularities of community zonal distribution in the Taimyr). In: Aleksandrova VD, Matveeva NV (eds) Arkticheskie tundry i poliarnye pustyni Taimyra. Nauka, Leningrad, pp 166–200

Chernov IuI, Matveeva NV (1983) Taksonomicheskyi sostav arkticheskoi flory i puti osvoeniia tsvetkovymi rasteniiami sredy tundrovoi zony. (Taxonomic composition of the arctic flora and ways of colonization by flowering plants in landscapes of the tundra zone). Zh Obshch Biol 44:187–201 (English Summary)

Chernov IuI, Matveeva NV (1986) Iuzhnye tundry v sisteme zonalnogo deleniia. (Southern tundras in the zonal division system). In: Chernov IuI, Matveeva NV (eds) Iuzhnye tundry Taimyra. Nauka, Leningrad, pp 192–204

Chernov IuI, Striganova BR, Anan'eva SI, Kuzmin LL (1979) Zhivotnyi mir polyarnoi pustnyi mysa Chelyuskin. (Animal world of polar desert at cape Chelyuskin). In: Aleksandrova VD, Matveeva NV (eds) Arkticheskie tundry i polyarnye pustnyi Taimyra. Nauka, Leningrad, pp 35–49

Chernov IuI, Medvedev LN, Khruleva OA (1993) Zhuki-listoedy (Coleoptera, Chrysomelidae) v Arktike [Leaf beetles (Coleoptera, Chrysomelidae) in the Arctic]. Zool. Zh 72:78–92 (English Summary)

Clarke AN (1973) The freshwater molluscs of the Canadian interior basin. Malacologia 13 (1–2):505

Danilov NN (1966) Puti prisposobleniia nazemnykh pozvonochnykh zhivotnykh v usloviiakh sushchestvovaniia v Subarktike. T. 2. Ptitsy. (The ways of adaptations of terrestrial vertebrates in subarctic environments, vol 2. Birds). Nauka, Sverdlovsk, 146 pp

Danks HV (1981) Arctic arthropods. Entomological Society of Canada, Ottawa, 608 pp

Danks HV (1990) Arctic insects: instructive diversity. In: Harington CR (ed) Canada's missing dimension: science and history in the Canadian Arctic islands. Can Mus Nature 2:444–470

Es'kov KIu (1985) Pauki tundrovoi zony SSSR (Spiders of the tundra zone of the USSR). In: Ovcharenko VI (ed) Fauna i ekologiia paukov SSSR. Nauka, Leningrad, pp 121–128

Kafanov AI (1987) Pravilo "age and area" J. Willisa i shirotnaia geterokhronnost' morskoi biot. (The rule "age and area" by J. Willis and latitudinal heterochroneity of marine fauna). Zh Obshch Biol 48: 105–114 (English Summary)

Kuz'min LL (1973) Fauna svobodno zhivushchikh nematod Zapadnogo Taimyra (Fauna of free living nematods of western Taimyr). In: Tikhomirov BA (ed) Biogeotsenozy taimyrskoi tundry i ikh produktivnost' vol 2. Nauka, Leningrad, pp 139–147 (English Summary)

Lantsov VI, Chernov IuI (1987) Tipuloidnye dvukrylye v tundrovoi zone (Tipuloid dipterans in tundra zone). Nauka, Moscow, 175 pp

Matveeva NV, Chernov IuI (1977) Arkticheskie tundry na severo-vostoke poluostrova Taimyr, 1. (Arctic tundras of the northeast Taimyr Peninsula, 1). Bot Zh. 62: 938–953 (English Summary)

Mayr E (1942) Systematics and origin of species. Columbia University Press, New York

Meyen SV (1986) Florogenez i evoliutsiia rastenii. (Florogenesis and plant evolution). Priroda 11: 47–57

Meyen SV (1987) Geografiia macroevoliutsii u vysshikh rastenii (Geography of macroevolution in higher plants). Zh Obshch Biol 48:291–309 (English Summary)

Piper SR, MacLean SF (1982) Enchytraeidae (Oligochaeta) from taiga and tundra habitats of northeastern USSR. Can J Zool 60:2594–2609

Ryabitsev VK (1993) Territorialnye otnosheniia i dinamika soobshchestv ptits v Subarktike (Area relationships and dynamics of bird communities in the subarctic). Nauka, Ekaterinburg

Schwarts SS (1963) Puti prisposobleniia nazemnykh pozvonochnykh zhivotnykh k usloviiam suchchestvovaniia v Subarktike. T.I. Mlekopitaiushchie. (The ways of adaptations of terrestrial vertebrates to environments in the subarctic. vol 1. Mammals). Nauka, Sverdlovsk

Schwarts SS, Ishchenko VG (1971) Puti prisposobleniia nazemnykh pozvonochnykh zhivotnykh k usloviiam sushchestvovaniia v Subarktike. T. III. Zemnovodnye. (The ways of adaptations of terrestrial vertebrates to environments in the subarctic, vol 3. Amphibians). Nauka, Sverdlovsk, pp 7–60

Stepanyan LC (1983) Nadvidy i vidy-dvoiniki v avifaune SSSR. (Superspecies and twin species in avifauna of the USSR). Nauka, Moscow

Stishov MS, Chernov IuI, Vronskii NV (1989) Fauna i naselenie ptits podzony arkticheskikh tundr. (Fauna and bird populations in the arctic tundra subzone). In: Chernov IuI (ed) Ptitsy v soobshchestvakh tundrovoi zony. Nauka, Moscow, pp 5–39

Whittaker RH (1977) Evolution of species diversity in land plant communities. Evol Biol 10:1–67

Zherikhin VV (1978) Razvitie i smeny melovykh i kainozoiskikh faunisticheskikh kompleksov. (The development and succession of Cretaceous and Cenozoic faunal complexes). Nauka, Moscow

7 Animal Diversity at High Altitudes in the Austrian Central Alps

E. MEYER and K. THALER

7.1 Introduction

In the central Alps of Austria, culminating in the Großglockner mts. 3798 m a.s.l. (Hohe Tauern range) and Wildspitze 3774 m (Ötztal Alps), there are two important borders for animal life in the high alpine environment: the timberline and the snow line. Above the timberline at ca 2000 m vegetation changes from dwarf shrub heath to steppe-like grasslands (sedge mats) of (1) the alpine zone (ca. 2000–2700 m), which gradually fade into the grassland fragments of (2) the subnival zone (2600–3000 m); (3) the nival zone above the snow line at ca. 2900–3100 m is characterized by open-cushion vegetation, mosses and lichens remaining at extreme sites. Decrease in temperature and the short period of growth, habitat fragmentation, the effect of wind and snow distribution are the most important ecological factors at high altitudes (see Körner, this vol.) Distribution and zonation of animals in the high Alps have been studied for more than 100 years (Bäbler 1910; Franz 1943; Holdhaus 1954; Janetschek 1956, 1993; Schmölzer 1962; Janetschek et al. 1987). A special fauna is found in the forefield of glaciers (Janetschek, 1949) and at the glacier surface itself. To survive under high alpine conditions, animals have evolved various morphological, physiological and behavioural adaptations (Sømme 1989). Our report concentrates on the regional situation in the Central Alps, with emphasis on invertebrates, but neglecting Protozoa (see Foissner 1987). Further regional information can be found in Franz (1981), Patzelt (1987), Cernusca (1989); see also results obtained at Munt La Schera in the Suisse National Parc (Matthey et al. 1981). The fauna of the Alps is also discussed in general overviews of high altitude biology presented by Mani (1968) and Franz (1979).

7.2 Altitudinal Zonation of the Main Groups of Animals

The sedge mats of the alpine zone harbour most of the main terrestrial orders present also at lower altitudes in the soil and at the soil surface (Meyer 1980). Absent are scorpions (only at the southern macroslope), woodlice and cockroaches, each with one species at the timberline, and Psocoptera. Almost absent

Institut für Zoologie, Technikerstrasse 25, 6020 Innsbruck, Austria

Ecological Studies, Vol. 113
Chapin/Körner (eds) Arctic and Alpine Biodiversity

are earwigs, Psyllina, Planipennia, and Mecoptera (*Boreus* excepted). Some main groups reach their upper limit of distribution within the alpine zone: among soil animals, earthworms, Pauropoda, Protura and Diplura; among surface dwellers, grasshoppers, Heteroptera, Cicadina, and ants. Amphibians and reptiles present at favourable localities are: *Salamandra atra* Laurenti, *Triturus alpestris* (Laurenti), *Rana temporaria* L., *Lacerta vivipara* Jacquin, *Vipera berus* L. Characteristic birds are ptarmigan [*Lagopus mutus* (Mont.)], alpine chough [*Pyrrhocorax graculus* (L.)], water pipit [*Anthus spinoletta* (L.)], alpine accentor (*Prunella collaris* Scop.), wheatear (*Oenanthe oenanthe* L.) and snow finch [*Montifringilla nivalis* (L.)]. The large vultures became extinct in the 19th century. Large herbivorous mammals of this zone are *Capra ibex* L., *Rupicapra rupicapra* (L.), marmot [*Marmota marmota* (L.)], snowshoe hare (*Lepus timidus* L.). For further information, see Stüber and Winding (1991). The small mammals (Insectivora and Rodentia) of this zone, even the snow vole [*Microtus nivalis* (Martins)], are infested by parasites: blood parasites (Sporozoa and trypanosomes), helminths, lice and fleas (Mahnert, Pfaller, various contributions 1970–1974).

In the sedge mat fragments of the subnival zone, further groups have their upper limit of distribution: gastropods, centipeds, millipedes, Symphyla, most beetles (Besuchet 1983) and Hymenoptera (Schedl 1976, 1982). Indigenous vertebrates are small mammals, esp. *Microtus nivalis* and the snow finch.

In the nival zone, microfauna (Protozoa, rotifers, nematodes, tardigrades) and mesofauna groups (Collembola, Acari) predominate, together with Enchytraeidae, Araneida (Thaler 1988), Diptera and Lepidoptera. Additional species belong to Opiliones, Pseudoscorpiones, Archaeognatha, Coccina and Aphidina. An der Lan (1958) even found some micro-Turbellaria in cushion plants from 3000–3200 m. In the nival zone, animal life concentrates near the soil surface. Food chains are very short and species interactions scarce. Admittedly, insect-flower relationships await further studies since an early overview in 1881.

7.3 Species Numbers of Invertebrates and Altitudinal Zonation

In the Central Alps, invertebrate groups decrease in species numbers with increase in altitude. Within the main life zones, this decline seems to be gradual (Fig. 1). At the main borders, it is stepwise (Fig. 2). Species loss at the timberline is evident for saprotrophic millipedes and for zootrophic spiders, it is dramatic for phytotrophic orders and is more drastic in sawflies than in butterflies and moths. From 364 species of sawflies present in North Tyrol, 55% still occur near the tree line, but only three species in the subnival zone (Schedl, pers. comm.). In beetles, with various trophic status, the main decline in species diversity apparently takes place below the timberline. More details on altitudinal zonation of spiders and beetles in North Tyrol can be seen in Table 1. Only eight species (< 2%) of spiders are found regularly within the nival zone, as also in Switzerland (Maurer and Hänggi 1990). Such a decrease in species numbers and percentage has also been

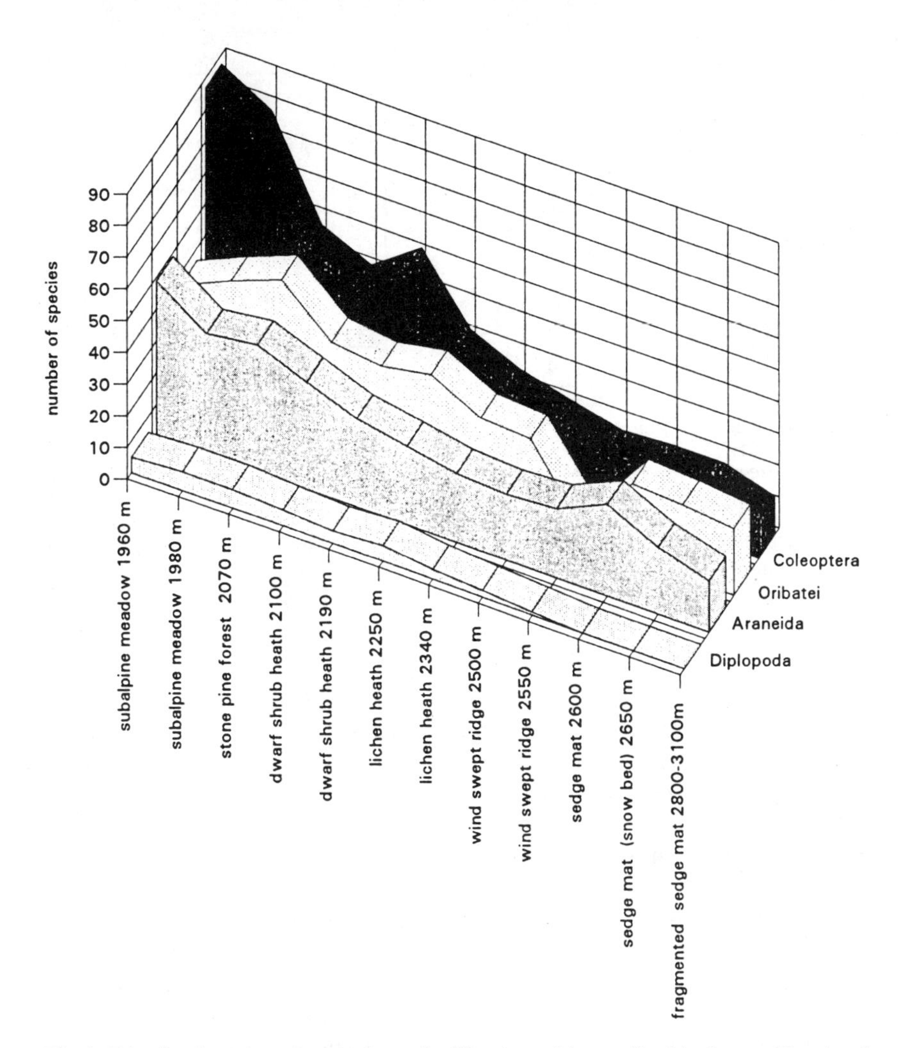

Fig. 1. Distribution of species numbers of millipedes, spiders, oribatid mites and beetles above the tree line in the Central Alps (Obergurgl, Austria): transect data from pitfall traps and soil samples along Festkogel slope, 1960–3100 m. Data from De Zordo (1979), Puntscher (1980), E. Meyer (unpubl.), Schatz (1978)

documented by Mani (1968) for different high mountain ecosystems of the world and by Danks (1981) for arctic North America.

In soil mesofauna as also in nematodes, diversity decreases less dramatically with altitude. Janetschek (1993) still found numerous species of oribatid mites (20) and springtails (24) in the nival zone of the Zillertal Alps above 3000 m. Lienhard (1980) mentions 41 Collembola species in grassland (*Caricetum firmae*) at 2500 m of the Swiss National Park. Species totals for North Tyrol are ca. 310 (mites) and 200 (springtails) respectively. In the high alpine and nival zones, the soil layer

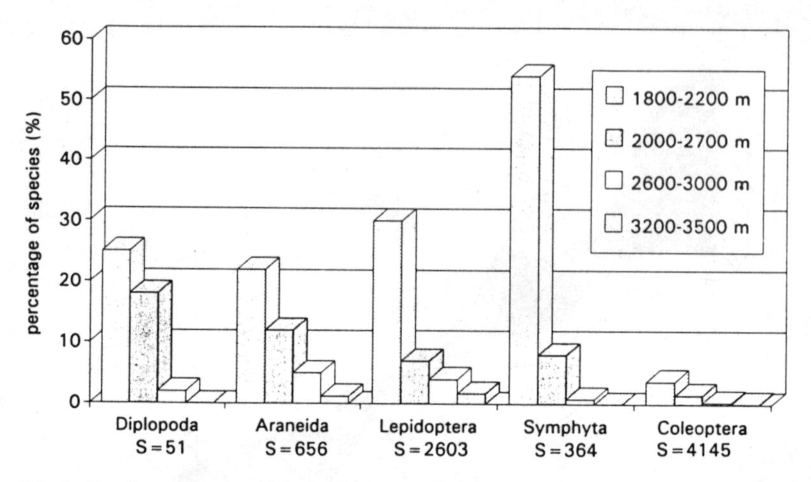

Fig. 2. Decline in species diversity in five macrofauna groups above the tree line in the Central Alps (Austria). *Columns* indicate the percentage of the number of species present in each life zone compared to the total in the whole area (North Tyrol). 1800–2200 m subalpine zone, 2000–2700 m alpine zone, 2600–3000 m subnival zone, 3200–3500 m nival zone. For reference, see Table 1, data for Lepidoptera from K. Burmann and G. Tarmann (pers. comm.), for Symphyta from W. Schedl (pers. comm.)

diminishes, but mesofauna diversity is still high due to the horizontal mosaic of various microhabitats.

7.4 Animal Communities Above the Timberline

This general decline in species diversity with altitude is due to a steady decrease in numbers of surviving euryoecious species, which are replaced by a minor number of species indigenous to the alpine and nival zones. There are very few euryzonal species, which occur over a wide range of life zones from lowlands < 500 m to mountain tops > 3000 m: one snail, *Arianta arbustorum* (L.) (Helicidae), one harvestman, *Mitopus morio* (F.) (Phalangiidae); various spiders, e.g. *Drassodes cupreus* (Blackwall) and *Haplodrassus signifer* (C.L. Koch) (Gnaphosidae), one millipede, *Ommatoiulus sabulosus* (L.) (Julidae). Only very few aspects of the ecology of these eurypotent species have been studied (Meyer 1985; Baur 1990). Naturally, some migrating species, called "tourists" by earlier authors, may casually and even regularly visit nival habitats, but they are not able to establish a lasting population. These are, among others, aeronautic money spiders (Lin. Erigoninae) and butterflies (Pieridae, Nymphalidae).

The percentage of indigenous species in high altitude zones has been assessed recently for some arthropod groups of the fauna of North Tyrol: Opiliones (2 out of a total of 38 spp.), false scorpions (1 from ca. 20), various spider families, Araneidae (1 from 32), Theridiidae (1 from 50), Agelenidae (2 from 20), Lycosidae

Table 1. Altitudinal distribution and numbers of species of Araneida and Coleoptera in the region of Obergurgl (Ötztal Alps)

Araneida[a] No. of species in North Tirol ( )	Subalpine forest, dwarf shrub heath 1800–2200 m	High alpine sedge mats 2000–2700 m	Sedge mat fragments 2600–3000 m	Mosses and lichens 3200–3500 m
Linyphiidae s.I. (255)	67	46	22	7
Lycosidae (55)	13	11	3	1
Gnaphosidae (50)	12	7	3	—
Thomisidae, Philodromidae (49)	9	3	3	—
Salticidae (49)	8	4	1	—
Theridiidae (50)	9	5	—	—
Clubionidae s.I. (34)	5	2	—	—
Araneidae s.I. (38)	9	1	—	—
Agelenidae (20)	5	—	—	—
Further 21 families (56)	5	2	—	—
Total number of species (656)	142	81	32	8
Percentage (100%)	22	12	5	1.2
Coleoptera[b] No. of species in North Tirol ( )				
Carabidae (323)	23	8	4	2
Curculionidae (514)	18	3	1	1
Staphylinidae (1165)	68	27	4	—
Scarabaeidae (83)	4	3	1	—
Byrrhidae (25)	3	1	1	—
Elateridae (89)	2	1	1	—
Chrysomelidae (324)	3	3	1	—
Cantharidae (66)	6	2	—	—
Further 76 families (1556)	26	7	—	—
Total number of species (4145)	153	62	13	3
Percentage (100%)	3.7	1.5	0.3	<0.1

[a] Data from Thaler (1979 and unpubl.), Puntscher (1980).
[b] Data from De Zordo (1979), Kahlen (1987), Thaler (1989).

(ca. 8 from 55), Gnaphosidae (ca. 10 from 50), millipedes (2 from 54) (K. Thaler, in prep.).

Within the alpine zone three main habitat types have been distinguished, each with a distinct community, grassland, snow beds as well as scree and rock sites. Within the nival zone only two types are found, forming a mosaic of sites with or without macroscopic vegetation (Franz, 1943; Schmölzer 1962). A more detailed typology of communities is difficult, since animals are mobile and microhabitats are closely interrelated. This can be seen from community tables for beetles

(De Zordo, 1979) and for spiders (Puntscher 1980; Thaler in Cernusca 1989) in alpine grasslands of the Ötztal Alps and the Tauern area. In these studies, the vertical zones again are well differentiated, whereas sites within a zone show a high degree of overlap in species occurrences. Problems of the micro-distribution of the soil mesofauna, which is much more minute, have been illustrated in Lienhard (1980). Of course, at the extreme sites of the nival zone, macrofauna groupings are much clearer (Fig. 3; see also Janetschek 1993).

Basic patterns of annual and daily periodicity of the epigeic macrofauna, as well as in insect emergence and flight, remain remarkably constant. They are manifold, with several successive peaks in the alpine zone, but patterns are simplified in nival habitats (Meyer 1980; Puntscher 1980; Stockner 1982; Troger et al. 1994).

Diversity values show that the structure of taxocoenoses is less diverse at extreme sites (Shannon index H′, ^{2}log). H′ values for spider communities investigated by pitfalls in the Ötztal and in the Tauern region respectively range from 2.3–3.1 (3.8–4.4) at timberline, 2.5–3.5 (2.7) in grassland, to 2.9 (1.7–2.3) at the subnival sites (Puntscher 1980; Thaler, in Cernusca 1989). According to biocoenotic principles, at the extreme sites, low species diversity may be outbalanced by high individual abundance.

7.5 Altitude-Related Change in Abundance and Biomass of the Soil Fauna

Vegetation type and litter quality are the key factors which determine composition, abundance and biomass of the soil fauna. In an altitudinal gradient of open habitats from the subalpine meadow to the high alpine *Carex* heath, there is a

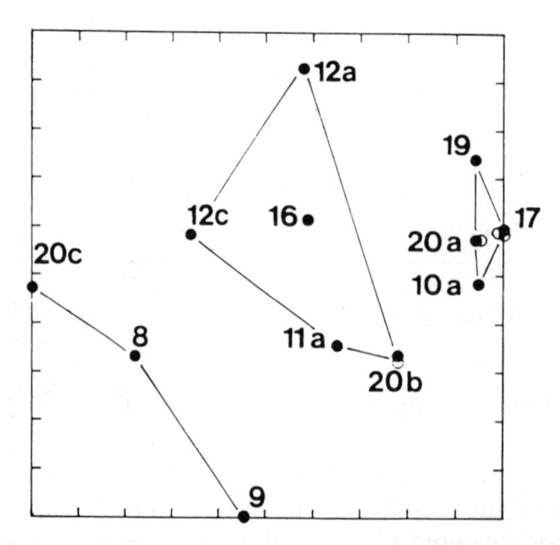

Fig. 3. Ordination of 16 spider communities in the nival zone of the Central Alps (Austria), 3000–3450 m. *Sites 9-8-20c* subnival (grassland fragments), *sites 10a-19-17* nival rocks with lichens, others intermediate. Note heterogeneity within subnival sites and close identity within the nival samples. (Data from Thaler (1981)

marked change in the soil fauna community (Meyer 1980; and in Franz 1981); (Fig. 4; Table 2). In a subalpine spruce forest, the soil fauna is dominated by mesofauna groups (mites, springtails), by enchytraeid worms as well as larvae of Diptera and Coleoptera. Larger soil animals, in particular Lumbricidae, are rare due to insufficient food and acid conditions. The mean dry mass of a macrofauna animal is only 0.17 mg, as calculated from Table 2. Therefore, the macrofauna biomass is very low. In contrast, at a subalpine meadow treated with organic manure, macrofauna biomass is more than ten times higher due to an enormous increase in earthworm biomass, mean dry mass per ind. 3.1 mg. In a subalpine pasture without manure, the increase in biomass is less drastic. In high alpine grassland (*Carex* heath), the soil fauna biomass diminishes gradually. Earthworms can be found mainly at sheep resting places, favourable for dung concentration. No earthworm was found in a *Carex* heath at 2600 m, where relatively large soil-living larvae of beetles and Diptera dominate, especially Tipulidae. Mean dry mass per ind. equals 0.23 mg. The importance of insect larvae in high

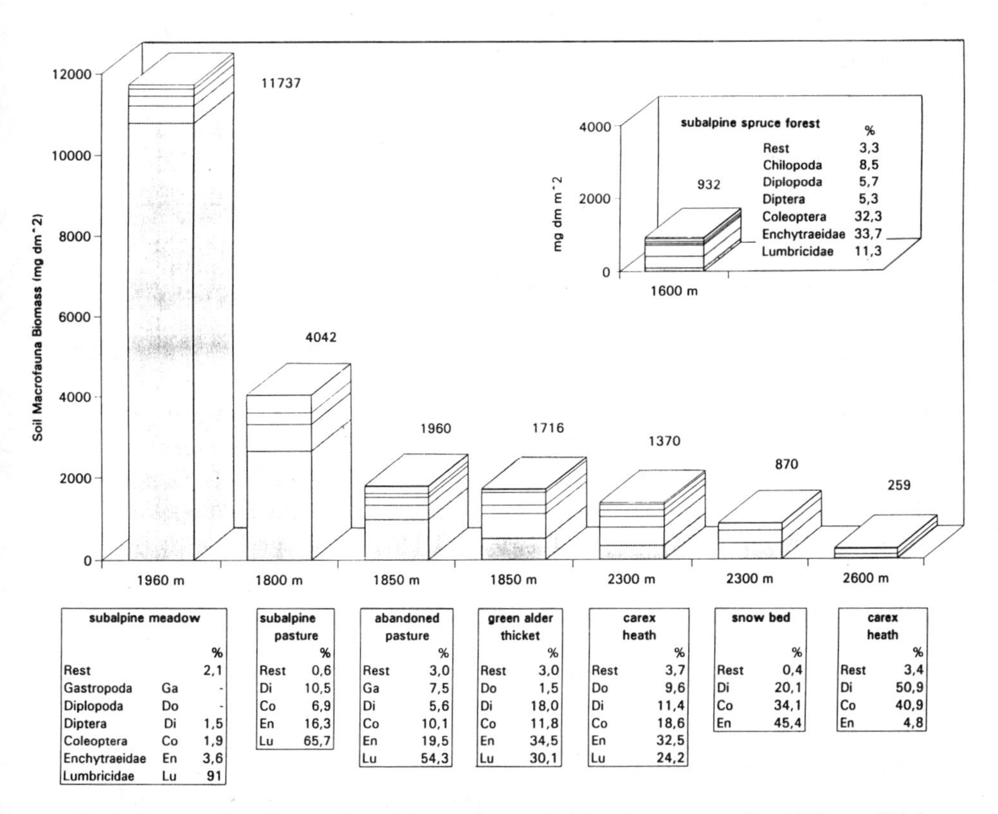

Fig. 4. Decrease in soil macrofauna biomass in open habitats between tree line 1960 m and high alpine *Carex* heath, 2600 m, Central Alps, Austria (Hohe Tauern and Ötztal Alps). The *number* above each column indicates total biomass. The situation in a subalpine spruce forest at 1600 m is given for comparison in inset. Data from Meyer (1980), Meyer (in Franz 1981), and unpubl

Table 2. Abundance and biomass of soil fauna and emerging insects at different altitudes in the Central Alps. Data from Meyer (1980 and unpubl.), Schatz (1981, Pedobiologia 22) and Troger et al. (1994). Emerging insects sampled with cloth-covered emergence traps. Days of capture: subalpine meadow 140, high alpine sedge mat 100

	Subalpine spruce forest 1600 m	Subalpine meadow 1960 m	High alpine sedge mat 2600 m
Abundance (ind. m^{-2})			
Macrofauna + Enchytraeidae	16 195	10 015	1 593
Mesofauna	329 500	—	27 300
Biomass (mg dry mass m^{-2})			
Macrofauna + Enchytraeidae	932	11 878	259
Mesofauna	1 523	—	307
Abundance (ind. $m^{-2}\,yr^{-1}$)			
Nematocera		623	885
Brachycera		627	370
Biomass ($mg\,m^{-2}\,yr^{-1}$)			
Nematocera		102	99
Brachycera		430	791

alpine soils is also shown in the high values of abundance and biomass of emerging Diptera.

7.6 Endemism and Vicariance in the Alpine Fauna

The high alpine fauna of the Alps is not only the result of present ecological conditions, but also of historical factors. The dramatic change in biota caused by the glaciations can still be recognized (Holdhaus 1954; Janetschek 1956). In the high glacial, the region was covered with an ice shield, separated from the northern glaciation by the ice-free corridor of central Europe. Isolation of species in ice-free mountain habitats (nunataks) and in the peripheral refugia during a high glacial led to distinct distributions and endemism of species. Accordingly, there are some paleoendemic species in nival sites (e.g. the mite *Mesoteneriffia steinboecki* Irk, Teneriffiidae; *Charimachilis relicta* Janetschek, Archaeognatha), as well as at the periphery (Maurer and Thaler 1988). Chemini (1991) has recently discovered a new cyphophthalmid species in the Bergamo Alps (Opiliones). Neoendemic species differentiated in glacial isolation and show as a rule marked vicariance. This was reported, among others for terrestrial leeches (genus *Xerobdella*), gastropods, millipedes, harvestmen, for some spider families (Thaler 1976; Maurer 1982), false scorpions, grasshoppers (Nadig 1987), beetles (Holdhaus 1954) and Lepidoptera. They abound in the favourable peripheral refugia, many actually under montane and subalpine conditions, most not even able to exist in nival situations. In the Central Alps, where nival habitats are interconnected, widespread endemics are present, some occurring even across the whole chain of the Alps and in the European mountain system. Local endemic species under nival conditions have been found in isolated mountain groups only, e.g. the Dolomites. Studies on vicariance and endemism in nival mesofauna groups are not available. Retreat of glaciation was followed by re-immigration from the main refugia (De Lattin 1957; Malicky et al. 1983). For re-immigration into alpine grassland, refugia at the eastern rather than at the southern border of the Alps might have been important. From the intermediate corridor of central Europe, arctic-alpine species came, known for Turbellaria, Enchytraeidae (?), snails, spiders, grasshoppers, and above all in beetles (Holdhaus and Lindroth 1939) and Lepidoptera. Other high alpine species evidently originated in the mountains and steppes of Asia. A general discussion about routes, intensity and success of post-glacial re-invasion of the fauna must be postponed. Distributions of some woodland spiders coincide with the ways of immigration of main tree species, beech, spruce and fir (Mayer 1974; Thaler 1980). Widely scattered distributions of some field species might be remnants of an extensive occurrence in the early post-glacial. These species have been named protocratic by Brinck (1966).

7.7 Conclusions

1. In the Central Alps, the general decrease in species diversity of invertebrates at the timberline is more drastic in phytotrophic than in zootrophic and saprotrophic orders. Most euedaphic and atmobiotic life-forms disappear within the subnival zone.
2. At the snow line the decline is more dramatic for the macrofauna than for the mesofauna. In the nival zone, animal life concentrates near the soil surface, under stones, in rock crevices and in the available plant cover. This is not only due to temperature conditions and short growing season, but also due to reduction of habitat structure and too little plant cover.
3. According to biocoenotic principles (Thienemann), there are only very few species at "the extreme" habitats, albeit sometimes in great numbers. Food chains in the nival zone are very short and species interactions are scarce. In the nival community, springtails and mites predominate together with spiders, midges and moths.
4. Basic patterns of annual and daily activity of the epigeic macrofauna, as well as in insect emergence and flight, remain remarkably constant with increasing altitude. They show several successive peaks in the alpine zone, but become more uniform in nival habitats. The short period of activity is balanced by a prolonged life cycle at least in Lycosidae and in millipedes.
5. Along an altitudinal gradient, the total soil fauna biomass decreases rapidly, largely due to the reduction in earthworms. In high alphine grassland, Enchytraeidae and larvae of beetles and Diptera predominate. The reduction in zoomass certainly reflects the overall decrease in primary production. Aerial import of arthropods from lower regions may substantially enrich the trophic situation in high-altitude ecosystems.

6. The recent fauna of the Central Alps is a result of glaciations. With increasing altitude rather widespread endemics and arctic-alpine species prevail. Even in the European Alps precise information about the number and distribution of species is not available for many animal groups.

References

An der Lan H (1958) Die ersten terricolen Turbellarien aus dem Ewigschneegebiet. Schlern-Schr (Innsbruck) 188: 161–166

Bäbler E (1910) Die wirbellose, terrestrische Fauna der nivalen Region. Rev Suisse Zool 18: 761–915

Baur B (1990) Seasonal changes in clutch size, egg size and mode of oviposition in *Arianta arbustorum* L. (Gastropoda) from alpine populations. Zool Anz 225: 253–265

Besuchet C (1983) Coleopteres des Alpes Suisses atteignant ou depassant l' altitude de 3000 m. Bull Romand Entomol 1: 167–176

Brinck P (1966) Animal invasion of glacial and late glacial terrestrial environments in Scandinavia. Oikos 17: 250–266

Cernusca A (ed) (1989) Struktur und Funktion von Graslandökosystemen im Nationalpark Hohe Tauern. Veröff österr MaB Progr 13: 1–625

Chemini C (1991) *Siro valleorum* n.sp. A new cyphophthalmid from the Italian Alps (Arachnida.. Sironidae). Riv Mus civ sci nat E Caffi Bergamo 14: 181–189

Danks HV (1981) Arctic arthropods. A review of systematics and ecology with particular reference to the North American fauna. Entomological Society of Canada, Ottawa

De Lattin G (1957) Die Ausbreitungszentren der holarktischen Landtierwelt. Verh Dtsch Zool Ges Hamburg Zool Anz Suppl 20: 380–410

De Zordo I (1979) Ökologische Untersuchungen an Wirbellosen des zentralalpinen Hochgebirges (Obergurgl, Tirol). III Lebenszyklen und Zönotik von Coleopteren. Veröff Univ Innsbruck 118 (Alpin-Biol Stud 11): 1–131

Foissner W (1987) Soil Protozoa: fundamental problems, ecological significance, adaptations in ciliates and testaceans, bioindicators, and guide to the literature. Prog Protist 2: 69–212

Franz H (1943) Die Landtierwelt der Mittleren und Hohen Tauern. Ein Beitrag zur tiergeographischen und -soziologischen Erforschung der Alpen. Denkschr Akad Wiss Wien Math-Naturw Kl 107: 1–552

Franz H (1979) Ökologie der Hochgebirge. Ulmer, Stuttgart

Franz H (ed) (1981) Bodenbiologische Untersuchungen in den Hohen Tauern 1974–1978. Veröff österr MaB-Hochgebirgsprogr Hohe Tauern 4: 1–300

Holdhaus K (1954) Die Spuren der Eiszeit in der Tierwelt Europas. Abh Zoolbot Ges Wien 18: 1–493

Holdhaus K, Lindroth CH (1939) Die europäischen Koleopteren mit boreoalpiner Verbreitung. Ann Naturhist Mus Wien 50: 123–293

Janetschek H (1949) Tierische Successionen auf hochalpinem Neuland. Nach Untersuchungen am Hintereis-, Niederjoch- und Gepatschferner in den Ötztaler Alpen, Ber Nat-Med Ver Innsbruck 48/49: 1–215

Janetschek H (1956) Das Problem der inneralpinen Eiszeitüberdauerung durch Tiere (Ein Beitrag zur Geschichte der Nivalfauna). Österr Zool Z 6: 421–506

Janetschek H (1993) Über Wirbellosen-Faunationen in Hochlagen der Zillertaler Alpen. Ber Nat-Med Ver Innsbruck 80: 121–165

Janetschek H, Meyer E, Schatz H, Schatz - De Zordo I (1987) Ökologische Untersuchungen an Wirbellosen im Raum Gurgl unter Berücksichtigung anthropogener Einflüsse. Veröff Österr MaB-Progr 10: 281–315

Kahlen M (1987) Nachtrag zur Käferfauna Tirols. Veröff Mus Ferdinandeum (Innsbruck), Beilageband 3: 1–288

Lienhard C (1980) Zur Kenntnis der Collembolen eines alpinen *Caricetum firmae* im Schweizerischen Nationalpark. Pedobiologia 20: 369–386

Malicky H, Ant H, Aspöck H, Dejong R, Thaler K, Varga Z (1983) Argumente zur Existenz und Chorologie mitteleuropäischer (extramediterran-europäischer) Faunen-Elemente. Entomol Gen 9: 101–119

Mani MS (1968) Ecology and biogeography of high altitude insects. Junk, The Hague

Matthey W, Dethier M, Galland P, Lienhard C, Rohrer N, Schiess T (1981) Étude écologique et biocénotique d' une pelouse alpine au parc national Suisse. Bull Ecol 12: 339–354

Maurer R (1982) Zur Kenntnis der Gattung *Coelotes* (Araneae, Agelenidae) in Alpenländern 1. Rev Suisse Zool 89: 313–336

Maurer R, Hänggi A (1990) Katalog der Schweizerischen Spinnen. Doc Faun Helv 12

Maurer R, Thaler K (1988) Über bemerkenswerte Spinnen des Parc National du Mercantour (F) und seiner Umgebung (Arachnida: Araneae). Rev Suisse Zool 95: 329–352

Mayer H (1974) Wälder des Ostalpenraumes. Fischer, Stuttgart

Meyer E (1980) Ökologische Untersuchungen an Wirbellosen des zentralalpinen Hochgebirges (Obergurgl, Tirol). IV. Aktivitätsdichte, Abundanz und Biomasse der Makrofauna. Veröff Univ Innsbruck 125 (Alpin-Biol Stud 13): 1–54

Meyer E (1985) Distribution, activity, life-history and standing crop of Julidae (Diplopoda, Myriapoda) in the Central High Alps (Tyrol, Austria). Holarct Ecol 18: 141–150

Meyer E, Steinberger KH (1994) Über die Bodenfauna in Wäldern Vorarlbergs (Österreich). Grundlegender Bestand und Auswirkungen von Gesteinsmehlapplikationen. Verh Ges Ökologie (Innsbruck) 23: 149–164

Nadig A (1987) Saltatoria (Insecta) der Süd- und Südostabdachung der Alpen zwischen der Provence im W, dem pannonischen Raum in NE und Istrien im SE .. 1: Laubheuschrecken (Tettigoniidae). Rev Suisse Zool 94: 257–356

Patzelt G (1987) MaB-Projekt Obergurgl. Veröff Österr MaB-Progr 10: 1–350

Puntscher S (1980) Ökologische Untersuchungen an Wirbellosen des zentralalpinen Hochgebirges (Obergurgl, Tirol). V Verteilung und Jahresrhythmik von Spinnen. Veröff Univ Innsbruck 129 (Alpin-Biol Stud 14): 1–106

Schatz H (1978) Oribatiden-Gemeinschaften (Acari: Oribatei) oberhalb der Waldgrenze im Raum Obergurgl (Tirol, Österreich). Ber Nat-med Ver Innsbruck 65: 55–72

Schatz H (1981) Abundanz, Biomasse und Respirationsrate der ArthropodenMesofauna im Hochgebirge (Obergurgl, Tiroler Zentralalpen). Pedobiologia 22: 52–70

Schedl W (1976) Untersuchungen an Pflanzenwespen (Hymenoptera: Symphyta) in der subalpinen bis alpinen Stufe der zentralen Ötztaler Alpen (Tirol, Österreich). Veröff Univ Innsbruck 103 (Alpin-Biol Stud 8): 1–85

Schedl W (1982) Über aculeate Hautflügler der zentralen Ötztaler Alpen (Tirol, Österreich) (Insecta: Hymenoptera). Ber Nat-Med Ver Innsbruck 69: 95–117

Schmölzer K (1962) Die Kleintierwelt der Nunatakker als Zeugen einer Eiszeit-Überdauerung. Ein Beitrag zum Problem der Prä- und Interglazialrelikte auf alpinen Nunatakkern. Mitt Zool Mus Berl 38: 171–400

Sømme L (1989) Adaptations of terrestrial arthropods to the alpine environment. Biol Rev 64: 367–407

Stockner J (1982) Ökologische Untersuchungen an Wirbellosen des Zentralalpinen Hochgebirges (Obergurgl Tirol). VII Flugaktivitat und Flugrhythmik von Insekten oberhalb der Waldgrenze. Veröff Univ Innsbruck 134 (Alpin-Biol Stud 16): 1–102

Stüber E, Winding N (1991) Die Tierwelt der Hohen Tauern, Wirbeltiere Z. Aufl. Carinthia, Klagenfurt

Thaler K (1976) Endemiten und arktoalpine Arten in der Spinnenfauna der Ostalpen (Arachnida: Araneae). Entomol Germ 3: 135–141

Thaler K (1979) Fragmenta Faunistica Tirolensia–IV (Arachnida .. Tipulidae). Veröff Mus Ferdinandeum Innsbruck 59: 49–83

Thaler K (1980) Die Spinnenfauna der Alpen: ein zoogeographischer Versuch. Veröff Int Kongr Arachnol 8 (Wien): 389–404

Thaler K (1981) Neue Arachnidenfunde in der nivalen Stufe der Zentralalpen Nordtirols (Österreich) (Aranei .. Pseudoscorpiones). Ber Nat-Med Ver Innsbruck 68: 99–105
Thaler K (1988) Arealformen in der nivalen Spinnenfauna der Ostalpen (Arachnida, Aranei). Zool Anz 220: 233–244
Thaler K (1989) Streufunde nivaler Arthropoden in den mittleren Ostalpen. Ber Nat-Med Ver Innsbruck 76: 99–106
Troger H, Janetschek H, Meyer E, Schatz W (1994) Schlüpfabundanz von Insekten (Diptera/Coleoptera/Hymenoptera) im zentralalpinen Hochgebirge (Tirol: Ötztal). Entomol Gen 18: 241–260

Part II
Past, Present, and Future Changes in Diversity

8 Arctic Tundra Biodiversity: A Temporal Perspective from Late Quaternary Pollen Records

L.B. Brubaker[1], P.M. Anderson[2], and F.S. Hu[1]

8.1 Introduction

During the Quaternary (the last ca. 2 million years), the earth's climate has oscillated between numerous predominantly warm or cold periods, causing dramatic changes in the distribution and composition of plant communities worldwide (Bradley 1885; Bartlein 1988; Bartlein and Prentice 1989). Arctic tundra has been particularly affected because large temperature fluctuations and the growth of continental ice sheets have extensively altered vegetation at high latitudes (Climap 1981; Bradley 1985). Since future warming is predicted to be greatest at high latitudes (e.g. Schlesinger and Mitchell 1987; Hansen et al. 1988), tundra will most likely continue to undergo significant change. Understanding the effects of potential climatic change on tundra biodiversity requires a variety of research approaches. In this effort, information on long-term responses of tundra to past climatic variations can complement results of shorter-term observations and experiments on contemporary landscapes.

Pollen in lake and bog sediments provides continuous records of the development of arctic tundra under different climatic conditions (e.g. Huntley and Birks 1983; Barnosky et al. 1987; Ritchie 1987; Lamb and Edwards 1988). In recent years, the spatial coverage of pollen records has increased throughout the Arctic, but fossil sites are still too sparse to provide a complete circumpolar history (Andrews and Brubaker 1991). Nevertheless, available records clearly demonstrate that the floristic and physiognomic composition of tundra varied over time and that the history of these changes differs among arctic regions (Ritchie 1987; Lamb and Edwards 1988; Anderson et al. 1989; Lozhkin et al. 1993). We summarize major trends in late Quaternary (ca. 20000 to 0 B.P.) pollen records from Alaska, with some comparisons to tundra history in other high latitudes of North America and Eurasia. The Alaskan pollen records are sufficiently detailed to address several questions about long-term patterns in the biodiversity of tundra ecosystems and to allow inferences about causes of those patterns (Barnosky et al. 1987; Anderson and Brubaker 1993, 1994). In particular, we explore three questions: What were the composition and regional patterns of tundra during periods of different late Quaternary climate (full glacial, glacial-interglacial transition, and present)? What were the patterns of shrub and tree invasions into

[1] College of Forest Resources, University of Washington, Seattle, WA, USA
[2] Quaternary Research Center, University of Washington, Seattle, WA, USA

Ecological Studies, Vol. 113
Chapin/Körner (eds) Arctic and Alpine Biodiversity

tundra when climate warmed at the end of the last glacial period? What major changes in ecosystem processes accompanied the conversion of shrub tundra to forested landscape at the end of the last glacial period? More detailed information about climate and vegetation history of this region can be found in recent reviews by Barnosky et al. (1987) and Anderson and Brubaker (1993, 1994).

8.2 Late Quaternary Climate History

During the Quaternary, variations in the earth's orbit changed the seasonal and latitudinal distribution of solar radiation, causing periodic advances and recessions of continental ice sheets (Climap 1981). In this chapter, we concentrate on the most recent fluctuation from full-glacial to interglacial climate. At the last glacial maximum (ca. 18 000 B.P.), thick ice sheets extended into temperate latitudes of North America and Europe. In addition, the Arctic Ocean and much of the North Atlantic Ocean were perennially ice-covered (Climap 1981; Denton and Hughes 1981). However, a great expanse of the Arctic from central Eurasia to northwestern Canada probably remained free of ice except for local alpine glaciers. Changes in global ice volumes had important secondary effects on the geography of some regions. For example, lowered sea level exposed a broad plain, called the Bering Land Bridge, connecting North America and Asia (Hopkins et al. 1982).

Late Quaternary climatic changes in the Arctic, as in other regions, are governed by variations in large-scale controls of atmospheric circulation, including the intensity and seasonality of insolation, ice volume and extent, sea surface temperature and atmospheric composition (e.g. Kutzbach and Guetter 1986). The influence of such variations on past climates has recently been evaluated by general circulation models (e.g. Manabe and Broccoli 1985; Kutzbach and Guetter 1986; Cohmap 1988; Mitchell et al. 1988; Wright et al. 1993), which simulate global circulation patterns under a given set of physical conditions. We summarize results for the Arctic from simulations by the National Center for Atmospheric Research Community Circulation Model over the past 18 000 years (Kutzbach and Guetter 1986; Cohmap 1988; Wright et al. 1993). These simulations highlight three distinctive periods of climate for examining arctic tundra biodiversity: full glacial (colder-drier than present), glacial-interglacial transition (including a period warmer than present in some areas), and modern.

During the full glacial (18 000 and 15 000 B.P. simulations), high global ice volume, lowered sea level, and reduced atmospheric CO_2 concentration resulted in cold, continental climates throughout the unglaciated Arctic. Strong high-pressure systems centered over continental ice sheets in North America and Europe and over the southern ice margin in Eurasia caused northeasterly airflows, which resulted in cold, dry climates south of the ice sheets. North-south temperature gradients were steep in these regions. During the glacial-interglacial transition (15 000 to 9000 B.P. simulations), continental ice sheets receded, and changes in the earth's orbital features amplified contrasts between winter and summer insolation. By 9000 B.P., summer radiation was near maximum values and temperatures were warmer than present in much of western North America, Europe, and Asia. In contrast, eastern North America did not experience maximum warming until 6000 B.P. because prior to that time the effect of the remaining ice sheet on atmospheric circulation dampened the effects of increased summer insolation. Climatic simulations for most arctic regions reached modern conditions between 6000 and 3000 B.P.

8.3 What Were the Composition and Regional Patterns of Tundra During Periods of Different Late Quaternary Climate?

Pollen records clearly show that late Quaternary climatic changes caused extensive reorganization rather than simple displacement of tundra communities (Lamb and Edwards 1988; Anderson et al. 1989; Anderson and Brubaker 1994). Although the composition of tundra communities probably varied at all spatial scales, only a few regions have been studied in sufficient detail to understand landscape-level patterns of variation. Several recent reviews describe the current understanding of tundra history in different regions of the Arctic (Ritchie 1984; Lamb and Edwards 1988; Anderson and Brubaker 1993; Peterson 1993; Ritchie and Harrison 1993). The history for Alaska summarized below is based on reviews by Anderson and Brubaker (1993, 1994) and Barnosky et al. (1987).

8.3.1 Full Glacial (ca. 20 000–14 000 B.P.)

An expansive herb tundra dominated unglaciated Eurasia (Sher 1974; Shilo 1987; Lozhkin et al. 1993) and far northwestern North America (Ritchie 1984; Anderson and Brubaker 1993, 1994) during the full glacial. Pollen assemblages from this period are dominated by Cyperaceae, Gramineae, and/or *Artemisia.* In contrast to this broad tundra zone, a narrow band of tundra or tundra-forest ecotone fringed the southern edge of ice sheets in most of North America and Europe (Huntley and Birks 1983; Whitlock 1992; Webb et al. 1993a). The diminished importance of tundra in these regions reflects the steep north-south temperature gradient at the margins of continental glaciers (Webb et al. 1993a, b). These continental-scale patterns in the pollen data suggest that the ranges of tundra species are more discontinuous in cold than in warm phases of the earth's climate. The effect of the full-glacial range discontinuities on the genetic diversity of tundra flora is discussed by Murray (this Vol.).

Full-glacial pollen assemblages in Alaska show relatively weak statistical similarity to modern pollen assemblages from the Canadian High Arctic and north coast of Alaska (Anderson et al. 1989). These differences reflect the greater importance of herbs and reduced importance of shrubs in full-glacial than in modern pollen assemblages. Plant cover in the full glacial was probably less continuous than in most modern Alaskan tundra communities because low pollen accumulation rates suggest that plant cover was sparse, pollen taxa indicate that open ground was present, and inorganic lake sediments suggest that rates of soil erosion or loess deposition were higher than at present.

The composition of full-glacial tundra communities varied at local to regional scales. Graminoid communities with moist herbaceous species (e.g. *Rubus chamaemorus*, *Polygonum amphidium*, and *Saxifraga hirculus*) were probably most common in lowlands and valleys of the western Brooks Range. However, vegetation cover may have been extremely sparse across most of eastern Alaska, because plant taxa that tolerate low moisture and frequent disturbances (e.g. Compositae, Cruciferae, Chenopodiaceae) were relatively common in pollen assemblages of this region. *Salix* was probably the only common shrub in the full-glacial tundra; dwarf *Salix* species may have been abundant on moist sites along seepages, drainages, and snowbeds. *Betula* and *Alnus* pollen was exceedingly rare or absent, indicating that low and tall shrub communities, which are common in modern Alaskan tundra, were not important components of the full-glacial vegetation.

8.3.2 Glacial-Interglacial Transition (ca. 14000–9000 B.P.)

Warming at the end of the last glaciation caused major changes in arctic tundra, but there is little similarity in the histories from different circumarctic regions (Huntley and Birks 1983; Ritchie 1987; Lamb and Edwards 1988; Anderson and Brubaker 1994). For example, although shrub tundra generally replaced herb tundra in the unglaciated Arctic, the period of shrub dominance varied greatly. In far northeastern Siberia, shrub tundra dominated the landscape 1000 to 2000 years before the expansion of *Larix* forests (Lozhkin et al. 1993), but in western Alaska, shrubs persisted as long as 6000 years before the invasion of *Picea* forests (Fig. 1). Both of these patterns differ from that of eastern North America, where a

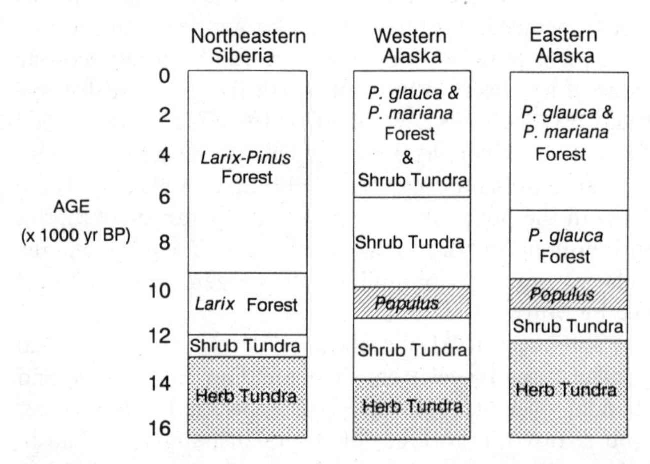

Fig. 1. Summary of major late Quaternary vegetation changes in far northeastern Siberia (Sosedne Lake, 62° 05′N 149° 30′E), western Alaska (Joe Lake, 66° 46′N 157° 13′W), and eastern Alaska (66° 02′N 147° 31W). Redrawn from Lozhkin et al. (1993)

transient band of shrub tundra moved northward close to receding ice and was rapidly replaced by forests (Gajewski et al. 1993; Ritchie and Harrison 1993; Webb et al. 1993a). The glacial-interglacial transition in Europe displayed yet another pattern, because the long-term warming trend was interrupted by a period of severe cold, called the Younger Dryas (also see Amman, this Vol.). Temperatures in western Europe declined sharply between ca. 11 000 and 10 000 B.P., when an influx of melt water from the North American ice sheet disrupted ocean circulation patterns in the North Atlantic Ocean, reducing the release of heat from ocean waters and cooling the European continent (Broecker and Denton 1990). In England, where open *Betula* forests had replaced tundra by ca. 12 500 B.P., colder temperatures caused the widespread reversion of forests to tundra on time scales of decades to centuries (Pennington 1986). Forests replaced tundra equally rapidly, when the more typical interglacial ocean circulation pattern was reestablished. This is a striking example from the fossil record that abrupt climatic changes can cause rapid, large-scale landscape conversions between forest and tundra.

In Alaska, the physiognomic character of the tundra changed profoundly at the beginning of the glacial-interglacial transition, because shrubs increased from being less common to much more common than today. Pollen assemblages were strongly dominated by *Betula*, and showed virtually no statistical similarity to modern pollen assemblages in North America (Anderson et al. 1989). The high percentages of *Betula* pollen suggest that shrublands covered extensive areas of Alaska. However, pollen accumulation rates indicate that herb pollen input to lakes did not decrease substantially with the dramatic increase in *Betula* pollen. *Betula* shrubs may have, therefore, invaded the sparsely vegetated full-glacial landscape without substantially altering population sizes of herbaceous taxa.

Differences in ecosystem processes between ancient and modern tundra landscapes can be inferred from studies of modern tundra communities on the northern foothills of the Brooks Range, Alaska. In this modern tundra, net primary productivity is higher in deciduous shrub communities (with *Betula*) than in other community types (Shaver and Chapin 1991). Although deciduous leaves are easily decomposed, decomposition rates and elemental fluxes are relatively low compared to the biomass of this community due to the slow decomposition of woody tissue. Organic matter decomposition in ancient shrub ecosystems may have been faster than suggested by these modern studies, however, since basal radiocarbon dates on organic soils in Alaska and northwest Canada suggest little net accumulation of organic matter during this period (Billings 1987; Ovenden 1990; Walker and Everett 1991; Marion and Oechel 1993). These results suggest that net primary productivity and elemental fluxes were higher in the shrubby tundra ecosystems of the glacial-interglacial transition than in their closest counterparts in modern tundra.

Populus cf. *balsamifera* pollen percentages reached peak values in Alaska during the middle of the glacial-interglacial transition. Spatial patterns in the pollen data suggest that *Populus balsamifera* formed extensive woodlands on both riparian and nonriparian sites at low elevations within *Betula* shrub tundra.

This scenario differs from the modern occurrence of *Populus* within Alaskan tundra, where small clonal populations are restricted to floodplains and south-facing slopes. Although glacial-interglacial conditions were probably unsimilar to projected warmer climates (Overpeck et al. 1992), the strongly deciduous, woody character of this period suggests that future warming may favor the expansion of deciduous trees/shrubs rather than conifers. This possibility is also suggested by vegetation modeling experiments for the Arctic, which project increases in deciduous shrublands or steppe in response to climate changes under doubled CO_2 conditions (Emanuel et al. 1985; Bonan et al. 1990).

8.3.3 Modern Tundra (ca. 6000 BP to Present)

In general, the modern composition and distribution of arctic tundra were established between 6000 and 3000 B.P. This conclusion is tentative, however, because the vegetation histories of large areas of the present arctic tundra (e.g. Canadian High Arctic) have not yet been documented (Andrews and Brubaker 1991). During this period, the southern limit of tundra contracted northward and then expanded southward in much of Canada (Payette et al. 1989; Gajewski et al. 1993; MacDonald et al. 1993) and in Iceland (Huntley and Prentice 1993), but not in Fennoscandia (Huntley and Prentice 1993) and Alaska (Anderson and Brubaker 1994). In northern Alaska, fossil pollen assemblages resembled modern by 6000 to 4000 B.P., showing nearly equivalent percentages of shrub (predominantly *Betula* and *Alnus*) and herb (predominantly Cyperaceae and Gramineae). This mixture of herb and shrub growth forms contrasts sharply with the herb-dominated pollen spectra of the full glacial and the shrub-dominated spectra of the glacial-interglacial transition.

8.4 What Were the Patterns of Shrub and Tree Invasions into Tundra When Climate Warmed at the End of the Last Glacial Period?

In glaciated regions, the factors controlling species invasions along receding ice sheets were probably complex, because the temporal and spatial patterns of ice recession, local processes related to soil development, as well as regional climatic conditions may have influenced plant population expansions. Pollen records from ice-free portions of the Arctic document plant invasions into tundra zones unaffected by recent deglaciation and are thus particularly relevant to questions of species movement into tundra under future climatic warming. Pollen data in north-central Alaska, the area roughly between the crests of the Alaska and Brooks Ranges (Fig. 2), are sufficiently detailed to examine late Quaternary shrub and tree invasions into such an unglaciated region. We describe invasion rates qualitatively (e.g. slow, fast), because additional well-dated sites are needed to estimate such rates with greater confidence.

Betula, *Ericaceae*, and *Alnus* shrubs spread from population centers in northwestern Alaska, but the timing and competitive context of their expansions

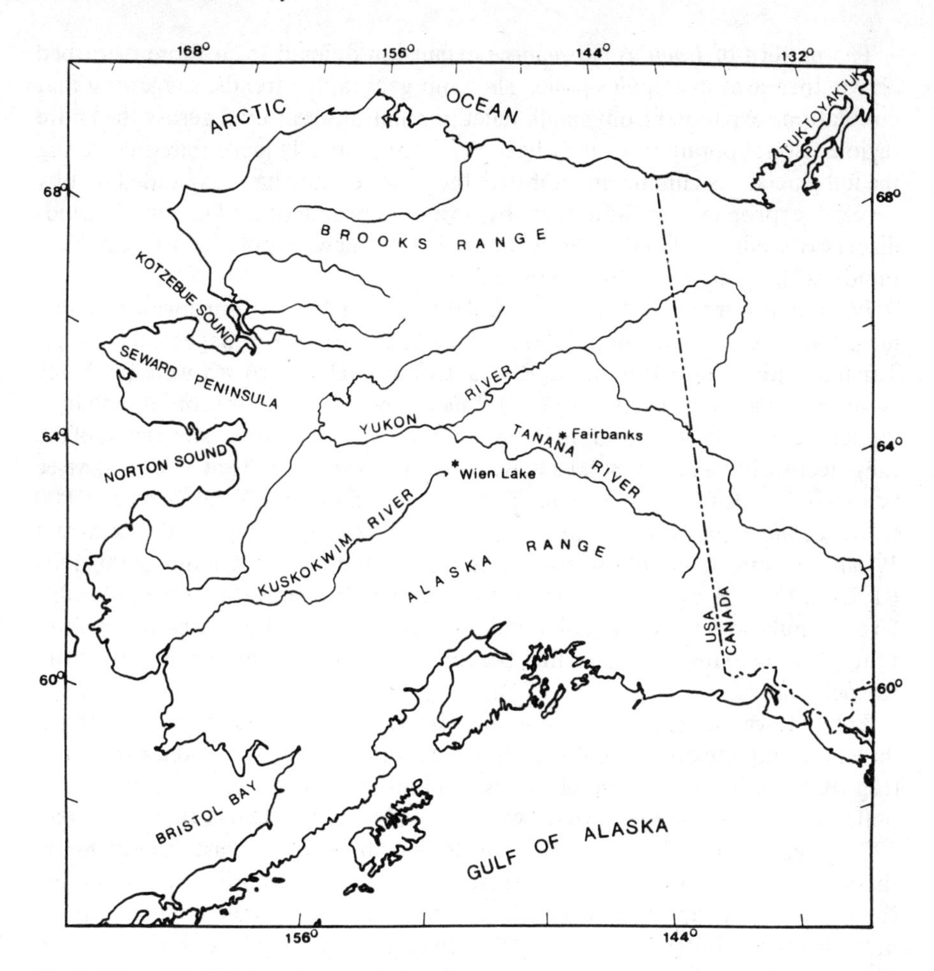

Fig. 2. Map of Alaska showing major geographic features and location of Wien Lake. (Hu et al. 1993)

differed. Although both *Betula* and Ericaceae populations initially expanded into herb tundra ca. 14000 B.P., *Betula* shrubs spread more rapidly, reaching north eastern Alaska by 12000 B.P. compared to 9000 B.P. for Ericaceae. *Alnus* shrubs initially spread slowly from the far western Brooks Range 10000 to 8000 B.P., but then expanded rapidly across the central and eastern Brooks Range and interior plateaus between 8000 and 7000 B.P. Population expansions 8000–7000 B.P. were so rapid that *Alnus* pollen increases are synchronous within the limits of radiocarbon dates. The predominant landscape ecosystems differed during these two expansion phases; *Alnus* invaded *Betula* shrub tundra of the western Brooks Range and *Picea* forested landscapes of central Alaska and the eastern Brooks Range. Its spread into boreal forest regions was particularly rapid, despite potential negative effects of competition with trees.

The pattern of *Populus balsamifera* expansion differed from those described above. Increases in *Populus* pollen show no geographic trends, suggesting that populations expanded from small, relict populations scattered across the entire region. Clonal populations may have persisted in locally favorable sites during the full glacial. As climate ameliorated, these clones may have expanded first by vegetative propagation and then by sexual reproduction. The small, wind-dispersed seeds of *Populus* would have allowed new populations to establish rapidly at long distances from parent clones.

Picea glauca invaded Alaska ca. 10 000 to 9000 B.P. from northwest Canada, most likely as riparian populations along major river drainages such as the Tanana, Yukon, and Porcupine Rivers. Conditions for growth were probably favorable on a wide variety of sites, as *P. glauca* pollen was more common than in modern sediments of the region. Populations decreased at the western front of the range (central Alaska) ca. 8000 B.P., but close interval sampling at Wien Lake (see below) indicates that populations sizes returned to predecline levels by ca. 7000 B.P. A second expansion of *P. glauca* began ca. 6000 B.P. when both *P. glauca* and *P. mariana* spread rapidly westward, reaching modern range limits 5000–4000 B.P. Like *Alnus*, *P. glauca* spread into different regional vegetation types. Initially, large populations invaded tall *Betula* shrub tundra and *Populus* woodlands. Later, *P. glauca* spread in small numbers into landscapes dominated by low shrub tussock tundra.

The patterns described above emphasize striking differences in the timing, direction, and rate of expansion of major shrub and tree taxa into Alaskan tundra (Fig. 3). Such individualistic behavior of plant species is a common feature of pollen records of late Quaternary vegetation change (e.g. Davis 1986; Brubaker 1988; Prentice 1992). Despite contrasting species histories, at least three common themes can be seen in the tree and shrub invasion patterns. First, species ranges can expand very rapidly. Second, the biological and ecosystem context of plant movements can differ greatly even for a given species. Third, migrating populations can be quite large, because peak pollen percentage and accumulation rates occur at the time of or soon after first arrival.

These observations indicate the difficulty of predicting how tundra will respond to future climate changes. Clearly, we should expect species responses to be individualistic. Species that presently occur together may follow very different population trajectories under different future climates. The fossil record indicates that plant movements can be very rapid and that individual species can establish in a variety of biological settings, suggesting that biological/ecosystem feedbacks are at times unimportant in controlling population expansions. Plant movements have been relatively slow at times, however, and may have been governed by internal controls or the spread in slowly improving environments. With the aid of geochemical sediment evidence (see below), it may be possible to identify some of the ecosystem changes associated with different types of invasion patterns. Overall, the fossil record of species expansion into tundra indicates that modern studies addressing potential consequences of climate change should focus on species-level as well as ecosystem-level

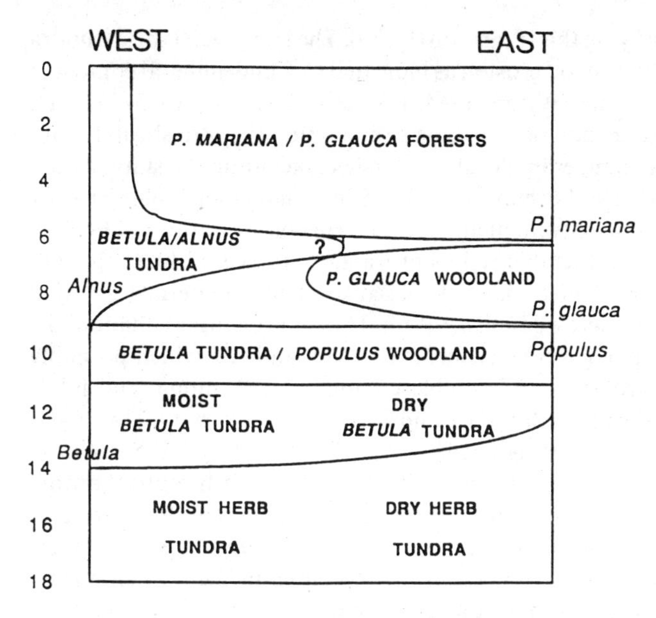

Fig. 3. Schematic diagram summarizing major temporal and geographic patterns in late Quaternary pollen data from north-central Alaska (Anderson and Brubaker 1994). On the horizontal axis, west corresponds to ca 62 °W and east to ca. 142°W longitude. On the vertical axis, numbers represent age × 10^3 B.P.

processes and should anticipate diverse responses of species to environmental changes.

8.5 What Major Changes in Ecosystem Processes Accompanied the Conversion of Tundra to Forest at the End of the Last Glacial Period?

To understand ecosystem processes associated with long-term vegetation changes, sediment constituents (e.g. elemental geochemistry, charcoal) other than pollen are needed. For example, geochemical investigations of lake sediments, when coupled with the pollen, can provide insights into the history of soil development and soil-vegetation interactions with in lake catchments (Mackereth 1966; Pennington et al. 1972; Engstrom and Wright 1984; Engstrom and Hansen 1985; Pennington 1986). Charcoal deposited in lake sediments provide information on the role of fires in vegetation changes (Patterson et al. 1987; Clark 1990). Unfortunately, few multiple proxy studies are available from arctic and subarctic regions (Ritchie 1984, 1987; Hu et al. 1993; MacDonald et al. 1993; Anderson and Brubaker 1994). The following discussion is primarily based on a record from Wien Lake in central Alaska (Fig. 2; Hu et al. 1993) with reference to relevant Canadian studies.

The Wien Lake (64° 20′ N, 152° 16′ W) record encompasses the transition from late-glacial shrub tundra to interglacial boreal forests and provides insights into soil processes accompanying this transition (Fig. 4). The late-glacial shrub tundra was characterized by intense soil erosion as indicated by a high mineral content of the sediments (Engstrom and Wright 1984; Likens and Moeller 1985). Around 10 500 B.P., the rapid expansion of *Populus-Salix* communities into shrub tundra induced major edaphic changes in the lake watershed, including soil stabilization and a marked increase in soil organic content. These changes are suggested by increases in sediment organic content and increases in concentrations of Fe, Mn, and Al within the acid-extractable sediment fraction. Increased litterfall from *Populus-Salix* communities most likely increased soil humic materials, elevated soil acidity, and possibly decreased soil aeration. These soil changes enhanced the solubility of redox- or pH-sensitive elements (Fe, Mn, and Al), which probably formed oxides, oxyhydroxides, and soluble complexes with humic and fulvic acids and entered the lake through surface and groundwater discharge (Jones and Bowser 1987; Engstrom and Wright 1984). Soils in the Wien Lake drainage probably changed from sparsely vegetated Entisol to forested Inceptisol during the period of *Populus-Salix* dominance. These soil changes might have, in turn, directly contributed to the disappearance of extensive *Populus-Salix* communities ca. 9500 B.P. (Cwynar 1982; Hu et al. 1993), resulting in the reversion from forest to tundra at some sites (Anderson et al. 1988).

The demise of *Populus-Salix* forest cover and the successive invasions and population fluctuations of *P. glauca* and *P. mariana* into the Wien Lake region did not result in changes in sediment geochemistry, suggesting that these vegetation shifts had no major consequences on soils in the watershed. These results contrast with observations in modern boreal forests of Alaska, where succession from

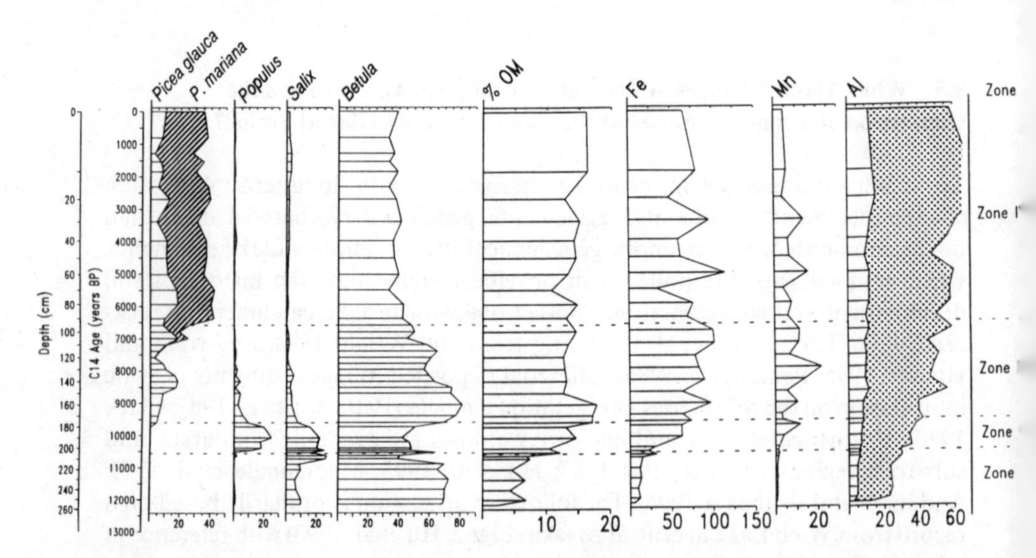

Fig. 4. Pollen percentages, sediment organic content (OM, % dry wt.), and elemental concentration (mg/g dry sediment) from Wien Lake, Alaska. *Stippled area* represents 10x exaggeration. *Dashed lines* are boundaries of pollen zones defined in Hu et al. (1993)

deciduous stands to coniferous stands and from *P. glauca* dominance to *P. mariana* dominance is typically accompanied by the accumulation of soil organic matter and the development of anaerobic, waterlogged soils (Van Cleve et al. 1983, 1986, 1991; Viereck et al. 1993).

Studies in central and eastern Canada show similar geochemical changes associated with the expansion of forest into shrub tundra, although the species of invading trees was not *P. balsamifera*, as in central Alaska. In central Canada, *P. mariana* populations expanded rapidly into shrub tundra during the middle Holocene (MacDonald et al. 1993), resulting in sharp increases in Mn, high sediment organic matter, and low concentrations of elements indicative of soil erosion (Ti, Na, Al, and Ca). As forest tundra reverted back to tundra ca. 4000 B.P., these sediment proxies resumed values similar to those prior to tree-line advance, suggesting reversion of soil conditions along with the vegetation. These geochemical changes might have also partially resulted from concurrent changes in the productivity and acidity of the lake systems (MacDonald et al. 1993). In eastern Canada, conversion of tundra to *P. glauca* woodland in the early Holocene and subsequent shifts from *P. glauca* to *Abies balsamea* to *P. mariana* corresponded to dramatic increases in sedimentary organic matter and concentrations of Fe and Mn (Engstrom and Hansen 1985), presumably because the solubility of these elements increases with the decrease in soil redox during the development of organic-rich, waterlogged soils (Mackereth 1966; Pennington et al. 1972; Engstrom and Wright 1984).

Sediment geochemistry records at each of the sites described above indicate that the establishment of forests in shrub tundra landscapes decreased erosion and accelerated losses of pH- and redox-sensitive elements from soils. Organic acids produced by the decomposition of increased leaf litter input and/or anaerobic conditions caused by soil paludification probably caused the losses of these elements from the watershed. Thus, tree invasions into tundra have major effects on soil development. The increased organic input to soils and the temperature increases, which ultimately caused tree invasions, were probably the predominant controls over these soil changes. Although the effects of differences in litter quality on soil forming processes is unclear, the invasion of deciduous vs evergreen conifer trees would most likely result in different rates of soil organic buildup. To investigate this possibility, fine-temporal studies are needed of sediment geochemistry in watersheds invaded by species with different litter quality.

8.6 Conclusion

The temporal perspective provided by late Quaternary pollen records clearly expands our ability to document arctic biodiversity from studies of contemporary landscapes. Differences between past and present tundra, both in terms of dominant taxa and physiognomy, suggest that modern investigations cannot include all possible states of arctic biodiversity and ecosystem function, because the large-scale boundary conditions of today's climate represent only a portion of the possible conditions that can result in tundra vegetation. Fossil records provide glimpses of tundra composition and function under different conditions. Just as past tundra environments differed from today, future arctic climates are projected to be substantially different from present. Although the fossil record does not provide clear predictions of future arctic ecosystems, it does provide evidence that such large-scale climate changes may result in major community reorganizations and new types of dominant ecosystems rather than simple changes in the areal extent of existing communities and ecosystems. Thus, our approaches to anticipating future changes should include experiments and manipulations under a wide array of conditions, some of which are not found in contemporary arctic tundra.

Acknowledgments. We thank F. Stuart Chapin and Christian Koerner for helpful comments on early drafts of the manuscript. The work was supported in part by National Science Foundation grants DPP-88922491 and ATM-9123449. This paper is Contribution #20 of the PALE (Paleoclimate from Arctic Lakes and Estuaries) project.

References

Anderson PM, Brubaker LB (1993) Holocene vegetation and climate histories of Alaska. In: Wright HE, Kutzbach JE, Webb T III, Ruddiman WF, Street-Perrott FA, Bartlein RJ (eds) Global climates since the glacial maximum. University of Minnesota Press, Minneapolis, pp 386–400

Anderson PM, Brubaker LB (1994) Vegetation history of northcentral Alaska: a mapped summary of late-Quaternary pollen data. Quat Sci Rev 13: 71–92

Anderson PM, Reanier RE, Brubaker LB (1988) Late-Quaternary vegetational history of the Black River region in northeastern Alaska. Can J Earth Sci 25: 84–94

Anderson PM, Bartlein PJ, Brubaker L B, Gajewski K, Ritchie JC (1989) Modern analogues of late-Quaternay pollen spectra from the western interior of North America. J Biogeogr 16: 573–596

Andrews JT, Brubaker LB (1991) Paleoclimate of arctic lakes and estuaries: science and implementation plan. Proc of a Steering Committee Meeting, March 16–17, 1991, University of Colorado, Boulder

Barnosky CW, Anderson PM, Bartlien PJ (1987) The northwestern U.S. during deglaciation: vegetation history and paleoclimatic implications. In: Ruddiman WF, Wright HE Jr (eds) North America and adjacent oceans during the last deglaciation. Geological Society of America, Boulder, pp 289–321

Bartlein PJ (1988) Late-Tertiary and Quaternary palaeoenviroments. In: Huntley B, Webb T III (eds) Vegetation history. Kluwer, Dordrecht, pp 113–152

Bartlein PJ, Prentice IC (1989) Orbital variations, climate and paleoecology. Trends Ecol Evol 4: 195–199

Billings WD (1987) Carbon balance of Alaskan tundra and taiga ecosystems, past, present and future. Quat Sci Rev 6: 165–177

Bonan GB, Shugart HH, Urban DL (1990) The sensitivity of some high-latitude boreal forests to climatic parameters. Clim Change 16: 9–29

Bradley RS (1985) Quaternary paleoclimatology. Allen and Unwin, Boston

Broeker WS, Denton GH (1990) What drives glacial cycles? Sci Am 256: 49–56

Brubaker LB (1988) Vegetation history and anticipating future vegetation change. In: Agee JK, Johnson DR (eds) Ecosystem management for parks and wilderness. University of Washington Press, Seattle, pp 41–61
Clark JS (1990) Fire and climate change during the last 750 yr in northwestern Minnesota. Ecol Monogr 60: 135–159
CLIMAP Project Members (1981) Seasonal reconstructions of the earth's surface at the last glacial maximum. Geol Soc Am Map Chart Ser MC-36
COHMAP Members (1988) Climatic changes of the last 18,000 years: observations and model simulations. Science 241: 1043–1052
Cwynar LC (1982) A late-Quaternary vegetation history from Hanging Lake, northern Yukon. Ecol Monogr 52: 1–24
Davis MB (1986) Climatic instability, time lags, and community disequilibrium. In: Diamond J, Case TJ (eds) Community ecology. Haprer and Row, New York, pp. 269–284
Denton GH, Hughes TJ (eds) (1981) The last great ice sheets. Wiley-Interscience, New York
Emanuel WR, Shugart HH, Stevenson MP (1985) Climate change and the broad-scale distribution of terrestrial ecosystem complexes. Clim Change 7: 29–43
Engstrom DR, Hansen BCS (1985) Postglacial vegetational change and soil development in southeastern Labrador as inferred from pollen and chemical stratigraphy. Can J Bot 63: 543–561
Engstrom DR, Wright HE Jr (1984) Chemical statigraphy of lake sediments as a record of environmental change. In: Harworth EY, Lund JWG (eds) Lake sediments and environmental history. Leicester University Pres, Leicester, 11–67
Gajewski K, Payette S, Ritchie JC (1993) Holocene vegetation history at the boreal forest- shrub tundra transition in northwestern Quebec. J Ecol 81: 433–444
Hansen JI, Fung A, Lacis A, Ring D, Russel G, Lebedeff S, Ruedy R, Stone P (1988) Global climate changes as forecast by the GISS 3-D model. J Geophy Res 93: 9341–9364
Hopkins DM, Matthews JV Jr, Schweger CE, Young SB (eds) (1982) Paleoecology of Beringia, Academic Press, New York.
Hu FS, Brubaker LB, Anderson PM (1993) A 12000 year record vegetation change and soil development from Wien Lake, central Alaska. Can J Bot 71: 1133–1142.
Huntley B, Birks HJB (1983) An atlas of past and present poolen maps for Europe: 0–13,000 years ago. Cambridge University Press, Cambridge
Huntley B, Prentice IC (1993) Holocene vegetation and climates of Europe. In: Wright HE, Kutzbach JE, Webb T III, Ruddiman WF, Street-Perrott FA, Bartlein RJ (eds) Global climates since the last glacial maximum. University of Minnesota Press, Minneapolis, pp 136–168
Jones BF, Bowser CJ (1987) The minerology and related chemistry of lake sediments. In: Lerman A (ed) Lakes, chemistry, geology, physics. Springer Berlin Heidelberg, New York, pp 179–235
Kutzbach JE, Guetter PJ (1986) The influence of changing orbital parameters and surface boundary conditions on climate simulations for the past 18,000 years. J Atmos Sci 43: 1726–1759
Lamb HF, Edwards ME (1988) The Arctic. In: Huntley B, Webb T III(eds) Vegetation history. Kluwer, Boston pp. 519–555
Likens GE, Moeller RE (1985) Chemistry. In: Likens GE (ed) An ecosystem approach to limnology: Mirror Lake and its watershed. Springer, Berling Heidelberg, New York, 392–409
Lozhkin AV, Anderson PM, Eisner WR, Ravako LG, Hopkins DM, Brubaker LB, Colinvaux PA, Millier MC (1993) Late Quaternary lacustrine pollen records from southwestern Beringia. Quat Res 39: 314–324
MacDonald GM, Edwards TWD, Moser KA, Pienitz R, Smol JP (1993) Rapid response of treeline vegetation and lakes to past climate warming. Nature (Lond) 361: 243–246
Mackereth FJH (1966) Some chemical observations on postglacial lake sediments. Philos Trans R Soc Lond B 250: 165–213
Manabe S, Brocolli AJ (1985) The influence of continental ice sheets on the climate of an ice age. J Geophy Res 90: 2167–2190

Marion GM, Oechel WC (1993) Mid- to late-Holocene carbon balance in Arctic Alaska and its implications for future global warming. Holocene 3: 193–200
Mitchell JFB Grahame NS, Needham KJ (1988) Climate simulations at 9000 years before present: seasonal variations and effects of Laurentine ice sheet. J Geophys Res 93: 8283–8303
Ovenden L (1990) Peat accumulation in northern peatlands. Quat Res 33: 377–386
Overpeck JT, Webb RS, Webb T III (1992) Mapping eastern North American vegetation change of the past 18 ka: no-analogs and the future. Geology 20: 1071–1074
Patterson WA, Edwards KJ, Maguire DJ (1987) Microscopic charcoal as a fossil indicator of fire. Quat Sci Rev 6: 3–23
Payette S, Filion L, Delwarde A, Begin C (1989) Reconstruction of treeline vegetation response to long-term change. Nature 341: 429–432
Pennington W (1986) Lags in adjustment of vegetation to climate caused by the pace of soil development: evidence from Britian. Vegetatio 67: 105–118
Pennington W, Haworth EY, Bonny AP, Lishman JP (1972) Lake sediments in northern Scotland. Philos Trans R Soc Lond B 264: 191–294
Peterson GM (1993) Vegetational and climate history of the western Former Soviet Union. In: Wright HE, Kutzbach JE, Webb T III, Ruddiman WF, Street-Perrott FA, Bartlein RJ (eds) Global climates since the last glacial maximum. University of Minnesota Press, Minneapolis, pp 169–193
Prentice IC (1992) Climate change and long-term vegetation dynamics. In: Glenn-Lewin DC, Peet RK, Veblen TT (eds) Plant succession: theory and prediction. Chapman and Hall, London, pp 293–339
Ritchie JC (1984) Past and present vegetation of the far northwest of Canada. University of Toronto Press, Toronto
Ritchie JC (1987) Postglacial vegetation of Canada. Cambridge University Press, Cambridge
Ritchie JC, Harrison SP (1993) Vegetation, lake levels, and climate in western Canada during the Holocene. In: Wright HE, Kutzbach JE, Webb T III, Ruddiman WF, Street-Perrott FA, Bartlein RJ (eds) Gloal climates since the last glacial maximum. University of Minnesota Press, Minneapolis, pp 401–414
Schlesinger ME, Mitchell JFB (1987) Model projection of the equilibrium climatic response to increased CO_2. Rev Geophy 25: 760–798
Shaver GR, Chapin FS III (1991) Production: biomass relationships and elemental cycling in contrasting arctic vegetation types. Ecol Monogr 61: 1–31
Sher AV (1974) Pleistocene mammals and stratigraphy of the far north east USSR and North America. Int Geol Rev 16: 1–283
Shilo NA (1987) Resolution: Interdisplinary Stratigraphic Congress for Quaternary Systems of Eastern USSR. Russian Ministry of Geology, USSR Academy of Science Far East Branch, Magadan
Van Cleve K, Dyrness CT, Viereck LA, Fox J, Chapin FS III, Oechel W (1983) Taiga ecosystems in interior Alaska, Bio Science 33: 39–44
Van Cleve K, Chapin FS III, Flanagan PW, Viereck LA, Dyrness CT (eds) (1986) Forest ecosystems in the Alaskan taiga: a synthesis of structure and function. Springer Berlin Heidelberg, New York
Van Cleve K, Chapin FS III, Dyrness CT. Viereck LA (1991) Element cycling in taiga forests: state-factor control. Bio Science 41: 78–88
Viereck LA, Dyrness CT, Foote MJ (1993) An overview of the vegetation and soils of the floodplain ecosystems of the Tanana River, interior Alaska. Can J For Res 23: 889–898
Walker DA, Everett KR (1991) Loess ecosystems of northern Alaska: regional gradient and toposequence at Prudhoe Bay. Ecol Monogr 61: 437–464
Webb T III, Bartlein PJ, Harrison SP, Anderson KH (1993a) Vegetation, lake levels and climate in eastern North America for the past 18,000 years. In:Wright HE, Kutzbach JE, Webb T III, Ruddiman WF, Street-Perrott FA, Bartlein RJ (eds) Global climates since the last glacial maximum. University of Minnesota Press, Minneapolis, pp 415–467
Webb T III, Ruddiman FA, Street-Perott FA, Markgraf V, Kutzbatch JE, Bartlein PJ, Wright HE Jr, Prell WL (1993b) Climatic changes during the past 18,000 years: regional syntheses, mechanisms, and causes. In: Wright HE, Kutzbach JE, Webb T III, Ruddiman WF,

Street-Perrott FA, Bartlein RJ (eds) Global climates since the last glacial maximum. University of Minnesota Press, Minneapolis, pp 514–535
Whitlock C (1992) Vegetational and climatic history of the Pacific Northwest during the past 20,000 years: implications for understanding present-day biodiversity. Northwest Environ J 8: 5–28
Wright HE, Kutzbach JE, Webb T III, Ruddiman WF, Street-Perrott FA, Bartlein RJ (eds) (1993) Global climates since the last glacial maximum. University of Minnesota Press, Minneapolis

9 Effects of Mammals on Ecosystem Change at the Pleistocene-Holocene Boundary

S.A. Zimov, V.I. Chuprynin, A.P. Oreshko, F.S. Chapin III, M.C. Chapin, and J.F. Reynolds

9.1 Introduction

The present vegetation-climate relationships have been used to deduce past climate based on paleoecological indicators of past vegetation (Grichuk et al. 1984; COHMAP) and to predict future vegegation distribution based on General Circulation Model projections of future climate (Pastor and Post 1988). The implicit assumption behind these climatic reconstructions is that climate is the major factor determining vegetation distribution. However, animals also strongly influence the species composition, structure, and processes of ecosystems (O'Neill 1976), including boreal forest (Pastor et al. 1988; Bryant et al. 1991) and tundra (Batzli et al. 1980). Consequently, changes in animal abundance can radically alter the structure and species composition of vegetation (Bond 1993), for example, converting grassland to shrubland or forest (Owen-Smith 1987; Schlesinger et al. 1990). Predicting the future structure and distribution of ecosystems (Pastor and Post 1988; IGBP 1990) may, therefore, require consideration of animal diversity and trophic interactions as well as direct effects of climate on vegetation.

One of the most dramatic vegetation changes of the last 20 000 years was the conversion of a vegetation mosaic dominated by semi-arid, grass steppe with dry soils and a well-developed grazing megafauna to a mosaic dominated by wet moss tundra (including forest tundra with scattered trees) without a large grazing fauna. This change occurred at the end of the Pleistocene 10 000–12 000 years before present (ka B.P.) and has generally been ascribed to climatic change (Hopkins 1967; Hopkins et al. 1982; Grichuk 1984). However, the extinction of Pleistocene megafauna (mammoths, bison, horses, etc.), which also occurred at the end of Pleistocene, could have contributed to this biome shift. Here, we address two questions: (1) Are the effects of grazing mammals on ecosystem processes sufficiently large to cause important changes in arctic soils and vegetation? (2) Could extinction of the Pleistocene megafauna have contributed to the Pleistocene-Holocene vegetation change?

Department of Integrative Biology, University of California, Berkeley, CA 94720, USA

Ecological Studies, Vol. 113
Chapin/Körner (eds) Arctic and Alpine Biodiversity

9.2 Mammalian Effects on Ecosystem Processes

In fertile grass-dominated meadows, a set of feedbacks promotes productivity and grazing (Fig. 1a; Chapin 1991). Growth of grasses in the current tundra environment is stimulated by nutrient inputs from fertilizers (McKendrick et al. 1978; Shaver and Chapin 1986), animal carcasses, and feces (Batzli et al. 1980; McKendrick et al. 1980), or intensive disturbance by humans (Chapin and Shaver 1981; Zimov 1990) or animals (Batzli and Sobaski 1980). Mowing stimulates the aboveground production of steppe grasslands five fold (Kucheruk 1985) and the aboveground production of sedge tundra three-fold (Peshkova and Andreiashchkina 1983), indicating that grazing stimulates aboveground productivity in these ecosystems. This grass-dominated vegetation has higher tissue nutrient concentrations than mosses or dwarf shrubs, is more digestible, and attracts grazers (White and Trudell 1980). Both the high litter quality and the rapid turnover of nutrients by grazers stimulate decomposition, mineralization, and nutrient turnover (Batzil et al. 1980; Chapin 1991) and reduce soil moisture (Chapin and Shaver 1985). The dense vascular litterfall shades out or physically smothers mosses, leading to a decline in their abundance (Chapin et al. 1994). In high-fertility sites many grasses are covered with snow while green and, therefore, maintain their high forage quality throughout the winter (Tishkov 1985), leading to intense overwinter microtine grazing (Batzli et al. 1980; Chapin and Shaver 1981). These feedbacks which maintain grasslands presently occur only in isolated locations of intensive lemming grazing in the far north, because there are no large grazing mammals that feed mainly on grasses.

Moss-dominated tundra, which predominates in the absence of grazing, has quite different effects on nutrient cycling. Mosses have a low nitrogen concentration and high concentrations of recalcitrant lignin-like polymers (Chapin et al. 1986) that retard decomposition (Coulson and Butterfield 1978; Johnson and Damman 1991). Moreover, the low thermal conductance of mosses reduces soil temperature and thaw depth during summer (Fig. 1b). Mosses have low rates of evapotranspiration, promoting development of waterlogged soils. Low soil temperature, increased acidity, waterlogged soils, and low litter quality combine to reduce rates of decomposition and nitrogen mineralization in moss-dominated ecosystems (Johnson and Damman, 1991: Van Cleve et al. 1991) and limit the productivity of vascular plants (Chapin and Shaver 1985). Moss growth is limited more strongly by water than by nutrients (Skre and Oechel 1979; Skre and Oechel 1981), so the moist, nutrient-deficient soil environment promotes moss growth. Eventually, as the mosses grow, thaw depth decreases to the point that only the moss layer itself thaws (Kriuchkov 1973; Chernov 1978; Van Cleve et al. 1983), and grasses, which require access to thawed mineral soil, are excluded from these communities. Because of these feedbacks (Fig. 1), moss ecosystems develop readily in the far north. For example, in northwestern Siberia, 60–70% of the forested area disappeared during historic time due to moss development and associated waterlogging of soils (Kriuchkov 1973), and Alaskan forests succeed to moss-dominated muskegs and spruce forests in the absence of fire (Van Cleve

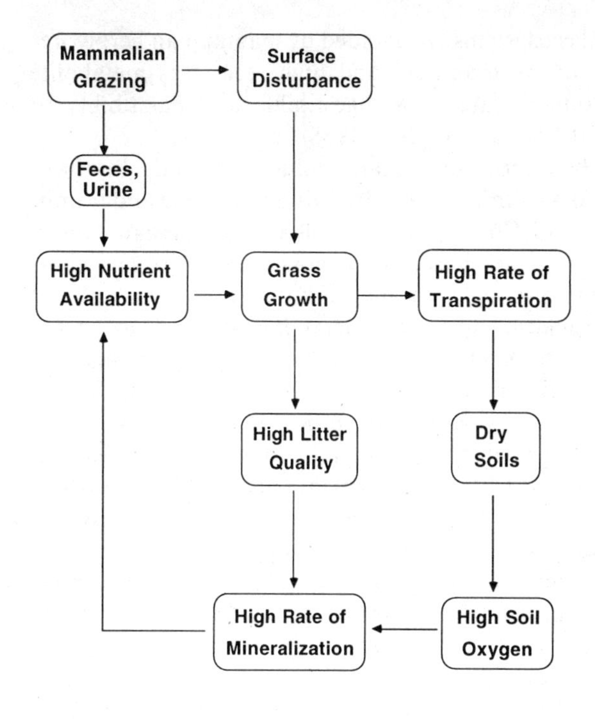

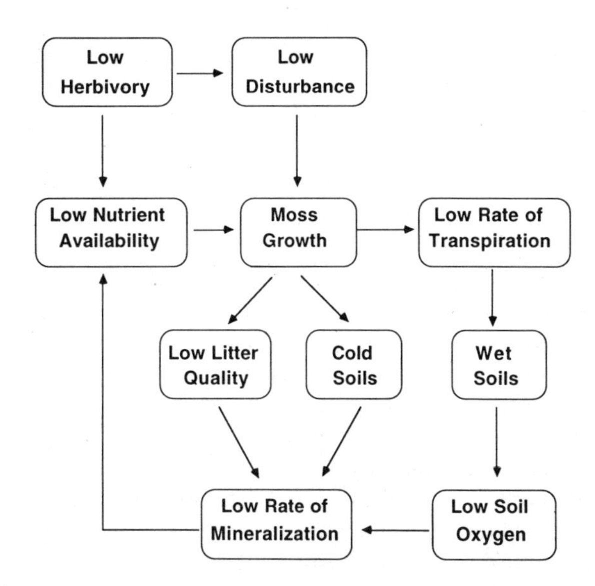

Fig. 1. Ecosystem processes in arctic tundra in **a** the presence and **b** absence of mammalian grazing

et al. 1991). Moss-dominated ecosystems are avoided by mammalian herbivores (Wolff 1980) because the low tissue-nutrient and high secondary-metabolite concentrations in the vegetations cause low digestibility and palatibility to animals (Bryant and Kuropat 1980; Pastor et al. 1988).

Mammalian grazers can be important in maintaining grasslands, but have large detrimental effect on moss-dominated tundra through grazing and trampling (Chernov 1980; Zimov and Chuprynin 1991). Areas of Russian tundra covered by deep snow, such as ravines, lack a dense moss cover. Here, lemmings congregate under the insulative snow cover and both destroy the mosses and fertilize the soil. During population outbreaks, lemmings expand into areas between ravines and eliminate mosses by grubbing (Chernov 1978). In this way, lemmings contribute to the maintenance of meadows, which are eight- to ten fold more productive than moss associations (Chernov et al. 1983). The high productivity of meadows is associated with an eight to ten fold increase in biomass of soil invertebrates and with rapid nutrient cycling (Chernov 1978; Chernov et al. 1983). Similarly, in Alaskan lowland polygonal tundra, troughs between ice-wedge polygons have deeper snow and support larger lemming populations than adjacent microhabitats (Batzli et al. 1980). The lemmings remove mosses by grubbing and stimulate nutrient cycling (Batzli et al. 1980). Reindeer grazing also causes expansion of graminoids and reduced cover of mosses and shrubs (Thing 1984). At a larger scale, where fire or human disturbance destroys the moss turf, meadow vegetation develops and becomes a focus for activity of mammalian herbivores (Fox 1978; Bryant and Chapin 1986; Field et al. 1992).

Differences in productivity between moss- and grass-dominated tundra could lead to differences in soil moisture. Transpiration rate correlates with photosynthesis at the leaf level because stomatal conductance is adjusted to match the biochemical potential for photosynthesis (Wolff 1980; Farquhar and Sharkey 1982) and at the canopy level because both processes correlate with leaf-area development (Collatz et al. 1991: Field et al. 1992). For these reasons, transpiration is proportional to photosynethic carbon gain (Chapin 1993). If primary productivity during the Late Pleistocene grass-dominated steppe was higher than that of the moss-dominated tundra of the Holocene, the Pleistocene soils could have been drier because of greater evapotranspiration rather than because of drier climate (Fig. 2). For example, addition of nutrients of tussock-tundra vegetation in Alaska increased productivity and reduced soil moisture (Chapin and Shaver 1985). The presence of large grazers increased the productivity of northern steppe vegetation and reduced soil moisture (Kucheruk 1985). Experiments with weighing lysimeters showed that arctic steppe vegetation with its deep roots and high transpiration rates reduced soil moisture more strongly than did moss-dominated tundra vegetation (Zimov and Chuprynin 1991).

In summary, grazing animals can have large effects on nutrient cycling and evapotranspiration through their effects on the balance between grass-vs moss-dominated tundra. How might this relate to extinction of megafauna and vegetation change at the end of the Pleistocene?

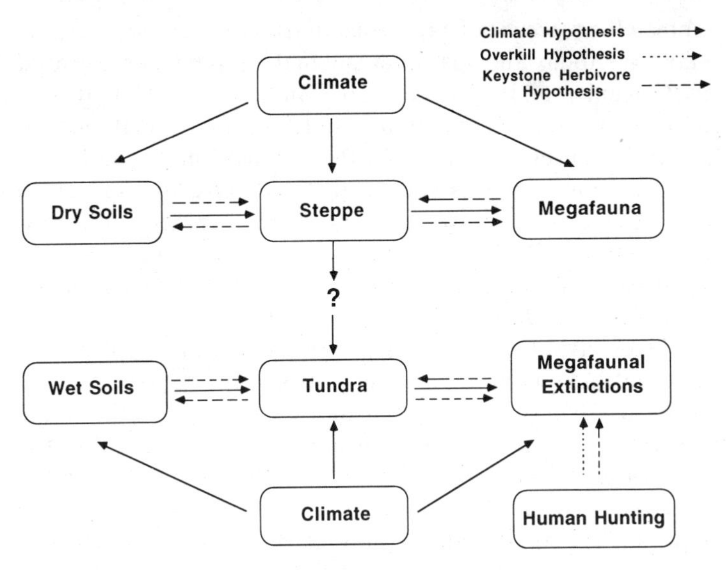

Fig. 2. Interaction among climate, Pleistocene megafauna, and soil moisture as described by the climate hypothesis, the overkill hypothesis, and the keystone-herbivore hypothesis

9.3 Hypotheses for the Steppe-Tundra Transition

The *Pleistocene overkill hypothesis* suggests that increased human hunting pressure caused or contributed to megafauna extinction at the end of the Pleistocene (Fig. 2; Hopkins 1967; Martin 1984). Early humans may have coexisted with steppe megafauna in the mid-Pleistocene, using sharpened bones and antlers as their hunting weapons. However, 10–15 ka B.P. Beringia became occupied by an anatomically modern people that hunted with stone microblades (stone chips) inserted into wooden or bone shafts (West 1981; Guthrie 1990) and fed extensively on the steppe megafauna (Haynes 1982). The spread of this modern microblade hunting technology through Beringia is *correlated* in time with the extinction of the steppe megafauna and with vegetation change from steppe to tundra.

Alternatively, the *climatic hypothesis* for megafauna extinction assumes that an arid, continental climate prevailed in Beringia during the Pleistocene, causing low summer precipitation and dry soils (Fig. 2; Yurtsev 1974; Hopkins et al. 1982; Sher 1988). Well-drained soils and warm summers promoted a productive steppe vegetation that suported populations of large grazers (mammoths, bison, and horses; Hopkins et al. 1982; Guthrie 1990). Dry soils facilitated animal movement. According to the climatic hypothesis, the climate became wetter during the Holocene: rainfall increased, a moss-lichen cover developed, the amount and productivity of the herbaceous vegetation decreased, and snow depth increased. Together these factors caused extinction of steppe megafauna (Tomirdiaro 1978; Guthrie 1990). The major evidence for the climatic hypothesis comes from widespread Pleistocene pollen and macrofossils of grasses, *Artemisia*, and other steppe taxa that today are most abundant in the arid regions of central Asia (Yurtsev 1974; Grichuk et al. 1984) reviewed by (Guthrie 1990). Thus, there is a strong *correlation* between presence of dry soils, steppe vegetation, and megafauna. It is unclear, however, how a cold Pleistocene climate could have allowed a steppe vegetation that was sufficiently productive to support an extensive fauna of large grazers. Moreover, reconstructions from general circulation models suggests that the Pleistocene soils of western Alaska and eastern Russia should have been wetter than today because of low evapotranspiration at low temperature (Manabe and Broccoli 1985; Rind 1987).

The keystone-herbivore hypothesis draws heavily on both the overkill and the climatic hypotheses and links them by assuming strong interactions between animal grazing and ecosystem processes. As with the climatic hypothesis, the keystone-herbivore hypothesis assumes that a productive steppe vegetation was necessary to support megaherbivores (Fig. 2). It extends the Pleistocene overkill hypothesis by suggesting that grazing by megaherbivores was essential to the maintenance of a productive grass-steppe vegetation, which produced dry soils through high rates of evapotranspiration. The keystone-herbivore hypothesis thus assumes that the presence of animal-induced steppe vegetation could have produced drier, more fertile soils than occur in the present tundra.

Given that both moss-dominated tundra and grass-dominated steppe ecosystems can create their own self-sustaining soil moisture and nutrient regimes (Fig. 1), what could have triggered a change in the relative abundance of steppe and tundra at the Pleistocene-Holocene boundary? We suggest that the loss of large mammalian grazers, perhaps triggered by the increase in human hunting pressure was responsible. Ungrazed, unfertilized meadows, resulting from fires, floods, and other disturbances, are the only grass-dominated ecosystems currently present in the north. They are unstable in the absence of herbivores or continued disturbance (Chernov 1978). When dry, uneaten litter accumulates, nutrients are sequestered in undecomposed organic matter, and nutrient turnover declines (Batzli 1977). Dry grass litter is highly reflective, reducing radiation gain, and is an effective heat insulator, minimizing heat input to soil during the short summer season, cooling the underlying soil, reducing depth of soil thaw, and promoting moss growth (Batzli 1977). Meadows degenerate and are replaced by shrubs and mosses that are more resistant to cold soil temperatures. A high density of grazers is necessary to maintain the productivity of natural tropical grasslands (McNaughton 1979).

9.4 Conclusions

1. Grazing promotes a series of postive feedbacks that favor grasses over mosses. In the presence of grazers, grasses are highly productive, have high transpiration rates, drying out soils, and a high litter quality, stimulating nutrient cycling and the maintenance of productive grassland.

2. In the absence of grazing, mosses predominate, insulating the soil and leading to cold waterlogged soils, where decomposition is slow due to both unfavorable soil environment and low litter quality.
3. We suggest that the Pleistocene megafaunal grazers contributed strongly to the maintenance of grassland during the Pleistocene and that the disappearance of these grazers at the end of Pleistocene, perhaps due to human hunting, contributed to the conversion of a productive grassland to a moss-dominated tundra.

We will never know the relative importance of the multiple causes of the transition from steppe to tundra at the end of the Pleistocene. However, the keystone-herbivore hypothesis provides a plausible explanation for the high productivity that must have been necessary to support megafauna in a Pleistocene climate that was colder than at present. Our analysis indicates that northern grazers have sufficiently large impacts on vegetation and ecosystem processes that their role must be considered in any effort to predict the pattern and productivity of past or future vegetation.

Acknowledgments. We thank M. Edwards, S. Armbruster, S. Brett-Harte, J. Bryant, S. Hobbie, M. Power, and H. Reynolds for critical review of the manuscript. The collaborative research was funded by the Russian Fund of Fundamental Research and the US Department Energy.

References

Batzli GO (1977) The influence of grazers on tundra vegetation and soils. Circumpolar Conf on Northern ecology, Proc 1: 215–225
Batzli GO, Sobaski S (1980) Distribution, abundance, and foraging patterns of ground squirrels near Atkasook, Alaska. Arct Alp Res 12: 501–510
Batzli GO, White RG, MacLean SF Jr, Pitelka FA, Collier BD (1980) The herbivore-based trophic system. In: Brown J, Miller PC, Tieszen, LL, Bunnel FL (eds) An arctic ecosystem: the coastal tundra at Barrow, Alaska. Dowden, Hutchinson and Ross, Stroudsburg, pp. 335–410
Bond WJ (1993) Keystone species. In: Schulze E-D, Mooney HA (eds) Ecosystem function and biodiversity. Springer Berlin Heidelberg New York, pp 237–253
Bryant JP Chapin FS III (1986) Browsing-woody plant interactions during boreal forest plant succession. In; Van Cleve K, Chapin FS III, Flanagan PW, Viereck LA, Dyrness CT (eds) Forest ecosystems in the Alaskan taiga; a synthesis of structure and function. Springer Berlin Heidelberg New York, pp. 213–225
Bryant JP, Kuropat PJ (1980) Selection of winter forage by subarctic browsing vertebrates: the role of plant chemistry Annu Rev Ecol Syst 11: 261–285
Bryant JP, Provenza FP, Pastor, J, Reichardt PB, Clausen TP, du Toit JT (1991) Interactions between woody plants and browsing mammals mediated by secondary metabolites. Annu Rev Ecol Syst 22: 431–446
Chapin FS III (1991) Effects of multiple environmental stresses on nutrient availability and use. In: Mooney HA, Winner WE, Pell EJ (eds) Response of plants to multiple stresses. Academic Press, San Diego, pp. 67–88
Chapin FS III(1993) Functional role of growth forms in ecosystem and global processes. In: Ehleringer JR, Field CB (eds) Scaling physiological processes: leaf to globe. Academic Press, San Diego, pp. 287–312

Chapin FS, III, Shaver GR (1981) Changes in soil properties and vegetation following disturbance in Alaskan arctic tundra. J Appl Ecol 18: 605–617
Chapin FS III, Shaver GR (1985) Individualistic growth response of tundra plant species to environmental manipulations in the field. Ecology 66: 564–576
Chapin FS III, McKendrick JD, Johnson DA (1986) Seasonal changes in carbon fractions in Alaskan tundra plants of differing growth form: implications for herbivores. J Ecol 74:707–731
Chapin FS III, Shaver GR, Gibbin AE, Nadelhoffer KG, Laundre JA (1994) Response of arctic tundra to experimental and observed changes in climate., Ecology (in press)
Chernov YI (1978) Struktura zhivotnogo naseleniia Subarktiki, Nauka, Moscow
Chernov YI (1980) Zhizn' tundry. Mysl', Moscow
Chernov YI, Matveeva N V, Zanokha LL (1983) Opyt izucheniia prirosta tsvetkovykh rastenii v coobshchestvakh Taimyra. Dokl Akad Nauk SSSR 272: 999–1002
COHMAP (1988) Climatic changes of the last 18,000 years: observations and model simulations. Science 241: 1043–1052
Collatz GJ, Ball JT,Grivet C, Berry JA (1991) Physiological and environmental regulation of stomatal conductance, photosynthesis and transpiration: a model that includes a laminar boundary layer. Agric For Meteorol 54: 107–136
Coulson JC, Butterfield J (1978) An investigation of the biotic factors determining the rates of plant decomposition on blanket bog. J Ecol 66: 631–650
Farquhar GD, Sharkey TD (1982) Stomatal conductance and photosynthesis. Annu Rev Plant Physiol 33: 317–345
Field C, Chapin FS III, Matson PA, Mooney H A (1992) Responses of terrestrial ecosystems to the changing atmosphere:a resource-based approach. Annu Rev Ecol Syst 23: 201–235
Fox JF (1978) Forest fires and the snowshoe hare-Canada lynx cycle. Oecologia 31: 349–374
Grichuk V P (1984) Late Pleistocene vegetation history. In: Velichko A A (eds) Late Quaternary environments of the Soviet Union. University of Minnesota Press, Minneapolis, pp. 155–178
Grichuk VP, Gurtovaya YY, Zelikson EM, Borisova OK (1984) Methods and results of Late Pleistocene paleoclimatic reconstructions. In: Velichko AA (eds) Late Quaternary environments of the Soviet Union. University of Minnesota Press, Minneapolis, pp 251–260
Guthrie RD (1990) Frozen fauna of the mammoth steppe: the story of Blue Babe. University of Chicago Press, Chicago
Haynes G (1982) Utilization and skeletal disturbances of North American prey carcasses. Arctic 35:266–281
Hopkins DM (1967) The Cenozoic history of Beringia–a synthesis. In: Hopkins DM (eds) The Bering land bridge. Stanford University Press, Stanford, pp 451–484
Hopkins DM, Matthews JJV, Schweger CE, Young SB (eds) (1982) Paleoecology of Beringia, Academic Press, New York
IGBP (1990) The international geosphere-biosphere programme: a study of global change. Rep No 12. IGBP Secretariat, Stockholm
Johnson LC, Damman AWH (1991) Species controlled *Sphagnum* decay on a South Swedish raised bog. Oikos 61:234–242
Kriuchkov VV (1973) Krainii Sever: problemy ratsional'nogo ispol'zovaniia prirodnykh resursov. Mysl', Moscow
Kucheruk VV (1985) Travoiadnye mlekopitaiushchie v aridnykh ekosisemakh vnetropicheskoi Evrasii. In: Sokolov BE (eds) Mlekopitaiushchie v nazemnykh ekosistemakh. Nauka, Moscow, pp 166–224
Manabe S, Broccoli AJ (1985) The influence of continental ice sheets on the climate of an ice age. J Geophys Res 90D:2167–2190
Martin PS (1984) Prehistoric overkill: the global model. In: Martin PS, Klein RG (eds) Quaternary extinctions. University of Arizona Press, Tucson, pp 354–403
McKendrick JD, Ott VJ, Mitchell GA (1978) Effects of nitrogen and phosphorus fertilization on carbohydrate and nutrient levels in *Dupontia fisheri* and *Arctagrostis latifolia*. In: Tieszen LL (eds) Vegetation and production ecology of an Alaskan arctic tundra. Springer, Berlin Heidelberg, New York, pp 509–537

McKendrick JD, Batzli GO, Everett KR, Swanson JC (1980) Some effects of mammalian herbivores and fertilization on tundra soils and vegetation. Arct Alp Res 12:565–578

McNaughton SJ (1979) Grazing as an optimization process: grass-ungulate relationships in the Serengeti.Am Nat 113:691–703

O'Neill RV (1976) Ecosystem persistence and heterotrophic regulation. Ecology 57:1244–1253

Owen-Smith RN (1987) Pleistocene extinctions: the pivotal role of megaherbivores. Paleobiology 13:351–362

Pastor J, Post WM (1988) Responses of northern forests to CO_2-induced climate change. Nature 334:55–58

Pastor J, Naimen RJ, Dewey B, McInnes P (1988) Moose, microbes, and the boreal forest. BioScience 38:770–777

Peshkova NV, Andreiashchkina NI (1983) Produktivnost' dvukh podvidov *Carex aquatilis* i ee izmenenie pod vliianiem mnogokratnogo otchuzhdeniia nadzemnoi biomassy. Ekologiia 4:30–35

Rind D (1987) Components of the ice age circulation. J Geophys Res 92D:4241–4281

Schlesinger WH, Reynolds JF, Cunningham GL, Huenneke LF, Jarrell WM, Virginia RA, Whitford WG (1990) Biological feedbacks in global desertification. Science 247:1043–1048

Shaver GR, Chapin FS III (1986) Effect of fertilizer on production and biomass of tussock tundra, Alaska, U.S.A. Arct Alp Res 18:261–268

Sher AV (1988) Sreda obitaniia plio-pleistotsenovykh mlekopitaiushchikh severo-vostochnoi Siberi. Stratigraphiia i korreliatsiia chetvertichnykh otlozhenii Azii i Tikhookeanskogo regiona. Akademii Nauk SSSR, Vladivostok, pp 78–79

Skre O, Oechel WC (1979) Moss production in a black spruce *Picea mariana* forest with permafrost near Fairbanks, Alaska, as compared with two permafrost-free stands. Holarct Ecol 2:249–254

Skre O, Oechel WC (1981) Moss functioning in different taiga ecosystems in interior Alaska. I. Seasonal, phenotypic, and drought effects on photosynthesis and response patterns. Oecologia 48:50–59

Thing H (1984) Feeding ecology of the West Greenland caribou (*Rangifer tarandus groenlandicus*) in the Sismiut Kangerlussuaq region. Dan Rev Game Biol 12:1–52

Tishkov AA (1985) Rastitel'noiadnye mlekopitaiushchie v ekosistemakh tundry. In: Sokolov VE (eds) Mlekopitaiushchie v nazemnykh ekosistemakh. Nauka, Moscow, pp 38–67

Tormirdiaro SV (1978) Natural process and development of territories of the permafrost zone. Nedra Press, Moscow

Van Cleve K, Dryness CT, Viereck LA, Fox J, Chapin FS III, Oechel WC (1983) Taiga ecosystems in interior Alaska. BioScience 41:78–88

Van Cleve K, Chapin FS III, Dryness CT, Viereck LA, (1991) Element cycling in taiga forest: state-factor control. BioScience 41:78–88

West FH (1981) The archaeology of Beringia. Columbia University Press, New York

White RG, Trudell J (1980) Habitat preference and forage consumption by reindeer and caribou near Atkasook, Alaska. Arct Alp Res 12:511–529

Wolff JO (1980) The role of habitat patchiness in the population dynamics of snowshoe hares. Ecol Monogr 50:111–129

Yurtsev BA (1974) Steppe communities in the Chukotka tundra and the Pleistocene "tundra-steppe". Bot Zh 59:484–501

Zimov SA (1990) Chelovek i priroda Severa: garmoniia protivopolozhnostei. Vestn Akad Nauk SSSR 2:118–133

Zimov SA, Chuprynin VI (1991) Ekosistemy: ustoichivost', konkurentsia tslenapravlennoye preobrazovanie. Nauka, Moscow

10 Paleorecords of Plant Biodiversity in the Alps

B. Ammann

10.1 Introduction

The study of past biodiversity allows us to investigate (1) types, (2) amplitudes, and (3) rate of changes that are different and possibly more rapid than those that can be observed even in long-term experiments. This may be partially caused by climatic changes that were stronger and/or faster than the ones in the instrumental meteorological record. For example, we may study the transition from the Late Glacial to the Holocene about 10 000 years ago, when the northern hemispheric and possibly the global temperatures rose as much as 7 °C in only about 50 years (Dansgaard et al. 1989). Therefore, ancient patterns of changes in biodiversity may add to our limited knowledge about how different species can respond to environmental changes, and how large the "plasticity of a species' autecology" may be.

High mountain ranges display steep gradients of environmental variables such as geology, soils, and climate producing a high diversity of habitats and thus a high species diversity (see Körner, this Vol.). In addition, mountain ranges have a complex history (see Agachanjanz and Breckle, this Vol.). For the Alps, classical plant-geographical literature discusses a threefold origin of the flora: survivors from the Tertiary, immigrants from three mountainous areas (inner-Asian, Scandinavian, and Mediterranean mountains), and the evolution of new species within the Alps during the late Tertiary and Quaternary (e.g. Merxmüller 1952, 1953, 1954; Favarger 1972, 1975; Küpfer 1974, Stebbins 1984, Ozenda 1985; Reisigl and Keller 1987).

Environmental history in high mountains therefore may have caused both speciation and extinction:

- tectonic uplift may have encouraged adaptive radiation;
- glaciations may have driven many Tertiary taxa to extinction, but also may have provided fragmentation of ranges and isolation of young species as well as refugia (Birks and Line 1993).

However I will concentrate the discussion on the period since the Last Glacial Maximum.

Institute for Geobotany, University of Bern, Altenbergrain 21, 3013 Bern, Switzerland

Ecological Studies, Vol. 113
Chapin/Körner (eds) Arctic and Alpine Biodiversity

10.2 Possibilities and Limitations of the Fossil Record

Besides high-resolution taxonomy (integrating morphology, molecular biology, and biogeography), the fossil record provides one of the few possibilites to reconstruct the biodiversity of the past.

Unfortunately, for many groups of organisms, virtually no fossils are found in Tertiary and Quaternary deposits, e.g. of birds, some groups of insects, or some algae. Pollen and spores of flowering plants and ferns, however, offer abundant identifiable fossils. Just as relevés of today's vegetation may give a rather rapid and precise description of an ecosystem, the record of fossil pollen–combined whenever possible with analysis of plant macrofossils–can serve as the best available indicator, albeit an approximation ("proxy"), of the past flora and vegetation. If some conditions are met, we can make inferences about species and habitat diversity of the past. Such conditions include (1) a high taxonomic resolution in pollen identification, (2) a refinement and completion of the list of taxa by the analysis of plant macrofossils (H. H. Birks 1993, 1994), (3) a fine time resolution of sampling, and (4) a time control by dating that is adequate for the processes of interest.

We also need to ask: What part of biodiversity can be addressed by studying the fossil record? Among the hierarchical levels of diversity (genetic, species, population, habitat, and landscape; Whittaker 1976), species or taxon diversity (consisting of richness and evenness) is the type of primary data we can collect by pollen analysis. In forested areas the pollen spectra of cores from very small hollows (Janssen 1986; Andersen 1988; Foster and Zebryk 1993) may be considered as a proxy for alpha- or within-habitat diversity. Cores from above the timberline resemble the more usual cores of forested areas from a lake or bog with larger pollen-source areas (Jacobson and Bradshaw 1981; Birks 1986, Fig. 1; Prentice 1988), which reflect the gamma or landscape diversity (Walker 1989; Birks and Line 1992).

10.3 Temporal Changes in Biodiversity in the Alps–Three Examples

Within the alpine mountain system from the Pyrenees to the Caucasus, the Alps provide the densest net of palynological sites, because of the long tradition of studies. But even so, we are still largely in the descriptive and narrative phases of collecting regional data, and we are only now beginning to explore the possibilities of developing an analytical phase (Ball 1980). Only by so doing can we again connect ecology and palaeoecology, as suggested (or even called for) by Walker (1990) and Birks (1993). The examples given below are therefore based on a few sites and cannot yet offer the great geographical coverage that one would desire. However, they are an attempt to develop our understanding of long-term ecological processes.

10.3.1 Species Richness Through Time

Only a few pollen studies in the Alps meet the three requirements needed for an estimation of changes in species richness through time: the situation of the coring site above timberline, the coverage of the entire Holocene, and high morphological detail in pollen identification (e.g. Bortenschlager 1970; de Beaulieu 1977; Welten 1982; Wick 1994).

Estimations of species richness derived from pollen records will primarily be of taxa richness (which depends partly on the abilities of identification). In many classical studies, the number of pollen types per sample (i.e. per stratigraphic horizons in a core) are given (e.g. Zoller 1960). However, because pollen sums counted for each level vary through a core, a standardization to the lowest number of grains counted per sample is needed. Birks and Line (1992) show how, among all the existing measures for diversity, rarefaction analysis is the most useful in palaeoecology. They present the theory, assumptions, and caveats of this method and summarize: "The technique of rarefaction analysis permits the estimation of the pollen richness ($E(T_n)$) that would be expected if all the pollen counts of the different samples had been the same size" and–after a summary of rarefaction used in ecology, palaeobiology, and palaeoecology–"In all cases the aim is to estimate richness for samples of different sizes when scaled down to a common size by considering the relative frequencies of individuals within categories". Thus, in $E(T_n)$, T is the number of pollen types in the counted sample and n the number of pollen grains chosen for standardization, i.e. the lowest pollen sum counted in the sequence.

For example, we chose Hopschensee on the Simplon Pass at 2017 m a.s.l (Welten 1982), and John Birks computed the first rarefaction analysis for the Alps. Figure 1 shows the stratigraphic plot through about the last 13 000 years for the estimated number of expected taxa. The thin curve gives the values resulting when long-distance-transported pollen is included, and the thick curve when it is excluded. We observe the following patterns:

1. A general trend towards increasing richness in pollen types through the Late Glacial and Holocene is apparent. We clearly need to test this trend in the Western and Eastern Alps. Küttel (1984) found a similar trend in the first half of the Holocene in the Swedish tundra, but he used a less appropriate diversity measure (Shannon index).
2. The values, including long-distance-transported pollen, are lower in the Late Glacial and higher in the Holocene than the values for local pollen; either because of the larger number of tree taxa in the valley bottom (especially after 9 000 BP) or because the number of pollen types was more stable at high altitudes.
3. The difference between our high-altitude site and the lowland sites is that we do not find the temporal pattern with high "protocratic", low "mesocratic", and high "*Homo sapiens* -phase" richness as demonstrated for lowland sites by

HOPSCHENSEE, 2017 m asl, Swiss Alps
Stratigraphic plot of estimated taxa diversity E(Tn)
Pollen analysis by M. WELTEN (1982)
Rarefaction by H.J.B. BIRKS

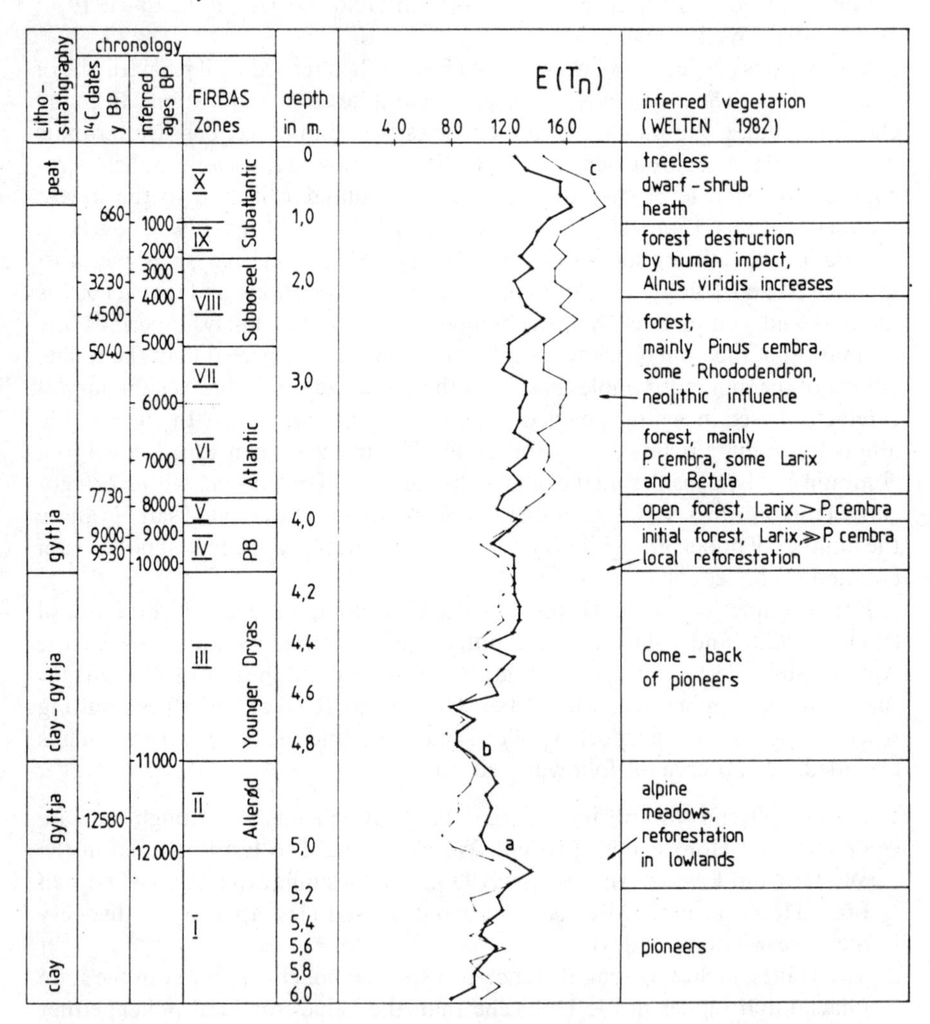

Fig. 1. Hopschensee on Simplon pass, 2017 m a.s.l Stratigraphic plot of estimated taxa diversity $E(T_n)$ according to the rarefaction analysis by H.J.B. Birks, pollen analysis, interpretation in terms of vegetation and dating by Welten (1982). *PB* Preboreal; V Boreal. Long-distance-transported pollen are included for the calculation of the thin-lined curve and excluded for the thick-lined curve. For the *thin curve* rarefaction is made to the lowest count (i.e. to the level with the lowest pollen sum) of 46 grains; for the *thick curve* to 52 grains. (In a third and a fourth run not shown in Fig. 1, the lowest three samples poor in pollen were deleted and the remaining smaples were rarefacted to basal counts of 185 grains and 126 grains, respectively; the results correspond very well with the ones including all samples). The three major decreases are labelled *a, b, c* (see text)

Birks and Line (1992). The protocratic phase at the beginning of an interglacial is characterized by basiphilous and heliophilous species of contrasting ecological and biogeographical affinities; in the mesocratic phase temperature deciduous forests have developed (see Iversen 1958; Birks 1986; Birks et al. 1988).
4. The local establishment of a subalpine forest of *Pinus cembra* and *Larix decidua* around 10 000 B.P. (Lang and Tobolski 1985) does not affect the richness in taxa, such as is observed at lowland sites for the period of reforestation around 13 000 to 12 700 B.P. (e.g. Gaillard 1984).
5. The generally cool period of the Younger Dryas consists of a first part poor in species and a second part rich in species. Welten (1982) described a bipartition of this period and concluded from several sites in the Central Alps that the first part of this period was rather oceanic, whereas the second part was rather continental.
6. The general increase in ($E(T_n)$) is interrupted by three major decreases: (a) the first one around 12 000 B.P. possibly explained by the reforestation in the valley for the thin-lined curve, and by the loss of pioneers for the local curve, (b) the second at the beginning of the cool phase of the Younger Dryas, and (c) the youngest one in the last 500 years (due either to overrepresentation of Cyperaceae after the infilling of the lake or to overgrazing). Further research is required to evaluate the potential causes such as competition, climate and human impact.

In addition, rapid changes in diversity can occur due to very local events. For example, Bortenschlager (1970, 1984) showed at Rotmoos (2260 m a.s.l.) that a local avalanche may produce a pollen record that resembles that of a climatic deterioration (but is not found at neighbouring sites and is associated with a different lithology). In both cases–regional climate and local slope process–the damages occur more rapidly than recovery. Stability over approx. the last 8000 years is also indicated by the most obvious vegetational limit, the timberline: where anthropo-zoogenic lowering can be excluded, the fluctuations occur only over about 150–200 m in altitude (Bortenschlager 1993; Lang 1993). Alpine grassland communities are characterized in part by what Klötzli (1992) calls “stability under unstable conditions”, i.e. traumatic impacts of weather and soil disturbances are frequent but natural “regeneration systems” are built in. The bog at Rofenberg (2760 m a.s.l; Bortenschlager 1993), the highest bog in the Eastern Alps, presents an impressive record of stability under harsh conditions. The patchiness of habitat diversity may buffer, to a certain extent, the vagaries of weather and slope processes, at least on the regional scale recorded in a pollen diagram. Two problems might develop under a future climate, a spatial and a temporal one: (1) For species on isolated mountains of intermediate altitude, the possibilities to escape from a changing climate may be very limited (see Grabherr et al. this vol.). (2) If future climate involves higher frequencies of extreme events, an acceleration of the disturbance regime may stress the “regeneration systems” more often often (and possibly too often) and push the ecosystems over a threshold “ beyond repair”.

10.3.2 Individualistic Response of Species: Leads and Lags

Long-term changes in species distribution in the alpine arch (e.g. Kral 1979; Welten 1982) prove to be just as individualistic as those on the broader continental scale (Firbas 1949; Huntley and Birks 1983). Individualistic in this context means that taxa co-occuring today may have very different histories. These may have included species having different lags between immigration and expansion. Forcing factors involve climate change, pedogenesis, and the mobility of species which depends, for example, on dispersal capacities and life cycles.

Figure 2A represents an example from the northern Alpine foreland (Ammann 1989) for two periods of rapid warming in the Late Glacial around 12 600 and 10 000 B.P (and a cooling at about 11 000 B.P; ages are expressed in uncalibrated radiocarbon years before present) as indicated by changes in oxygen-isotope ratios. In Fig. 2B (alpha), the July mean temperatures are reconstructed from fossil insect assemblages based on today's biogeographic ranges of insects (C and E). The changes in stable isotopes are thought to record climatic warming and cooling with little or no time lag (Wright 1984; Siegenthaler and Eicher 1986). The water plant *Typha latifolia* and the beetle *Donacia cinerea* respond quickly to the warming around 12 600 B.P; (exact dating is impeded by a plateau of constant age in the radiocarbon time scale; Ammann and Lotter 1989; Lotter et al. 1992). Insects are mobile and have short generation times (Coope 1977), and water plants can disperse rapidly, e.g. by water fowl, and are thought to be independent of pedogenesis (Iversen 1964). The hazel *Corylus avellana*, however, in spite of its range being similar to that of *Donacia cinerea*, needs about 3500 years more to arrive in southern-central Europe. Reconstruction of summer temperature based on the immigration of woody plants only could thus produce an error of several thousand years. After arrival, the expansion (population growth) of hazel is extremely rapid (e.g. Gaillard 1984; Clark et al. 1989). A number of hypotheses have been discussed to explain this: the late arrival may be caused by migrational lag and/or climate (including high seasonality with cold winters and late frosts) between 12 000 and 10 000 B.P; for the subsequent high rate of migration of *Corylus* through Europe, Huntley (1993) discusses seven groups of hypotheses. The behaviour of *Corylus* in Europe is a well-studied example that illustrates unexpectedly late arrival and then unexpectedly fast expansion. For herbaceous taxa at high-altitude sites, similar studies are in progress (Wick 1994 and in prep.) that illustrates individualistic behaviour of species to climatic change.

10.3.3 Temporal Changes over Altitudes

Overall vegetational change through time is reflected by palynological change in pollen diagrams. The amount of this change can be quantified by applying the numerical techniques of detrended correspondence analysis (DCA, Hill and Gauch 1980) and canonical correspondence analysis (CCA, ter Braak 1986), used widely in modern ecology and applied to the sequence of pollen samples. Figure 3 presents the results of a CCA of the Late Glacial (from deglaciation to 10 000 B.P)

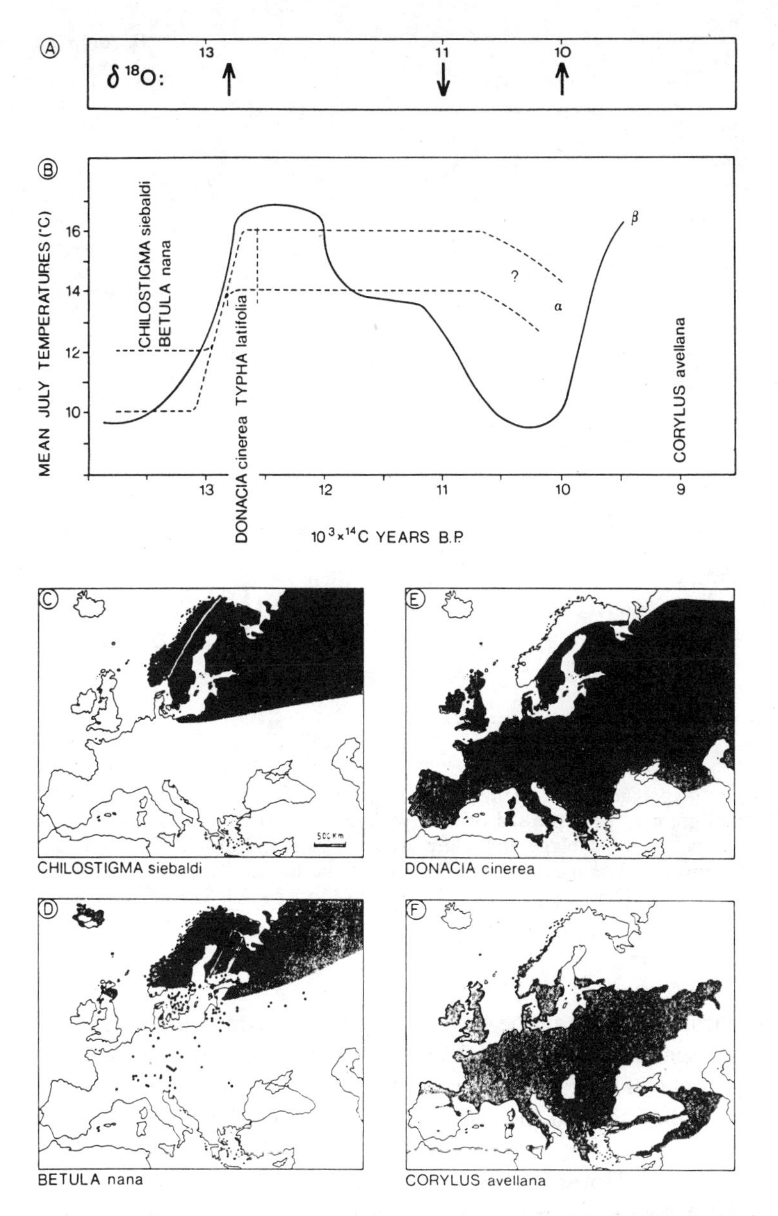

Fig. 2. A Shifts in oxygen isotopes (measured on lake marl) indicate periods of rapid warming around 12 700 and 10 000 B.P. and a cooling aroung 11 000 B.P. The cool phase of the Younger Dryas corresponds to the period between about 11 000 and 10 000 B.P. when expressed as radiocarbon years (after Eicher and Siegenthaler 1983). **B** Mean July temperature as inferred from insects (the band *alpha* for Lobsigensee, the curve *beta* for central England for comparison with a strong signal between 11 000 and 10 000 B.P. (after Elias and Wilkinson 1983) and Late Glacial occurrences of plant and insect species. **C–F** Today's distribution of some insects and plants used for palaeoecology and palaeotemperatures. **A–F** The caddis fly *Chilostigma siebaldii* and the dwarf birch *Betula nana* with a similar range today co-occurred before 13 000 B.P. The rapid warming indicated around 12 000 B.P. by the oxygen isotopes is rapidly answered by the beetle *Donacia cinerea* and by the water plant *Typha latifolia. Corylus avellana*, however, arrives about 3500 years later and then expands very rapidly (for alternative explanations, see text)

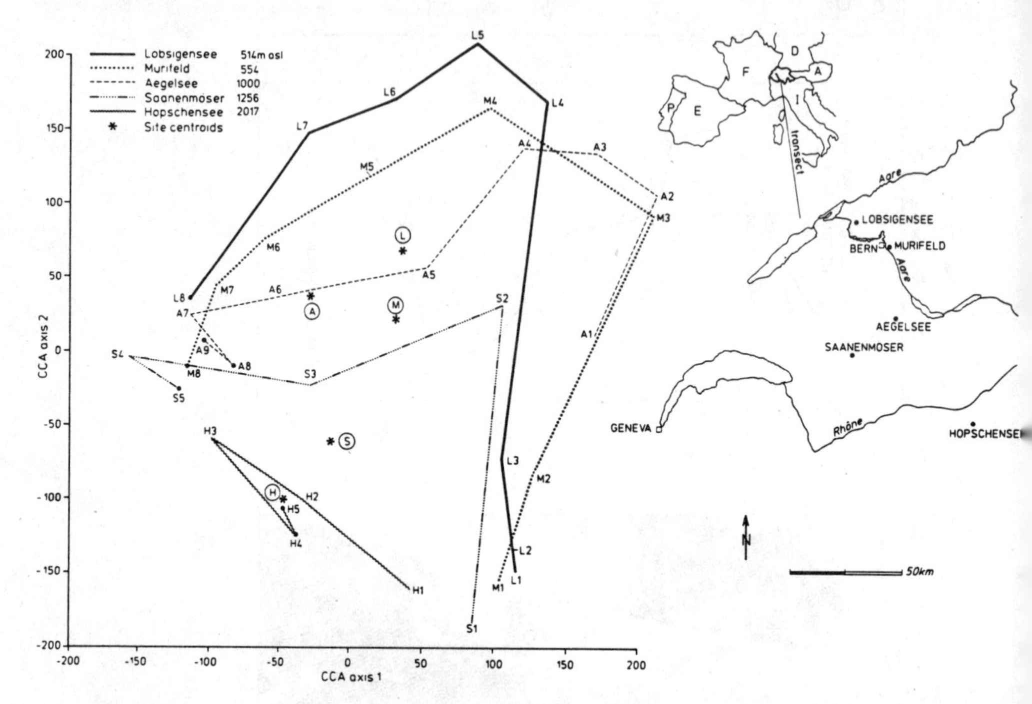

Fig. 3. The amount of palynological change in late glacial pollen records from five sites along an altitudinal gradient as expressed on the two axes of a canonical correspondance analysis. The local pollen assemblage zones are labelled from the oldest one as 1 (e.g. Hopschensee H1) upwards. The *black dots* at the end of the trajectories mark the pollen assemblages of the Younger Dryas (L7 to L8, M7 to M8, A7 to A8+A9, S4 to S5 and H3 to H4+H5)

at five sites on a transect through the northern Swiss Alps (Ammann et al. 1993). The local pollen assemblage zones are labelled for each site from the oldest to the youngest, for example for Hopschensee from H1 to H5. The more important the change from one pollen zone to the next, the longer the gradient length on the CCA plot.

Among these five sites, Hopschensee is the only one at timberline, which today is at ca. 1900–2000 m a.s.l. Before anthropogenic lowering to create pastures, it was at about 2200–2250 m a.s.l (see Fig. 1). Some main results presented in Fig. 3 are: (1) no two sites have identical pollen stratigraphies. (2) The major palynological differences correspond to the site location and altitude. Thus, the sequences (of both the trajectories and their "summaries", which are the means (or centroids) in the pollen stratigraphical space of CCA correspond to the sequence in geographical space. (3) At the sites at low and middle altitudes (500 to 1500 m a.s.l.), the early, large changes (transitions of local pollen zones L3 to L4, M2 to M3, S1 to S2) reflect reforestation in the area. (Around Hopschensee the trees arrived only after 10000 B.P; see Fig. 1). (4) The climatic cooling at the transition from the

Allerød to the Younger Dryas is weak around 500 m, but strong at middle and high altitudes, where it appears as a reversal towards pollen assemblages of treeless vegetation; this is usually interpreted as a lowering of the timberline during the cool phase. Points 1, 2 and 4 illustrate that not only did species migrate individualistically in response to climatic change (Sec. 3.2), but that each site showed an individualistic pattern of vegetation development. This probably resulted from site differences in local conditions such as geology, altitude, slope, aspect and vegetational inertia and thresholds (Smith 1965).

10.4 Needs for Future Research

10.4.1 High resolution

Ecological processes vary over vast changes of spatial and temporal scales (Birks 1986, Fig. 1; Körner 1993, Fig. 1). The selection of the scale appropriate to an ecological process also calls for adequate spatial and temporal resolution. In most cases, this can be done by site selection, sampling design, and high-quality dating. In addition to high resolution in space and time high taxonomic resolution is needed for studies in biodiversity. The "top 20 taxa", most of which are woody plants, will not suffice. For research on alpine diversity, priorities for refinement of pollen morphology should be given to plant families that (1) play important roles in communities above or beyond the timberline and (2) promise successful identifications based on their various morphological groups, e.g. Apiaceae, Fabaceae, Caryophyllaceae, Rosaceae.

10.4.2 Interdisciplinarity

In order to understand the leads and lags in species response to climatic change, we need to improve knowledge of several groups of organisms. The useful combination of bio- and lithostratigraphies to record changes in aquatic and terrestrial ecosystems calls for cooperation on selected sites between biologists (e.g. working with pollen, plant macrofossils, diatoms, chrysophytes, cladocera, ostracods, beetles, chironomids), geochemists (analyzing subfossil pigments of algae and bacteria or the composition of the minerogenic matrix or the pollution history), physicists (for stable isotopes and dating), as well as archaeologists and historians for reconstructing the land-use history (e.g. Berglund 1991; Gaillard et al. 1991, 1992). This "multi-proxy" approach opens possibilities either to reconstruct various parameters of past climate or to study responses of groups of organisms to climate change traced by independent lines of evidence, e.g. stable isotopes (Lemdahl 1991; Lotter et al. 1992; Ponel et al. 1992). By combining high resolution and interdisciplinary studies with powerful statistical techniques, the contributions of palaeoecology may be of significance to research on communities and biodiversity (Birks 1986, 1993; Walker 1990; Ritchie 1991; Schoonmaker and Foster 1991, Birks and Line 1992).

10.5 Conclusions

Environmental history may be one of the most important factors controlling biodiversity because it embraces time scales ranging from ecological time to evolutionary time (sensu Simberloff et al. 1981; Delcourt et al. 1982; Birks 1986; Ricklefs 1987). Processes involved include climatic change, disturbance regimes, pedogenesis, and human impact, all of which can control succession and floristic and vegetational composition. The few studies above the timberline available today suggest:

1. A high stability on a regional scale, the so-called stability under unstable conditions
2. Damages occurs faster than recovery.
3. Individualistic responses of species to changes in climate and human impact (see also Grabherr et al. this vol.).
4. High destabilizing frequencies of extreme events may disrupt existing plant communities.

The combination of points 2, 3, and 4 could result in new types of plant associations. For future research on biodiversity, this means that we need to overcome what Kingsland (1985) has called "the eclipse of history" (Ricklefs 1987) and be aware of the several time scales involved.

Acknowledgments. My cordial thanks go to John Birks for the rarefaction analysis and for improving earlier drafts of the manuscript, to Margrit Kummer and Sandrine Zaugg for drawing the figures and to Christian Körner, F. Stuart Chapin and Herbert E. Wright for suggestions and discussions.

References

Ammann B (1989) Late-Quaternary palynology at Lobsigensee. Regional vegetation history and local lake development. Diss Bot 137: 1–157

Ammann B, Lotter AF (1989) Late-Glacial radiocarbon- and palynostratigraphy on the Swiss Plateau. Boreas 18: 109–126

Ammann B, Wick L (1993) Analysis of fossil stomata of conifers as indicators of the alpine tree-line fluctuations during the Holocene. In: Frenzel B, Eronen M, Vorren K-D, Gläser B (eds) Oscillations of the alpine and polar tree limits in the Holocene. Paläoklimaforschung 9: 175–186

Ammann B, Birks HJB, Drescher-Schneider R, Juggins S, Lang G, Lotter AF (1993) Patterns of variation in Late-Glacial pollen stratigraphy along a north-west–south east transect through Switzerland – a numerical analysis. Quat Sci Rev 12:277–286

Andersen STh (1988) Changes in agricultural practices in the Holocene indicated in a pollen diagram from a small hollow in Denmark. In: Birks HH, Birks HJB, Kaland PE, Moe D (eds) The cultural landscape – past, present and future. Cambridge University Press, Cambridge pp 395–408

Ball IR (1980) The status of historical biogeography. Actis XVII Congr Int Ornithol 1283–1288

Berglund BE (ed) (1991) The cultural landscape during 6000 years in southern Sweden. Ecol Bull 41:1–495

Birks HH (1993) The importance of plant macrofossils in late-glacial climatic reconstructions in western Norway. Quat Sci Rev 12:719–726
Birks HH (1994) Plant macrofossils and the Nunatak theory of per-glacial survival. Diss Bot 234:129–143
Birks HJB (1986) Late-Quaternary biotic changes in terrestrial and lacustrine environments, with particular reference to north-west Europe. In: Berglund BE (ed) Handbook of Holocene palaeoecology and palaeohydrology. Wiley, Chichester, pp 3–65
Birks HJB (1993) Quaternary palaeoecology and vegetation science – current contributions and possible future developments. Rev Palaeobot Palynol 79:153–177
Birks HJB, Line JM (1992) The use of rarefaction analysis for estimating palynological richness from Quarternary pollen-analytical data. Holocene 2:1–10
Birks HJB, Line JM (1993) Glacial refugia of European trees – a matter of chance? Diss Bot 196:283–291
Birks HJB, Line JM, Person T (1988) Quantitative estimation of human impact on cultural landscape development. In: Birks HH, Birks HJB, Kaland PE, Moe D (eds) The cultural landscape – past, present and future. Cambridege University Press, Cambridge, pp 229–240
Bortenschlager S (1970) Waldgrenz- und Klimaschwankungen im pollenanalytischen Bild des Gurgler Rotmooses. Mitt Ostalp-din Ges Vegetkde 11: 19–26
Bortenschlager S (1984) Beiträge zur Vegetationsgeschichte Tirols I. Inneres Oetztal und unteres Inntal. Ber Nat-Med Ver Innsbruck 71:19–56
Bortenschlager S (1993) Das höchstgelegene Moor der Ostalpen "Moor am Rofenberg" 2760 m. Diss Bot 196:329–334
Clark JS, Merkt J, Müller H (1989) Post-glacial fire, vegetation, and human history on the northern alpine forelands, south-western Germany. J Ecol 77:897–925
Coope GR (1977) Fossil coelopteran assemblages as sensitive indicators of climatic changes during the Devensian (Last) cold stage. Philos Trans R Soc Lond B 280:313–340
Dansgaard W. White JWC, Johnsen SJ (1989) The abrupt termination of the Younger Dryas climate event. Nature 339: 532–534
de Beaulie J-L (1977) Contribution palynologique à l'histoire tardiglaciaire et holocène de la végétation des Alpes méridionales francaises. Thèse Marseille III, 358 pp
Delcourt HR, Delcourt PA, Webb T III (1982) Dynamic plant ecology: the spectrum of vegetational change in space and time. Quat Sci Rev 1: 153–175
Eicher U, Siegenthaler U (1983) Stable isotopes in lake marl and mollusc shells from Lobsigensee (Swiss Plateau). Rev Paléobiol 2: 217–220
Elias SA, Wilkinson B (1983) Late glacial insect fossil assemblages from Lobsigensee (Swiss Plateau). Rev Paléobiol 2: 189–204
Favarger C (1972) Endemism in the mountain floras of Europe. In: Valentine DH (ed) Taxonomy, phytogeography and evolution. Academic Press, New York, pp 191–204
Favarger C (1975) Cytotaxonomie et histore de la flore orophile des Alpes et de quelques autres massifs montagneux d'Europe. Lejeunia 77: 1–45
Firbas F (1949) Spät- und nacheiszeitliche Waldgeschichte Mitteleuropas nördlich der Alpen. Fischer, Jena, 480 pp
Foster DR, Zebryk TM (1993) Long-term vegetation dynamics and disturbance history of a TSUGA-dominated forest in New England. Ecology 74: 982–998
Gaillard MJ (1984) Etude Palynologique de l'Evolution Tardi- et Postglaciaire de la Végétation du Moyen-Pays Romand (Suisse). Diss Bot 77: 1–346
Gaillard MJ, Dearing JA EI-Daoushy F, Enell M, Håkonsson H (1991) A late Holocene record of land-use history, soil erosion, lake trophy and lake-level fluctuations at Bjäresjön (South Sweden). J Paleolimnol 6: 51–81
Gaillard MJ, Birks HJB, Emanuelsson U, Berglund BE (1992) Modern pollen/land-use relationships as an aid in the reconstruction of past land-uses and cultural landscapes: an example from south Sweden. Veg Hist Archaeobot 1: 3–17
Hill MO, Gauch HG (1980) Detrended correspondence analysis: an improved ordination technique. Vegetation 42: 47–58
Huntley B (1993) Rapid early-Holocene migration and high abundance of hazel (*Corylus avellana* L.): alternative hypotheses. In: Chambers FM (ed) Climate change and human

impact on the landscape. Studies in palaeoecology and environmental archaeology. Chapmen & Hall, London, pp 205–215
Huntley B, Birks HJB (1983) An atlas of past and present pollen maps for Europe: 0–13,000 years ago. Cambridge University Press, Cambridge
Iversen J (1958) The bearing of glacial and interglacial epochs on the formation and extinction of plant taxa. Uppsala Univ Aarskrift 6: 210–215
Iversen J (1964) Plant indicators of climate, soil, and other factors during the Quaternary. INQUA Rep of the VI Congr in Warsaw 2: 421–428
Iversen J (1973) The development of Denmarks's nature since the last glacial. Dan Geol Unders V 7C: 126
Jacobson GL, Bradshaw RHW (1981) The selection of sites for paleovegetational studies. Quat Res 16: 80–96
Janssen CR (1986) The use of local pollen indicators and the contrast between regional and local pollen values in the assessment of the human impact on vegetation. In: Behre K-E (ed) Anthropogenic Indicators in pollen diagrams. Balkema, Rotterdam, pp 203–208
Kingsland SE (1985) Modeling nature, episodes in the history of population ecology. University of Chicago Press, Chicago
Klötzli F (1992) Alpine Vegetation: stabil und natürlich? Publ Schweiz Akad Naturwiss 5: 70–83
Körner Ch (1993) Scaling from species to vegetation: the usefulness of functional groups. In: Schulze E-D, Mooney HA (eds) Biodiversity and ecosystem function. Ecological Studies 99. Springer, Berlin Heidelbey New York, pp 117–140
Kral F (1979) Spät- und postglaziale Waldgeschichte det Alpen auf Grund der bisherigen Pollenanalyse. Veröff Inst f Waldbau Univ Bodenkultur, Wien
Küpfer Ph (1974) Recherches sur les liens de parenté entre la flore des Alpes et celle des Pyrénées. Boissiera 23: 1–322
Küttel M (1984) Veränderung von Diversität und Eveness der Tundra, aufgezeichnet im Pollendiagramm des Vuolep Allakasjaure. Bot Helv 94: 279–283
Lan G (1993) Holozäne Veränderungen der Waldgrenze in den Schweizer Alpen – Methodische Ansätze und gegenwärtiges Kenntnisstand. Diss Bot 196: 317–327
Lang G, Tobolski T (1985) Hobschensee – late-glacial and holocene environments of a lake at the timberline in the central Swiss Alps. Diss Bot 87: 209–228
Lemdahl G (1991) A rapid climatic change at the end of the Younger Dryas in south Sweden – palaeoclimatic and palaeoenvironmental reconstructions based on fossil insect assemblages. Palaeogeogr, Palaeoclimatol Palaeoecol 83: 313–331
Lotter AF, Eicher U, Siegenthaler U, Birks HJB (1992) Late-glacial climatic oscillations as recorded in Swiss lake sediments. J Quat Sci 7: 187–204
Merxmüller H (1952, 1953, 1954) Untersuchungen zur Sippengliederung und Arealbildung in den Alpen. Ver zum Schutz d Alpenpflanzen und -tiere, No 17, 18, 19, München
Ozenda P (1985) La végétation de la chaîne alpine dans l'espace montagnard européen. Masson, Paris
Ponel Ph, de Beaulieu J-L, Tobolski K (1992) Holocene palaeoenvironments at the timberline in the Taillefer Massif, French Alps: a study of pollen, plant macrofossils and fossil insects. Holocene 2: 117–130
Prentice IC (1988) Records of vegetation in time and space: the principles of pollen analysis. In: Huntley B, Webb T (eds) Vegetation history. Kluwer, Dordrecht, pp 17–42
Reisigl H, Keller R (1987) Alpenpflanzen im Lebensraum. Alpine Rasen, Schutt- und Felsvegetation. Fischer, Stuttgart
Ricklefs RE (1987) Community diversity: relative roles of local and regional processes. Science 235: 167–171
Ritchie JC (1986) Climate change and vegetation response. Vegetation 67: 65–74
Ritchie JC (1991) Palaeoecology: status and prospect. In: Shane LCK, Cushing EJ (eds) Quaternary landscapes. University of Minnesota Press, Minneapolis, pp 113–128
Schoonmaker PK, Foster DR (1991) Some implications of palaeoecology for contemporary ecology. Bot Rev 57: 204–245

Siegenthaler U, Eicher U (1986) Stable oxygen and carbon isotope analyses. In: Berglund BE (ed) Handbook of Holocene palaeoecology and palaeohydrology, Wiley, Chichester, pp 407–422

Simberloff D, Heck KL, McCoy ED, Connor EF (1981) There have been no statistical tests of cladistic biogeographical hypotheses. In: Nelson G, Rosen DE (eds) Vicariance biogeography: a critique. Columbia University Press, New York, pp 40–63

Smith AG (1965) Problems of inertia and thresholds related to post-glacial habitat changes. Proc R Soc B 161: 331–342

Stebbins GL (1984) Polyploidy and the distribution of arctic-alpine flora: new evidence and a new approach. Bot Helv 94: 1–13

ter Braak CJF (1986) Canonical correspondence analysis: a new eigenvector technique for multivariate direct gradient analysis. Ecology 67:1167–1179

Walker D (1990) Purpose and method in Quaternary palynology. Rev Palaeobot Palynol 64: 13–27

Welten M (1982) Vegetationsgeschichtliche Untersuchungen in den westlichen Schweizeralpen: Bern – Wallis. Denksch Schweiz Naturforsch Ges 95: 104 + 37 Diagr

Whittaker RH (1976) Evolution of species diversity in land communities. Evol Bio 10: 1–67

Wick L (1994) Early-Holocene reforestation and vegetation changes at a lake near the Alpine forest limit: Lago Basso (2250 m asl), N-Italy. Diss :Bot 234: 555–563

Wright HE (1984) Sensitivity and response time of natural systems to climatic change in the Late Quaternary. Quat Sci Rev 3: 91–131

Zoller H (1960) Pollenanalytische Untersuchungen zur Vegetationsgeschichte der insubrischen Schweiz. Denkschr Schweiz Naturforsch Ges 83: 45–156

11 Implications for Changes in Arctic Plant Biodiversity from Environmental Manipulation Experiments

T. V. CALLAGHAN[1] and S. JONASSON[2]

11.1 Introduction

The importance of the Arctic both as a modifier of global climate, through greenhouse gas emissions from its soils, and as a possible early indicator of climate change, for example atmospheric warming, has been emphasised several times (e.g. Callaghan et al. 1992; Chapin et al. 1992). Arctic biota might be particularly sensitive to such changes because life in the Arctic is strongly regulated by climatic constraints. Any change in climatic conditions is, therefore, likely to affect the arctic biota more than in more benign environments where biological regulation may be more important (Callaghan and Emanuelsson 1985).

There are several ways of quantifying the impacts of climate change on arctic ecosystems. Inferences can be made about the likely relationships between future biota and a warmer climate by seeking analogues in warmer times in past historical periods or in warmer geographical regions at the present time (e.g. Emanuel et al. 1985). Assessments made using these methods give static equilibrium models. These fail to identify the changing opportunities and barriers to migration of biota and implicitly assume that whole communities will move en bloc (Melillo et al. 1990). More mechanistic models of the responses of biomes to climate change are now being developed (e.g. Melillo et al. 1993), but again, predictions of climate change impacts on the biota during the transition period are generally lacking although the transitional conditions are likely to be more important than those at an equilibrium which may never be reached.

In contrast to considering climate change impacts on biota at the biome level, studies of physiological processes, although precise, have not yet been successfully scaled up to even population level. A method to quantify the impacts of climate change on arctic ecosystems intermediate between the biome level and physiological level studies is to simulate some of the factors involved in climate change under field conditions and then to monitor the responses. This approach has been applied with success to one community in arctic Alaska (Chapin and Shaver 1985; Tissue and Oechel 1987; Grulke et al. 1990) and along latitudinal

[1] Centre for Arctic Biology, School of Biological Sciences, University of Manchester M13 9PT, Manchester, UK
[2] Botanical Institute, Department of Plant Ecology, University of Copenhagen, Øster Farimagsgade 2D, 1353 Copenhagen K, Denmark

Ecological Studies, Vol. 113
Chapin/Körner (eds) Arctic and Alpine Biodiversity

and altitudinal gradients ranging from a Scandinavian subarctic birch forest heath community to a high arctic polar semi-desert community in Svalbard (Havström et al. 1993; Jonasson et al. 1993; Wookey et al. 1993, 1994; Parsons et al. 1994).

Such an approach permits direct observations of concomitant responses of several variables that can be integrated to form a basis for predictions of likely events at an ecosystem level if the climate changes. This allows integration with the few studies in the Arctic which directly monitor and model population and growth responses to climate variables either in real time (Carlsson and Callaghan 1990) or retrospectively (Callaghan et al. 1989). Thus, the impacts of natural interannual variations in climate can be compared with those generated by manipulations of climate.

In this chapter, we concentrate on the results from various experiments which use the approach of manipulating arctic ecosystems to simulate aspects of climate change. We discuss likely longer-term trends in biodiversity based on demographic responses of arctic plants to environmental manipulations and changes in other components of the ecosystems. Two concepts are given priority. Firstly, comparisons along an environmental severity gradient are used to test the generality of results from any one site in the context of the wider Arctic and to acknowledge that responses to the same change in climate should differ between different arctic ecosystems. Secondly, we conceive that individual species may respond differently to environmental change in different parts of their altitudinal/latitudinal geographical ranges and that they might be particularly responsive to climate change at their boundaries.

11.2 Mechanisms of Change in Biodiversity

Changes in biodiversity reflect the net balance between immigration, expansion of existing species/genotypes and species/genotype loss. Such changes result from the dynamics of population processes: e.g. species loss is achieved by mortality exceeding natality resulting in negative population growth. This may respond directly to the impact of an environmental factor on survival or fecundity, or it may result indirectly from changed competitive interactions. Loss of genotypic diversity may occur under conditions of positive population growth but when that growth is achieved by the proliferation of vegetative meristems rather than recruitment from sexual reproduction (Takamiya and Tanaka 1982). Loss may also result from changes in competitive interactions between genotypes when population growth is increasing. For biodiversity to increase, a population/individual of a species formerly absent must become established. This can either occur by recruitment from a seed/propagule bank of a species not represented as established plants, by the local encroachment of clonal plants or by the immigration of seeds/propagules from outside during the process of migration. Thus, within the life cycle of a plant, there may be a series of sensitive stages and constraints (Callaghan and Carlsson 1995). In the high Arctic, the constraint can be the seed producing stage (Wookey et al. 1993), whereas in the low Arctic it might be seedling establishment. In periods of rapid change, the constraint is likely to be dispersal and speed of migration.

In arctic areas, long horizontal distances between vegetation types present obstacles for migration which are not present in alpine regions where vegetation zones are compressed into relatively short distances. More likely migrations in the Arctic may occur from refugia, i.e. from islands of species and genotypes/ecotypes which form point sources of "inocula". Examples are arctic oases where trees are found at a considerable distance beyond the tree line, riparian habitats where shrubs are often found, and nunataks surrounded by ice. This latter example may have played a significant role in the revegetation of the Arctic during the warming following the last ice age (Dahl 1987). More locally, ecotypes restricted to cold, moist depressions, for example, might displace those dominant on warmer, dry ridges if the climate becomes warmer and wetter. Such a mechanism has been proposed for *Saxifraga oppositifolia* on Svalbard (Crawford et al. 1993).

The type of population process which dominates is often related to the characteristics of the vegetation which, in turn, are related to environment. For example, vegetative proliferation is generally associated with the closed vegetation of the low Arctic, whereas seedling recruitment is important for maintaining discontinuous distributions of populations in the high Arctic (Bell and Bliss 1980; Callaghan and Emanuelsson 1985). Along such a gradient of environmental severity, the impact of any particular environmental factor will vary. For example, on a small scale, the biomass, shoot density and height of *Salix herbacea* varied along a snow bed gradient (Wijk 1986). At one extreme, short growing seasons mainly decreased biomass and shoot height, but at the other extreme competitive interactions mainly decreased density, whereas height growth increased.

11.3 Responses of Soils to Environmental Manipulation: Implications for Plant Nutrition and Biodiversity

11.3.1 Implications for Biodiversity from Changes in Soil Moisture

Soil moisture limitation is normally not considered to exert a direct control on plant growth in the Arctic, except in the dry polar deserts. This is supported by observations of low or no short-term response in growth or reproduction to water additions to subarctic (Karlsson 1985; Parsons et al. 1994), low arctic (Oberbauer et al. 1989) and high arctic (Henry et al. 1986; Wookey et al. 1993) vegetation. However, in a polar semi-desert, the growth of the ericaceous dwarf shrub *Cassiope tetragona* is usually positively correlated with precipitation as snow in spring in the high Arctic (Callaghan et al. 1989) and stable isotope analyses have recently suggested that this is due to dependence of growth on melting snow (Welker et al. 1995).

The general lack of response of plant growth to increased precipitation conflicts with data on plant productivity and the long-term differentiation of arctic vegetation types which are strongly correlated with moisture regime (e.g. Webber 1978; Oberbauer and Dawson 1992). This correlation in polar deserts could be a direct effect of water availability (Tenhunen et al. 1992). At the moister lower latitudes, however, the correlation between plant productivity, vegetation types and water availability is more likely to be indirect because of increased availability of nutrients through enhanced nutrient diffusion in moist tundra on level ground or mass flow of nutrients past roots of vascular plants in sloping terrain (Oberbauer et al. 1989).

In terms of biodiversity, any change to a moister or drier Arctic will probably cause a slow succession of vegetation types with an ultimate change in species composition and productivity. Seed germination and seedling survival, which are reduced by drought stress in dry tundras (Bliss 1958), are likely to increase if moisture conditions improve (Bell and Bliss 1980) with an associated increase in biodiversity. This is likely to have the most pronounced effects in the high Arctic and in the polar deserts where recruitment from seeds is more important than in the low Arctic (Callaghan and Emanuelsson 1985). In extreme cases, changes in soil moisture, could, however, directly induce more rapid changes, even with massive mortality, due to waterlogging with associated development of anoxic conditions or, in cases of decreased moisture, mortality due to increased drought stress. Indeed, death by drought and decrease in biodiversity would be particularly severe in some arctic areas of low precipitation where retreating permafrost will allow drainage from currently moist soils (Callaghan et al. 1992).

11.3.2 Relationship Between Air and Soil Warming and Implications for Nutrient Availability

The effects of increasing temperature on stimulating decomposer activity and the consequent rate of nutrient mineralisation and release of plant nutrients may be particularly important to plant growth and carbon storage in arctic ecosystems where plant-available nutrients, generally N or P, often severely limit plant growth (Shaver and Chapin 1980; Nadelhoffer et al. 1992). Changes in nutrient availability in these ecosystems through increased temperature – or other factors (e.g. increased N deposition, thermokarst formation, etc.) – are likely to affect biodiversity through changes in the competitive balance between species with different abilities to take up and utilise soil nutrients (Berendse and Jonasson 1992).

In several experiments, which increased air temperature, soil temperature was raised considerably less than air temperature, and could even decline (Table 1). The lower degree of warming of soils compared with air agrees with predictions from soil heat-flux models (Kane et al. 1992). The low degree of soil warming did not increase net nutrient mineralisation at four sites from the sub- and high Arctic (sites 1, 2, 3, and 5, Table 1; Jonasson et al. 1993; Robinson et al. submitted), suggesting that increased air temperature may not cause any pronounced release of extra nutrients. More surprisingly, net nutrient mineralisation was also unaffected when soil temperature was increased by 4–5 °C after transplantation

Table 1. Air and soil temperatures, and changes in temperature induced by greenhouse treatments at a range of experimental sites along an environmental severity gradient[a]

Site and location	Dominant vegetation	Mean temperatures + enhancement (°C)	
		Air	Soil
1. Subarctic forest understorey, Abisko, Sweden, 68°21′N 18°49′E; 400 m a.s.l.	*Vaccinium/Empetrum* heath under *Betula pubescens* ssp. *tortuosa*	11.8 + 2.7	5.3 − 0.3
2. Subarctic heath above tree line, Abisko, Sweden, 68°19′N 18°61′E; 450 m a.s.l.	*Cassiope tetragona, Vaccinium, Empetrum, Rhododendron, Salix* spp.	11.0 + 3.9	9.1 + 1.0
3. Subarctic Fell-field, Abisko, Sweden, 68°20′N 18°41′E; 1150 m a.s.l.	*Cassiope tetragona, Salix polaris, S. herbacea, Aulacomium* sp.	6.9 + 4.8	7.9 + 1.9
4. High arctic heath, Ny Ålesund, Svalbard, 78°56′N 11°50′E; 10 ml a.s.l.	*Cassiope tetragona, S. polaris, Racomitrium lanuginosum*	8.0 + 2.5	4.8 + 0.3
5. High arctic polar semi-desert, Ny Ålesund, Svalbard, 78°56′N 11°50′E; 22 ml a.s.l.	*Dryas octopetala, S. polaris, Saxifraga oppositifolia*	5.4 + 3.5	6.1 + 0.7

[a] See Havström et al. 1993 (sites 2, 3, 4), Wookey et al. 1993 (sites 1, 5) and Coulson et al. 1993 (site 4) for further details. Soil temperatures are at root depth (about 5 cm), and are daytime (07.00–19.00 h) temperatures measured during the 18- and 21-day period at sites 2 and 3 respectively. Elsewhere, they are derived from mean hourly temperatures measured throughout the growing season.

of soil from a cold to a warmer site, probably because soil microorganisms immobilised any extra nutrients that were released from the soil organic matter at the higher temperature (Jonasson et al. 1993). Furthermore, both under ambient conditions and after soil warming, elemental immobilisation was site-specific: one of the elements (N or P) was generally immobilised, whereas the other was not (Fig. 1). This suggests that either N or P can limit microbial growth in these ecosystems (Jonasson et al. submitted). If the limiting nutrient is taken up efficiently by soil microorganisms, any increased decomposition rate in a warmer climate may not lead to any pronounced increase in the nutrients allocated to plants.

11.3.3 Implications for Soil Nutrient Availability from Increasing Atmospheric CO_2 Concentrations and Enhanced UV-B Flux

Microbial decomposition of organic matter which may release nutrients for plant uptake depends upon the resource quality of this organic material in addition to the physical environment discussed above (Swift et al. 1979). It is well known that increased atmospheric concentrations of CO_2 often reduce the nitrogen content of leaves relative to carbon (Fajer et al. 1989). The resulting decrease in leaf litter

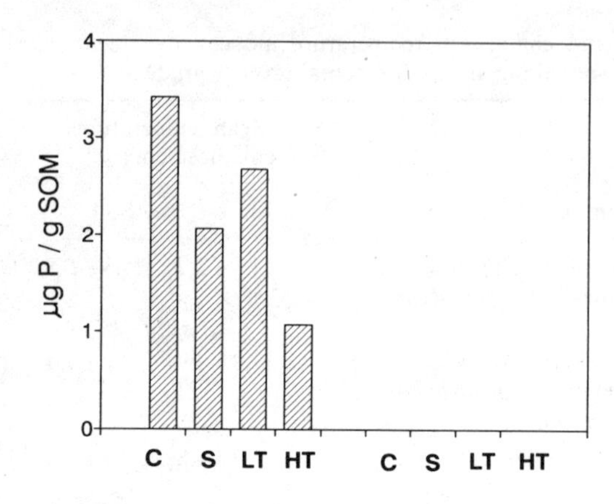

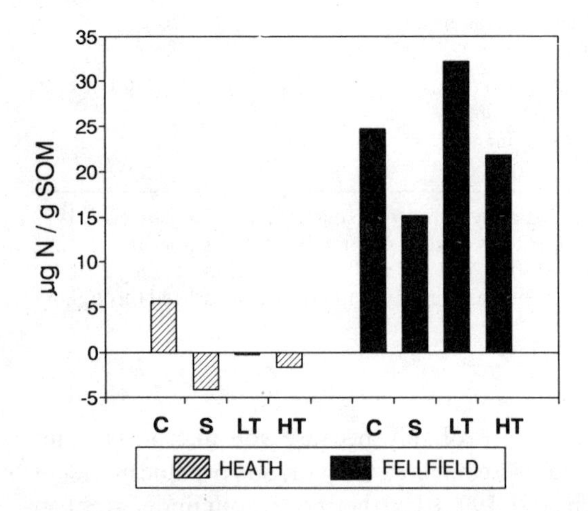

Fig. 1. Net mineralisation of nitrogen and phosphorus per g soil organic matter (*SOM*) in in situ buried bags at a low altitude heath (450 m a.s.l.) and high altitude fell-field (1150 m a.s.l.), Abisko, Swedish Lapland. Data represent net mineralisation from late June to early Septem-ber in control (*C*), shaded (*S*), low temperature (2–3 °C; *LT*) and high temperature (4–5 °C; *HT*) greenhouse treatments. Phosphorus was not possible to measure with confidence at the fell-field because of extremely low concentrations both before and after incubation. (Adapted from Jonasson et al. 1993)

"quality" as a resource for microbial decomposers has been shown to affect invertebrates (e.g. Fajer et al. 1989) and may reduce the growth and abundance of those involved in the early stages of litter breakdown. In addition, initial decomposition (over 3 months) of leaf litter by various groups of soil fauna and flora is retarded by the impact of high atmospheric CO_2 on leaf litter quality (Couteaux et al. 1991).

UV-B radiation which is increasing as a result of anthropogenic destruction of stratospheric ozone, particularly at high latitudes (Stolarski et al. 1992) also affects decomposition of leaf litter. In areas where litter is exposed to sunlight (e.g. with an open or intermittent vegetation canopy), enhanced UV-B radiation may accelerate the breakdown of lignocelluloses in physico-chemical reactions (Moorhead and Callaghan 1994). In other areas of closed vegetation, as for example in subarctic heathlands, enhanced UV-B radiation has been shown to retard decomposition because it changes plant tissue chemistry and the quality of resources for microorganisms, and also becaue of its direct effect in reducing or eliminating some species of soil fungi (Gehrke et al. 1995).

The combined effects of enhanced UV-B radiation and elevated atmospheric concentrations of CO_2 are likely to reduce the availability of nutrients in the Arctic and to moderate any increase in nutrient availability due to warming. The degree of soil warming likely to occur and the fate of extra nutrients mineralised are uncertain, as discussed above, and the balance between the opposing impacts of warming and increasing CO_2 and UV-B is totally unknown.

11.3.4 Implications for Biodiversity from Changes in Nutrient Availability

In terms of biodiversity, changes in air and soil temperatures together with changes in nutrient availability are unlikely to directly kill plants. Changes in biodiversity are more likely to be generated by altered competitive interactions between existing species in response to the changes in environment.

An air temperature increase combined with restricted heat flux to deeper soil layers would favour nutrient uptake by plants with shallow roots, but carbon lost through respiration would be greater than in those species with a large belowground biomass deeper in the soil where temperature increases may be very small. For instance, several ericaceous species, which usually have shallow roots, would probably be at a nutritional advantage in comparison with plants with deep roots. With an increased gradient of temperature between air and soil, aboveground processes such as photosynthesis, transpiration, shoot growth and flowering (Havström et al. 1993, 1995) would be favoured in relation to belowground processes such as nutrient mineralisation, root growth and nutrient uptake. Greater impacts on shoots compared with roots could lead to increased carbon and/or tannin to nutrient ratios which occur during both experimental warming and naturally occurring warm years (Jonasson et al. 1986; Laine and Henttonen 1987). This effect would be particularly pronounced during increases in atmospheric CO_2 concentrations and UV-B and would lead to a reduction of the quality of plant tissues as perceived by herbivores and decomposers. Such changes in plant tissue quality are likely to be species-specific and will differentially alter the growth rates and population dynamics of herbivores and decomposers, thereby changing competitive interactions within several trophic levels and altering biodiversity. For example, grazing pressure on plants with either decreased nutrient content or decreased biomass would probably increase to compensate for the general decline of the food quality and quantity, thereby selecting against these plant species.

Any increase in nutrient mineralisation rates resulting from warmer soils, or other factors, would increase the growth of plants (dry weight and height). In addition, however, it could directly affect demographic processes by, for example, increasing tillering rates in graminoids (Chapin and Shaver 1985) and increasing seed output (Havström et al. 1994; Wookey et al. 1994; Carlsson and Callaghan 1995). Enhanced proliferation of growing points together with increased weight and height would have local effects on the relative abundance of genets and individuals of a species, whereas increased output of seed has the potential to maintain or increase genets at greater distances via dispersal and establishment.

The outcome of the changes in the relative abundance of species which respond individualistically to changes in environment (Chapin and Shaver 1985) has been shown experimentally to favour an increase in graminoids initially (Jonasson 1992; F.S. Chapin and G.R. Shaver, pers. comm.; Parsons et al. 1994) and dwarf shrubs in the longer term (F.S. Chapin and G.R. Shaver, pers. comm.) in arctic dwarf shrub heaths and tussock tundra. Although changes in species composition have only been recorded from the open vegetation of the high Arctic polar semi-desert (C.H. Robinson et al. pers comm.), changes in the relative abundance of genotypes/ecotypes are likely to be the first changes in biodiversity responding to changes in soil processes (Crawford et al. 1993).

11.4 Direct Responses of Plants to Environmental Perturbations and Implications for Changes in Populations, Communities and Biodiversity

11.4.1 Reproduction

In the high Arctic, dramatic responses to environmental perturbations were recorded in enhanced phenological development and flowering of *Dryas octopetala* (Wookey et al. 1993) and also in seed germinability (Wookey et al. in press). A 3.5 °C temperature increase was sufficient to allow seed set to occur, whereas this was almost absent under ambient temperature conditions (Fig. 2a). Nutrient and water additions had no significant effect on flowering and seed set at this site. The reproductive development and output from *Polygonum viviparum*, growing at the same high arctic site, were increased by higher temperatures, but increased nutrients also had a significant effect (Wookey et al. 1994; Fig. 2b).

In the Subarctic, in contrast, the berry production of *Empetrum hermaphroditum* was totally nutrient-dependent with no significant effect of temperature (Fig. 2c; Wookey et al. 1993).

Growth differences of a species between the Subarctic and the high Arctic provide the background against which responses to environmental manipulations at a high arctic site can be placed in context. The increased bulbil production, which is particularly important in the high Arctic (Bell and Bliss 1980), in response to warming parallels the general trends obtained when comparing the reproductive output of *Polygonum* from sites along a gradient of environmental severity reaching into temperate alpine latitudes (Callaghan and

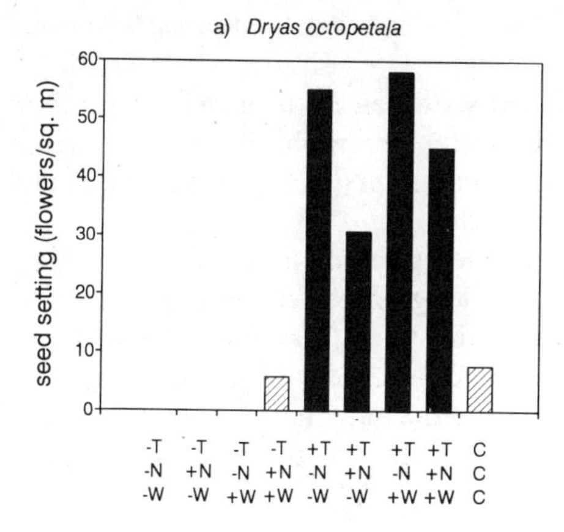

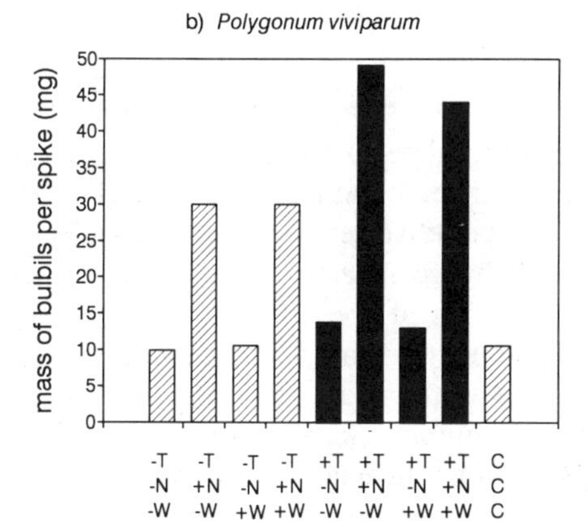

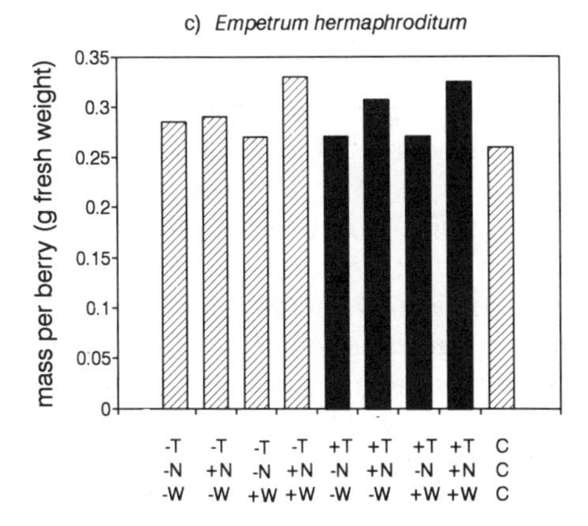

Fig. 2 a–c. Response of reproductive parameters to environmental perturbations at high and subarctic sites. *C* Absolute control; *T* temperature enhancement; *N* nutrient addition; *W* water addition. **a** Flowers setting seed in *Dryas octopetala* growing at a high arctic polar semidesert site on Svalbard (Wookey et al. 1993). **b** bulbil mass in *Polygonum viviparum* growing at the high arctic polar simidesert site on Svalbard (Wookey et al. 1994). **c** Fresh mass of berries of *Empetrum hermaphroditum* growing at a subarctic forest understorey site in Swedish Lapland. (Wookey et al. 1993)

Collins 1981). Similarly, the increased seed-setting in *Dryas* under temperature enhancement, corresponds approximately to the geographical/climatic distance between the Subarctic where seed setting is common (Kjellberg et al. 1982) and the high Arctic where it appears to be sporadic.

Climate change variables other than temperature and nutrients affect reproduction in the Arctic. High CO_2 can have a long-term effect on populations by increasing tillering rather than growth of existing tillers (Tissue and Oechel 1987). High CO_2 may also affect seed production and the time to sexual maturity. This last parameter may be important in increasing the potential rate of migration of long-lived species which normally take many years to reach sexual maturity (Callaghan and Carlsson 1995).

11.4.2 Vegetative Growth

Responses in vegetative growth to environmental perturbations along gradients of environmental severity show that, in general, temperature effects on vegetative growth are not as significant as those on reproductive development. In a study of the responses of the ericaceous dwarf shrub *Cassiope tetragona* to enhanced temperature at three locations (subarctic heath at tree line, subarctic alpine fell-field and high arctic heath), total leaf length, mass and numbers were increased most where ambient temperatures were lowest, i.e. at the high arctic and high altitude fell-field sites, whereas leaf number, length and biomass were increased more by nutrients at the warmest subarctic heath site (Havström et al. 1993). Similar responses of growth to nutrient additions at low arctic sites have been reported for some time (Shaver and Chapin 1980, 1986) but in the high Arctic, increased nutrients may even lead to the death of some species (Henry et al. 1986).

Under ambient conditions, leaf size may decrease as temperature increases along a gradient of environmental severity (Callaghan et al. 1989; Havström et al 1993; Fig. 3a). Similarly, increases in temperature, but particularly when combined with increases in water, decreased leaf size in *Dryas* subjected to manipulation experiments at a polar semi-desert site (Fig. 3b; Welker et al. 1993).

In the high Arctic, nutrient additions significantly increase the vegetative growth of *Polygonum viviparum*, but temperature does not affect growth (Wookey et al. 1994; Fig. 3c); also, there was no significant response in net photosynthesis to any perturbation (Wookey et al. 1994). The responses in mass, architecture and allometry of species in a subarctic dwarf shrub community of a forest understorey were also mainly generated by nutrient addition (Parsons et al. 1994). Temperature and nutrient enhancement were frequently synergistic as in the study by Chapin and Shaver (1985). In the forest understorey community, increased growth of the grass *Calamagrostis lapponica* was dramatic (Parsons et al. 1994) and similar to the responses of grasses to nutrients in previous experiments (Jonasson 1992). These findings also agree with those of Chapin and Shaver (F. S. Chapin and G. R. Shaver, pers. comm.) from Alaska which show that graminoids

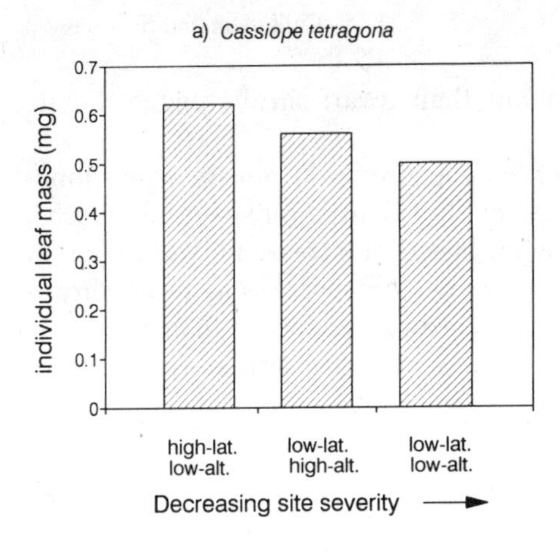

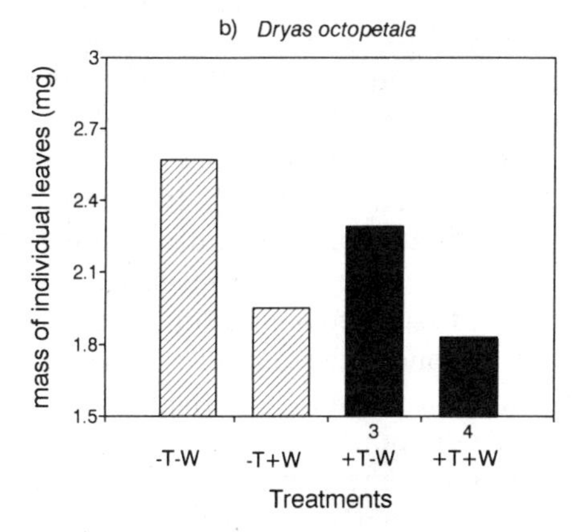

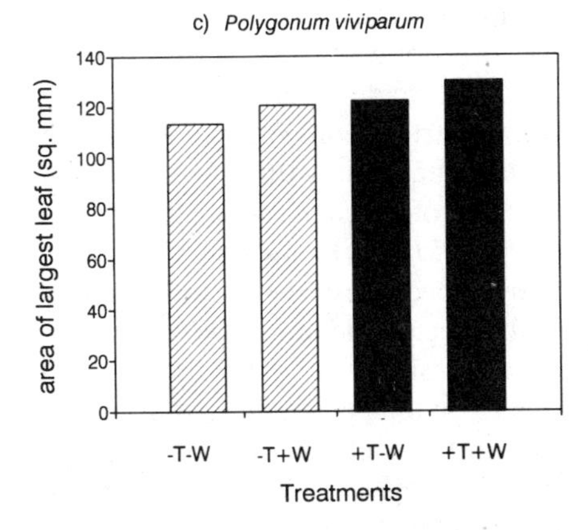

Fig. 3 a–c. Responses of leaf size to increases in temperature and/or precipitation illustrating how leaf size can decrease as temperature and/or precipitation increase and how co-occurring species may show different trends. **a** *Cassiope tetragona* sampled across a latitudinal/altitudinal severity gradient (Havström et al. 1993). **b** *Dryas octopetala* from the high arctic polar semi-desert site (Welker et al. 1993) **c** *Polygonum viviparum* from the same site as *Dryas*. (Wookey et al. 1994)

initially became relatively more abundant than dwarf shrubs under climate warming.

Increasing levels of atmospheric CO_2 in the Arctic do not have any long-term effects on growth due to acclimation of photosynthesis (Tissue and Oechel 1987). In contrast, recent experiments on the effects of increased UV-B radiation on a subarctic dwarf shrub community show that evergreen dwarf shrubs accumulate damage and show greater annual growth reductions than deciduous dwarf shrubs (Johanson et al. 1994). The potential for long-term restructuring of the community is obvious.

11.4.3 Biodiversity

In the perturbation experiments reviewed here, no individual plants were directly killed by increased temperature as suggested by Gauslaa (1984), but populations of some litter-dwelling fungi were absent or reduced in high UV-B radiation (Gehrke et al. 1995). Although no plants were killed directly, some species showed reduced growth, e.g. leaf size in *Dryas* declined as temperature increased and annual growth in evergreen dwarf shrubs decreased as UV-B increased (Johanson et al. 1995). Such species are likely to be at a competitive disadvantage compared with those in which growth does not decrease, e.g. *Polygonum.* In the longer term, loss of biodiversity could occur due to such poor competitiveness. In cases where different species in a community respond in the same direction to an environmental perturbation, it is the difference in magnitude of the responses which will change competitive balances with the potential to lead to species exclusions and loss in biodiversity. Differences in the responsiveness to climate change of currently co-existing life forms, such as cryptogams and graminoids, may lead to the exclusion of one form and a loss in biodiversity (Jonasson 1992). In general, shade exerts significant effects on arctic vegetation (Chapin and Shaver 1985; Havström et al. 1995) and increased shade developed by a particularly responsive species/life form may be an important competitive mechanism.

Increases in biodiversity are most likely to occur initially through the recruitment of new genotypes of existing species because of the constraints on migration and dispersal (Crawford et al. 1993). Genotypes of *Dryas octopetala* will increase in the high Arctic as a result of climate warming, because of the positive response of flowering (Wookey et al. 1993) and viable seed production (Wookey et al. 1995) to warming, if seedling establishment is successful. Thus, the present limitation on subspecific biodiversity seems to be lack of viable seed at present but seedling establishment may become the limiting factor under climate warming. This illustrates the significance of identifying sensitive stages of life cycles to population growth (Carlsson and Callaghan 1995).

Increases in biodiversity are most likely to occur in those situations where the constraints on establishment from sexual reproduction are least. In disturbed areas or in polar deserts with large proportions of bare ground, the constraint to establishment of a closed and competitive vegetation canopy is missing. Response of colonisation of bare ground by cryptogams to an increase in temperature in the Antarctic can increase cover from 5 to 75%(Wyn Williams 1990), while four higher plant species invaded bare ground in the high Arctic following nutrient addition (C. H. Robinson, pers. comm.). In the closed vegetation of the Subarctic, increases in biodiversity are likely to occur only in the long term because of the buffering against change provided by a continuous cover of vegetation (Jonasson 1994), unless there is disturbance.

11.5 Conclusions

1. Relatively few environmental changes, apart from extremes of soil moisture, are likely to affect biodiversity by directly killing plants.
2. Biodiversity is most likely to respond in the long term to changing environmental conditions as a result of altered competitive interactions resulting from species-specific reactions: in manipulation experiments, short-term responses to an individual treatment varied between co-occurring species implying that community structure will change.
3. Competitive interactions leading to changes in biodiversity can be generated by changes in soil nutrient availability, temperature, atmospheric CO_2 and UV-B.
4. Effects of soil fertility in determining biodiversity will be particularly important:
   a) when effected through vegetative growth;
   b) in subarctic vegetation;
   but net nutrient mineralisation and corresponding increases in soil fertility could be lower than anticipated because air warming is likely to be followed initially by a proportionally lower warming of the soil, nutrients can be differentially immobilised by soil microorganisms and substrate "quality" can be reduced during increases in temperature, atmospheric CO_2 and UV-B radiation.
5. Direct temperature effects on plant biodiversity will be particularly important:
   a) when effected through reproductive processes;
   b) in vegetation at the highest latitude and altitude;
   but increases in temperature (and nutrients) can reduce plant growth.
6. The constraint on increasing biodiversity varies from seedling recruitment in the Subarctic to seed production in the high Arctic; dispersal distances and migration rates are overall constraints.
7. Initial changes in biodiversity are likely to be subtle with changes only in abundance of existing species or changes in genotypes/ecotypes but more dramatic changes will occur following disturbances or in areas where colonisable bare ground is present i.e. in the high Arctic.

Acknowledgments. Many colleagues have allowed access to material in press and in preparation. We are particularly grateful to: C. Gehrke, M. Havström,

U. Johanson, Prof. J. A. Lee, Dr. A. Parsons, Dr. M. C. Press, Dr. C. Robinson, Dr. J. M. Welker and Dr. P. A. Wookey. For hospitality and logistic support, we thank Prof. M. Sonesson and Nils Åke Andersson (Abisko) and Nick Cox and the KBKC company, Ny Ålesund, Svalbard. T. V. Callaghan wishes to thank the UK Natural Environment Resource Council for support.

References

Bell KL, Bliss LC (1980) Plant reproduction in a high arctic environment. Arct Alp Res 12: 1–10

Berendse F, Jonasson S (1992) Nutrient use and nutrient cycling in northern ecosystems. In: Chapin FS III, Jefferies RL, Reynolds JF, Shaver GR, Svoboda J (eds) Arctic ecosystems in a changing climate. An ecophysiological perspective. Academic Press, San Diego, pp 337–356.

Bliss LC (1958) Seed germination in arctic and alpine species. Arctic 11: 180–188

Callaghan TV, Collins NJ (1981) Life cycles, population dynamics and the growth of tundra plants. In: Bliss LC, Heal OW, Moore JJ (eds) Tundra ecosystems: a comparative analysis. International Biological Programme 25, Cambridge University Press, Cambridge, pp 257–284

Callaghan TV, Carlsson BÅ (1995) Impacts of climate change on demographic processes and population dynamics in arctic plants. In: Oechel WC, Callaghan TV, Gilmanov T, Holten JI, Maxwell B, Molau U, Sveinbjörnsson B (eds) Global change and arctic terrestrial ecosystems. Proceedings of invited and plenary papers from the international conference, 21–26 Aug 1993, Oppdal, Norway. Springer, Berlin Heidelberg New York (in press)

Callaghan TV, Emanuelsson U (1985) Population structure and processes of tundra plants and vegetation. In: White J (ed) The population structure of vegetation. Junk, Dordrecht, pp 399–439

Callaghan TV, Carlsson BÅ, Tyler NJC (1989) Historical records of climate related growth in *Cassiope tetragona* from the Arctic. J Ecol 77: 823–837

Callaghan TV, Sonesson M, Sφmme L (1992) Responses of terrestrial plants and invertebrates to environmental change at high latitudes. Philos Trans R Soc Lond B 338: 279–288

Carlsson BÅ, Callaghan TV (1990) Effects of flowering on the shoot dynamics of *Carex bigelowii* along an altitudinal gradient in Swedish Lapland. J. Ecol 78: 152–165

Carlsson BÅ, Callaghan TV (1995) Simulated effect of climate change on the population dynamics of *Carex bigelowii*. Ecography (in press)

Chapin FS III, Shaver GR (1985) Individualistic growth responses of tundra plant species to environmental manipulations in the field. Ecology 66: 564–576

Chapin FS III, Jefferies RL, Reynolds JF, Shaver GR, Svoboda J (1992) Arctic ecosystems in a changing climate. An ecological perspective. Academic Press, San Diego, 469 pp

Coulson S, Hodkinson ID, Strathdee A, Bale JS, Block W, Worland MR, Webb NR (1993) Simulated climate change: the interaction between vegetation type and microhabitat temperatures at Ny Ålesund, Svalbard. Polar Biol 13: 67–70

Couteaux MM, Mousseau M, Celerier ML, Bottner P (1991) Increased atmospheric CO_2 and litter quality: decomposition of sweet chestnut leaf litter with animal food webs of different complexities. Oikos 61: 54–64

Crawford RMM, Chapman HM, Abbott RJ, Balfour J (1993) Potential impact of climatic warming on Arctic vegetation. Flora 188: 367–381

Dahl E (1987) The nunatak theory reconsidered. Ecol Bull 38: 77–94

Emanuel WH, Shugart HH, Stevenson MP (1985) Climate change and the broad-scale distribution of terrestrial ecosystem complexes. Clim Change 7: 29–43

Fajer ED, Bowers MD, Bazzaz FA (1989) The effects of enriched carbon dioxide atmospheres on plant-insect herbivore interactions. Science 243: 1198–1200

Gauslaa Y (1984) Heat resistance and energy budget in different Scandinavian plants. Holarct Ecol 7: 1–78

Gehrke C, Johanson U, Callaghan TV, Chadwick D, Robinson CH (1995) The impact of enhanced ultraviolet-B radiation on litter quality and decomposition processes in *Vaccinium* leaves from the subarctic. Oikos (in press)
Grulke NE, Reichers GH, Oechel WC, Hjelm U, Jaeger C (1990) Carbon balance in tussock tundra under ambient and elevated CO_2. Oecologia 83: 485–494
Havström M, Callaghan TV, Jonasson S (1993) Differential growth responses of *Cassiope tetragona*, an arctic dwarf shrub, to environmental perturbations among three contrasting high- and subarctic sites. Oikos 66: 389–402
Havström M, Callaghan TV, Jonasson S (1995) Effects of simulated climate change on the sexual reproductive effort of *Cassiope tetragona*. In: Callaghan TV, Oechel WC, Gilmanov T, Holten JI, Maxwell B, Molau U, Sveinbjörnsson B, Tyson M (eds) Global change and arctic terrestrial ecosystems. Proceedings of contributed poster papers from the international conference, 21–26 Aug. 1993, Oppdal, Norway. Commission of the European Communities Ecosystems Research Report, Brussels (in press)
Henry GHR, Freedman B, Svoboda J (1986) Effects of fertilization on three tundra plant communities of a polar desert oasis. Can J Bot 64: 2502–2507
Johanson U, Gehrke C, Björn LO, Callaghan TV, Sonesson S (1995) The effects of enhanced UV-B radiation on a subarctic heath ecosystem. Ambio (in press)
Jonasson S (1992) Growth responses to fertilisation and species removal in tundra related to community structure and clonality. Oikos 63: 420–429
Jonasson S (1995) Buffering of arctic plant responses in a changing climate. In: Oechel WC, Callaghan TV, Gilmanov T, Holten JI, Maxwell B, Molau U, Sveinbjörnsson B (eds) Global change and arctic terrestrial ecosystems. Proceedings of invited and plenary papers from the international conference, 21–26 Aug 1993, Oppdal, Norway. Springer, Berlin Heidelberg New York (in press)
Jonasson S, Bryant JP, Chapin FS III, Andersson M (1986) Plant phenols and nutrients in relation to variations in climate and rodent grazing. Am Nat 128: 394–408
Jonasson S, Havström M, Jensen M, Callaghan TV (1993) In situ mineralisation of nitrogen and phosphorus of arctic soils after perturbations simulating climate change. Oecologia 95: 179–186
Jonasson S, Michelsen A, Schmidt IK, Nielsen EV, Callaghan TV, Regulation of nutrients in two arctic soils by microbes: implications for plant nutrient uptake. Oikos (submitted)
Kane DL, Hinzman LD, Ming-ko Woo, Everett KR (1992) Arctic hydrology and climate change. In: Chapin FS III, Jefferies RL, Reynolds JF, Shaver GR, Svoboda J (eds) Arctic ecosystems in a changing climate. An ecophysiological perspective. Academic Press, San Diego, 0035–57
Karlsson PS (1985) Effects of water and mineral nutrient supply on a deciduous and an evergreen dwarf shrub: *Vaccinium uliginosum L* and *V. vitis-idaea* L. Holarct Ecol 8: 1–8
Kjellberg B, Karlsson PS, Kerstensson I (1982) Effects of heliotrophic movements of flowers of *Dryas octopetala* L. on gynoecium temperature and seed development. Oecologia 54: 10–13
Laine K, Henttonen H (1987) Phenolics/nitrogen ratios in blueberry *Vaccinium myrtillus* in relation to temperature and microtine density in Finnish Lapland. Oikos 50: 389–395
Melillo JM, Callaghan TV, Woodward FI, Salati E, Sinha SK (1990) Effects on ecosystmes. In: Houghton JT, Jenkins GJ, Ephraums JJ (eds) Climate change, the IPCC scientific assessment. Cambridge University Press, Cambridge, pp 282–310
Melillo JM, McGuire AD, Kicklighter DW, Moore B III, Vorosmarty CJ, Schloss AL (1993) Global climate change and terrestrial net primary production. Nature 363: 234–240
Moorhead D, Callaghan TV (1994) Effects of increasing UV-B radiation on decomposition and soil organic matter dynamics: a synthesis and modelling study. Biol Fertil Soils 18: 19–26
Nadelhoffer KJ, Giblin AE, Shaver GR, Linkins AE (1992) Microbial processes and plant nutrient availability in arctic soils. In Chapin FS III, Jefferies RL, Reynolds JF, Shaver GR, Svoboda J (eds) Arctic ecosystems in a changing climate. An ecophysiological perspective. Academic Press, San Diego, pp 281–300
Oberbauer SF, Dawson TE (1992) Water relations of arctic plants. In: Chapin FS III, Jefferies RL, Reynolds JF, Shaver GR, Svoboda J (eds) Arctic ecosystems in a changing climate. An ecophysiological perspective. Academic Press, San Diego, pp 259–279

Oberbauer SF, Hastings SJ, Beyers JL, Oechel WC (1989) Comparative effects of downslope water and nutrient movement on plant nutrition, photosynthesis, and growth in Alaskan tundra. Holarct Ecol 12: 324–334

Parsons AN, Welker JM, Wookey PA, Press MC, Callaghan TV, Lee JA (1994) Growth responses of four sub-arctic dwarf shrubs to simulated climate change. J Ecol 82: 307–318

Robinson CH, Wookey PA, Parsons AP, Welker JM, Callaghan TV, Press MC, Lee JA (1995) Responses of plant litter decomposition, nutrient concentrations in soil solution and nitrogen mineralisation to simulated environmental change in a high arctic polar semi desert and a subarctic dwarf shrub heath. Oikos (submitted)

Shaver GR, Chapin FS III (1980) Response to fertilisation by various plant growth forms in an Alaskan tundra: nutrient accumulation and growth. Ecology 61: 662–675

Shaver GR, Chapin FS III (1986) Effects of fertilizer on production and biomass of tussock tundra, Alaska, U.S.A. Arct Alp Res 18: 261–268

Stolarski R, Bojkov R, Bishop L, Zerefos C, Staehelin J, Zawodny J (1992) Measured trends in stratospheric ozone. Science 256: 342–349

Swift MJ, Heal OW, Anderson JM (1979) Decomposition in terrestrial ecosystems. Blackwell, Oxford

Takamiya M, Tanaka AR (1982) Polyploid cytotypes and their habitat preferences in *Lycopodium clavatum*. Bot Mag 95: 419–434

Tenhunen JD, Lange OL, Halm S, Siegwolf R, Oberbauer SF (1992) The ecosystem role of poikilohydric tundra plants. In: Chapin FS III, Jefferies RL, Reynolds JF, Shaver GR, Svoboda J (eds) Arctic ecosystems in a changing climate. An ecophysiological perspective. Academic Press, San Diego, pp213–237

Tissue DT, Oechel WC (1987) Responses of *Eriophorum vaginatum* to elevated CO_2 and temperature in the Alaskan arctic tundra. Ecology 68: 401–410

Webber PJ (1978) Spatial and temporal variation of the vegetation and its productivity, Barrow, Alaska. In: Tieszen LL (ed) Vegetation and production ecology of an Alaskan arctic tundra. Springer Berlin Heidelberg New York, pp37–112

Welker JM, Wookey PA, Parsons AN, Press MC, Callaghan TV, Lee JA (1993) Leaf carbon isotope discrimination and vegetative responses of *Dryas octopetala* to temperature and water manipulations in a high arctic polar semi desert, Svalbard. Oecologia 95: 463–469

Welker JM, Heaton THE, Spiro B, Callaghan TV (1995) Indirect effects of winter climate on the d^{13} C and dD characteristics of annual growth segments in the long-lived, arctic plant *Cassiope tetragona*: a preliminary analysis. In: Proc Eur Science Foundation Worksh on Problems of stable isotopes in tree rings, lake sediments and peat bogs as climatic evidence for the Holocene, Bern, Switzerland, April 1993, European Science Foundation, European Paleoclimate Programme (in press)

Wijk S (1986) Performance of *Salix herbacea* in an alpine snow-bed gradient. J. Ecol 74: 675–684

Wookey PA, Parsons AN, Welker JM, Potter J, Callaghan TV, Lee JA, Press MC (1993) Comparative responses of phenology and reproductive development to simulated environmental change in sub-arctic and high arctic plants. Oikos 67: 490–502

Wookey PA, Welker JM, Press MC, Callaghan TV, Lee JA (1994) Differential growth, allocation and photosynthetic responses of *Polygonum viviparum* L. to simulated environmental change at a high arctic polar semi-desert. Oikos 67: 490–502

Wookey PA, Robinson CH, Parsons AN, Welker JM, Press MC, Callaghan TV, Lee JA, Environmental constraints on the growth and performance of *Dryas octopetala* ssp. *Octopetala* at a high arctic polar semi-desert. Oecologia (in press)

Wyn Williams D (1990) Microbial colonization processes in Antarctic fellfield soils – an experimental overview. Proc Natl Inst Polar Res Symp Polar Biol 3: 164–178

12 Patterns and Current Changes in Alpine Plant Diversity

G. GRABHERR, M. GOTTFRIED, A. GRUBER, and H. PAULI

12.1 Introduction

Global warming, resulting from increased concentrations of greenhouse gases, may affect ecosystems in different ways and to various extents (Emanuel et al. 1985; Bolin et al. 1986; Solomon and Shugart 1993, etc.). Coral reefs, mangroves, the arctic tundra, and high mountain ecosystems are particularly vulnerable (Markham et al. 1993).

Based on a temperature decrease of 0.5–0.7 °C per 100 m and the projected warming of the atmosphere of 3 °C by the middle of the next century, it is possible that vegetation belts on mountain slopes will show an upward migration trend of 400–600 m (Nilsson and Pitt 1991; Ozenda and Borel 1991). In danger of extinction are alpine tundras on moderately high mountains due to the encroachment of today's subalpine forests (Boer et al. 1990; Ozenda and Borel 1991; Holten 1993). Thus in high mountains areas an extensive reduction in biodiversity may occur.

However, this calculation ignores the dynamics of migration. The processes of competition and upward migration depend on the potential expansion rates of particular species-specific plant populations, which are for the vast majority of alpine plants unknown. However, for most species, these expansion rates may not be adequate to keep pace with climate warming. Furthermore, the small-scale relief of high mountains offers many protected sites to migrating flora and fauna, which may survive far beyond climatically defined belts. The sites that can be reached depend on the barriers or migration corridors.

In this chapter, in order to present a more detailed picture, some features of biodiversity in high mountains are discussed, based on data from the literature as well as our own, mostly unpublished, observations from the Central Alps. Taking into consideration the altitudinal extension and differentiation of the alpine life zone viewed in a global perspective, altitudinal biodiversity patterns at different spatial scales are described for the Central Alps. Finally, evidence for upward movement of summit vegetation in the Alps is presented along with empirical findings regarding rates of movement. Underlying physiological responses of alpine plants to the changing climate are discussed in more detail by Körner (1992, and this Vol.).

Institute of Plant Physiology, Department of Vegetation Ecology and Conservation Biology, University of Vienna, Althanstraße 14, 1090 Vienna, Austria

Ecological Studies, Vol. 113
Chapin/Körner (eds) Arctic and Alpine Biodiversity

12.2 The Altitudinal Limits of Plant Life

High mountain environments which exclude the growth of trees and tall shrubs are conventionally considered as "alpine", regardless of striking differences in climate, flora, and vegetation. The upper limit of the alpine zone is assumed to coincide with the climatic snow line, i.e. the altitude at which snow does not completely melt. Favorable microhabitats above the snow line support scattered vegetation, which gradually diminishes up to point where no further plant life exists. This is the "nival zone".

While primitive plants such as bacteria, soil algae, and fungi survive at the highest summits of the Himalayas, the upper limit of higher plant forms is confined to lower altitudes, e.g., 6000 m in subtropical and 4000 m in temperate mountains (Table 1). It is unclear whether higher plants have the physiological potential to grow at even higher elevations at tropical latitudes. On the one hand, individuals of Afro-alpine species occur close to the summits of Mt. Kenya (at 5190 m), Mt. Kilimanjaro (at 5760 m; Beck et al. 1983; Beck 1988), and at the highest peak of the Ruwenzoris (5119 m; pers. observ.). On the other hand, in the tropical Andes (Chimborazo) higher plants do not grow beyond 5100 m (Halloy 1991).

Below the uppermost, isolated outposts of higher plants, species numbers increase with decreasing altitude. In both tropical and temperate mountains, the altitudinal interval between the uppermost occurrences and the uppermost closed vegetation may approach 1000 m (Table 1). From this elevation, it may be an additional 1000 m down the mountain to the timber-line. Generally, the growth belt of higher plant life in the alpine/nival zone on high mountains may extend over 2000 m. However, only the lower half will favor the development of closed plant communities such as heathlands and grasslands.

12.3 Patterns of Diversity

12.3.1 Large-Scale Pattern

Species richness decreases with increasing altitude, as illustrated in the Montafon mountains of Austria (Fig. 1), varying widely in the lower alpine zone with an average of about 100 species/25-ha plot. This value does not change significantly with elevation until a sudden drop to 55 species at 2500 m, indicating the lower limit of the upper alpine zone. Again, the average number of species remains constant over an altitudinal interval of at least 300 m, at which point species richness decreases exponentially. At the summit, i.e. Piz Buin (3313 m), eight to ten species can still be found (not shown in Fig. 1). A similar sigmoidal decrease in species richness of higher plants was also found in the arctic tundra along a latitudinal gradient in the Taymyr Peninsula, Russia (Matveeva 1993).

The exponential decrease in species in the nival zone is also reflected in the literature. Figure 2 presents all the species of the Alps known to grow at altitudes above 3000 m, 3100 m, etc.

Table 1. Altitudinal limits of plant life in tropical, subtropical and temperate mountains

Uppermost recorded species or communities	Elevation (m)	Reference
Tropical mountains		
Mt. Kenya		
summit	5190	
uppermost vascular plants	5190	Rehder et al. (1988)
uppermost giant rosettes	5000	Rehder et al. (1988)
uppermost closed stands of giant rosette plants	4400	Rehder et al. (1988)
Kilimanjaro		
summit	5896	
mosses and lichens	5896	Beck (1988)
uppermost vascular plants	5760	Lind and Morrison (1974)
uppermost communities	5700	Beck (1988)
spring water community	4500	Beck et al. (1983)
closed shrub community	4300	Beck et al. (1983)
Ruwenzori		
summit	5119	
uppermost vascular plants	5119	Gottfried and Pauli (pers. observ.)
uppermost giant rosettes	4600	G. Grabherr (pers. observ.)
uppermost closed communites	4500	G. Grabherr (pers. observ.)
Chimborazo		
summit	6310	
uppermost mosses	5730	Halloy (1991)
uppermost vascular plants	5100	Halloy (1991)
continuous vegetation	4600	Halloy (1991)
Subtropical mountains		
Himalaya		
highest summit	8848	
soil bacteria/fungi	8400	Miehe (1991)
uppermost lichens	7400	Kunavar (cited in Miehe 1989)
uppermost vascular plants		
Saussurea gnaphalodes	6400	Miehe (1991)
Ermania himalayensis	6300	Miehe (1991)
Arenaria bryophylla	6180	Wollaston (1921; cited in Polunin and Stainton 1985)
uppermost community (9 higher plant species)	5960	Miehe (1989)
uppermost closed swards	5500	Miehe (1991)
Andes		
highest summit considered	7084	Halloy (1991)
uppermost liches	6700	Halloy (1991)
uppermost mosses	6060	Halloy (1991)
uppermost vascular plants	5800	Halloy (1991)
continuous vegetation incl. vascular plants	4600	Halloy (1991)
Temperate mountains		
Alps		
highest summit	4807	
uppermost mosses and lichens	4634	Vaccari (1906)
uppermost vascular plants		
Saxifraga biflora	4450	Anchisi (1986)
Ranunculus glacialis	4270	Heer (1885)
3 higher plant species	4000–4270	Heer (1885)
11 higher plant species	3800–3969	Vaccari (1911)
uppermost closed swards	3480	M. Gottfried and H. Pauli (pers. observ.)

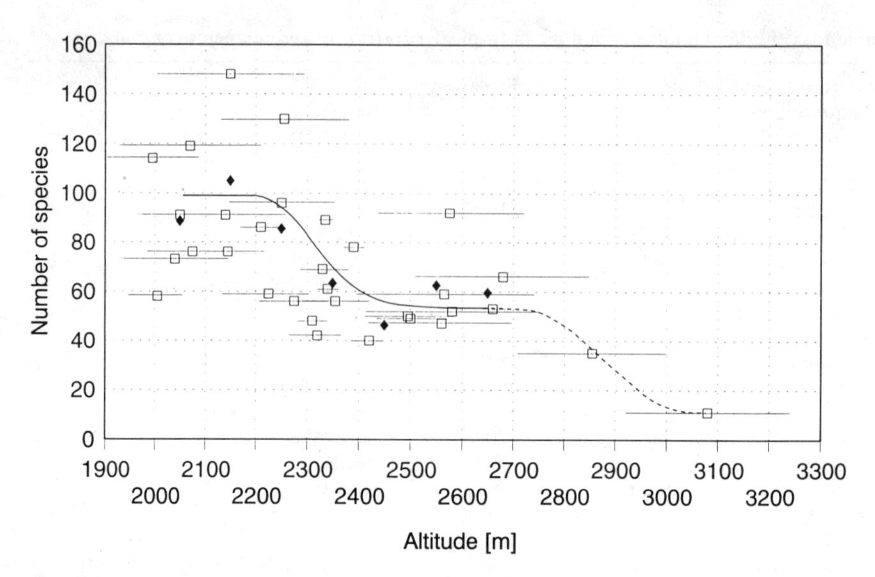

Fig. 1. Decrease in species richness along an altitudinal gradient from the timberline to the highest summits in the Montafon mountains, Austria. Data represent species richness of randomly selected 25-ha plots in the alpine/nival zone. *Horizontal lines* indicate the range in altitude for the particular plot; *squares* represent the mean; diamonds denote altitude of particular summits. Data from Grabherr (1985, unpubl.)

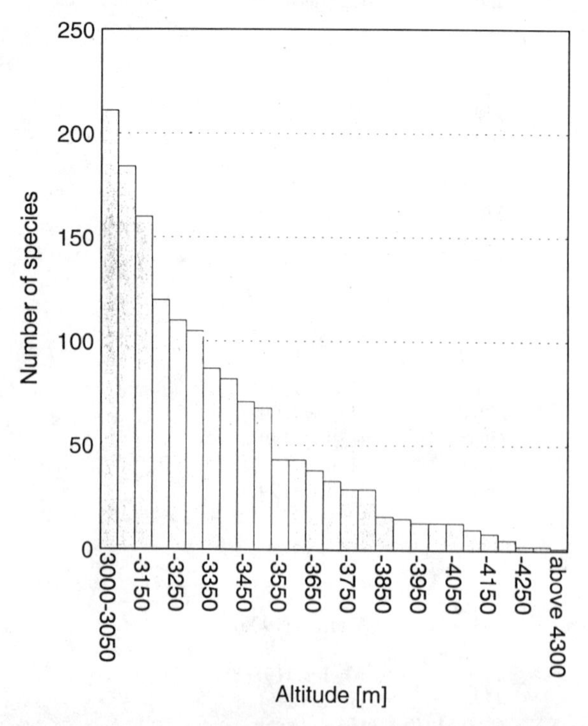

Fig. 2. Decrease in species richness for the nival zone only (above 3000 m) in the Alps. *Columns* represent number of species at this or higher altitudes. (After Reisigl and Pitschmann 1958; updated by H. Pauli, unpubl.)

Examination of the Alps as a whole shows that more than 200 species – about one-third of the alpine/nival flora – occurs at altitudes above 3000 m. At 3000 m upwards, however, there is a significant decrease in species richness, with a gradual reduction of about 120 species at 3200 m to 70 at 3500 m. Above this, about 30–40 species may be found up to 3800 m. Slightly more than 20 species can be found at altitudes above 4000 m; the greatest recorded altitude is from Mt. Dom in Switzerland at 4450 m (Saxifraga biflora; Anchisi 1986).

Despite the clear exponential decrease in species richness in the nival zone, some discontinuities (e.g. at 3550 m) are notable (Fig. 2). These may be caused by cryoturbation in the soil, which becomes more and more important with increasing altitude. At a certain point – as a result of frequent night frosts in summer – conditions might be quite similar to the diurnal climate in the tropics. Only "solifluction acrobats" (Miehe 1989) may survive. Threshold limits at which cryoturbation can be expected regularly may coincide with the above-mentioned discontinuities.

The discontinuities below the nival zone, however, especially between the lower and upper alpine zone (Fig. 1), are more striking. These may be linked to greater habitat variety, a pronounced higher altitudinal variance of the lower alpine plots (see Fig. 1), and the altitudinal distribution of key species (sect. 12.3.3). Regardless of the explanation, it is evident that situations exist in which biodiversity does not change gradually with altitude. Thus, the working hypothesis of a continuous change must be rejected. The different altitudinal zones are separated from each other by true discontinuities.

12.3.2 Small-Scale Pattern

Plots of 25 ha include a mosaic of different communities and, therefore, represent vegetation complexes. On Montafon, a single plot contains 10–15 community types in the alpine zone. In most plots, zonal types such as dwarf shrub heath, alpine meadows and pastures, *Carex curvula* grassland, and related types occupy most of the area. Interspersed are azonal communities including spring communities, fens, rock and scree communities, pioneer vegetaion on moraines, snow beds, and tall forb communities in avalanche paths.

Within-community species richness varies from 3–4 higher plant species in wetland communities to more than 40 species in some grassland communities. Table 2 presents some values for species richness of zonal communities which are compared with figures from Afro-alpine sites. The high species richness of alpine grasslands in the Alps is obvious, although only a few key species like the sedges or grasses dominate. However, zonal communities are not always rich in species, as indicated by the values for dwarf communities and open nival cushion vegetation (Table 2).

Low community diversity is typical for Afro-alpine communities, especially for the giant rosette communities. In addition, fewer alpine community types occur on tropical than on temperate mountains. Rehder et al. (1981) distinguished 16 different communities using floristic criteria for the alpine/nival zone at Mt.

Table 2. Species richness of zonal alpine plant communites (= climax communities) along altitudinal gradients in the Alps and in East-African mountains

Mountain system and type of plant community	Plot[a] size (m²)	Within-community species richness Mean	Min	Max	No. of samples
Mt. Kenya (Rehder et al. 1981)					
lower alpine *Senecio-Lobelietum*	49	10	9	10	3
alpine *Dendrosenecio woodlands*	49	4	3	4	2
alpine tussock grasslands	49	13	7	18	4
upper alpine *Lobelietum*	100	14	.	.	1
nival herb community	?	6	.	.	1
Mt. Kilimanjaro (Beck et al. 1983)					
lower alpine *Alchemilla* shrubs	25	15	10	18	9
upper alpine *Helychrysum* shrubs	25	11	5	17	5
subnival open tussock vegetation	100	10	9	10	2
Alps (G. Grabherr, unpubl.; acid soil vegetation)					
lower alpine dwarf shrubs (*Loiseleurio-Cetrarietum*)	9 16	18	13	24	11
lower alpine grassland community (*Caricetum sempervirentis*)	16–25	37	25	54	18
upper apline grassland community (*Caricetum curvulae*)	16–25	23	16	36	17
nival open cushion vegetation (Androsacetum alpinae)	> 100	7	3	13	7

[a] Plot size is the minimum area that includes 90% of all species within a certain plant community.

Kenya. When considering communities on acid soils only, the equivalent value is about 40 for the above-cited Montafon area in the Alps. This difference is due simply to the lack of some habitat types in the tropics (e.g. snow beds, avalanche paths, wind ridges blown free from snow, etc.).

The various habitats along with their communities contribute differently to the overall species diversity. At least one-third of the alpine/nival flora (vascular plants) of the Alps is restricted to azonal habitats, especially rocks, screes, and snow beds. Habitat, and therefore community diversity, are certainly key factors affecting the diversity of high mountains (see Körner, this Vol.). Different habitats might be variously influenced by global warming. Snow beds, for example, may disappear long before essential changes in the composition of alpine grasslands can be observed. Predictions regarding the future of diversity must take this fact into consideration.

12.3.3 Altitudinal Distribution of Key Species

In many plant communities, including alpine communities, both community processes and biodiversity are strongly controlled by dominating key species

(Grabherr 1989). They provide specific niches for associated species or exclude species as a result of competition for nutrients and water, or, simply, lack of space. The altitudinal distribution of such key species from alpine heathlands and grasslands may be responsible for the above-mentioned discontinuities.

Loiseleuria procumbens and *Carex sempervirens* (Fig. 3), as indicator species for the lower alpine zone, have an optimum extension across 300 m. The distribution curve is clearly not Gaussian. Between 2300 and 2400 m the frequencies decline rapidly to a low level. In particular, the *Carex sempervirens* grasslands are very rich in species (Table 2). Thus the overall decrease in biodiversity at this altitude (Fig. 1) might be, among other factors, due to the disappearance of this key species. The distribution of the associated species, *Phytheuma hemisphaericum*, may support this view; *P. hemisphaericum* has a distribution similar to its "host" (Fig. 3).

The third dominating species on acid soils, i.e. *Carex curvula*, ocurs over a wide range (Fig. 3). In contrast to the above-mentioned species, *C. curvula* obviously survives well in the upper alpine zone. The associates (*Oreochloa disticha*, a tiny grass species, and *Phytheuma globulariifolia*, a tiny rosette plant) occupy the gaps between the small clusters of this sedge. In the taller *Carex sempervirens* community, these species cannot survive.

In contrast, the distribution of *Ranunculus glacialis* (Fig. 3) exemplifies plants growing at the limits of plant life, however, they may also occur far lower. The low altitude habitat of *R. glacialis* is a recently formed glacier forefield. The absence of competition seems to be the precondition for the establishment of this "alpine ruderal".

These few examples demonstrate that a complex pattern of interferences may determine vegetation changes on a small scale. Concentrating on key species may also allow predictions for associated species. However, ruderals such as *Ranunculus glacialis* may react very individualistically.

12.4 Effects of Global Warming on Diversity

12.4.1 Biodiversity Disasters

As stated previously, vegetation zones may shift upwards as the climate warms (Boer et al. 1990; Nilsson and Pitt 1991; Ozenda and Borel 1991; Holten 1993). In areas like the Snowy Mountains of SE Australia, this could eliminate a unique flora (Costin et al. 1982); the same may become true for many other isolated "island mountains" with a distinct flora. Larger mountain systems, like the interzonal mountain system of the Alps or long chains such as the Urals, the Rocky Mountains, or Andes, might be less prone to such biodiversity disasters. However, deep valleys split these mountain systems into parts and further subdivisions; such migration barriers may facilitate extinction processes.

For example, the endemic flora of the Alps is concentrated in specific refugias like the northernmost, southeast, central-south, and southwesternmost Alps (Pawlowski 1970), where most summits do not exceed 2500 m (Fig. 4). Species

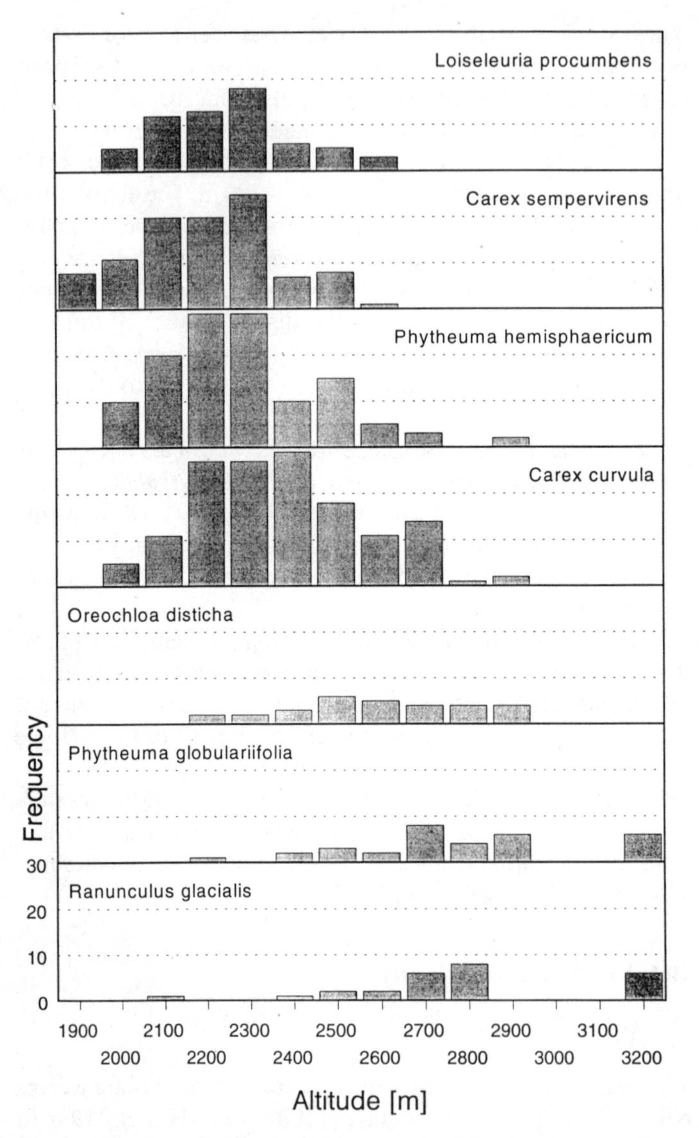

Fig. 3. Altitudinal distribution of key species (*Loiseleuria procumbens*, *Carex sempervirens*, *Carex curvula*) of alpine climax communities in the Montafon mountains, Austria (dwarf shrub heath, alpine grasslands) together with associated species (*Phytheuma hemisphaericum*, *P. globulariifolium*, *Oreochloa disticha*). The distribution of *Ranunculus glacialis* represents an example for a nival plant. Frequency is given as percentage occurrence in a set of sampling plots; plot size: 0.25 ha. (Grabherr 1985, unpubl.)

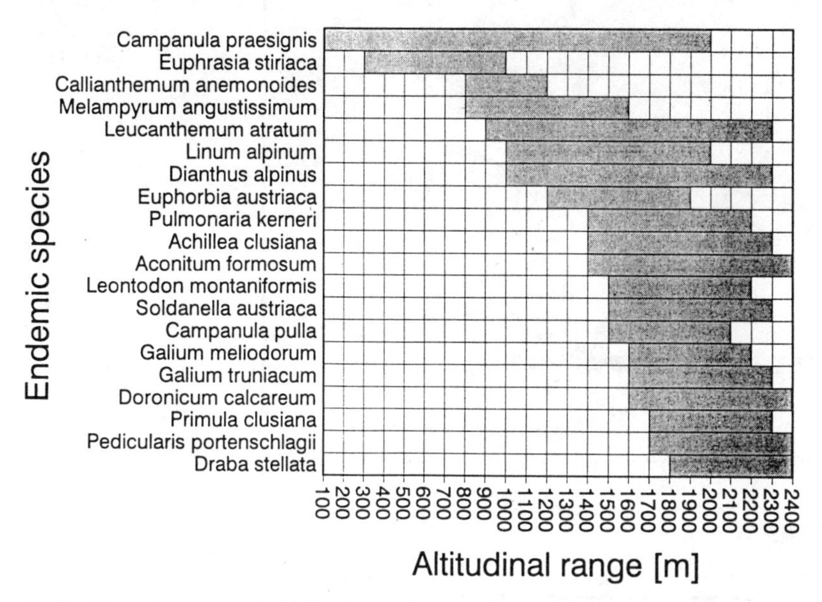

Fig. 4. Altitudinal range for the endemic vascular plant species of the nothernmost Alps, one of the refugias during glaciation. Each *column* represents an altitudinal interval of 100 m. The summits are about 2000 m high; thus, upward movement induced by warming may lead to the extinction of about one-half of the species. Data have been compiled from standard floras.

extinction may occur when there are no higher, adjoining mountains to provide an area to re-establish.

12.4.2 Filling Versus Moving

A more precise picture of vegetation migration is seen on Mt. Patscherkofel, the "island mountain" (Tyrol, Austria; Fig. 5; A. Gruber, unpubl.). It is characterized by well-developed krummholz up to 2100 m followed by lower alpine vegetation and fragments of typical upper alpine communities at its peak (2234 m; Fig. 5). The altitudinal range can be classified into two groups: the alpine grassland group close to the summit (Fig. 5, above), and the subalpine tree species group (Fig. 5, below); these species grow everywhere except in some uppermost plots. Towards the top, the tree species grow as krummholz.

Because the timberline is close to the summit, an increase in the annual temperature of only 1–2 °C would cause the extinction of 13% of the present flora. In addition, isolated outpost populations of "lowland species" would expand as a result of warming. At Mt. Patscherkofel, the uppermost forest trees and understory species will show increased growth and regenerate more effectively. Thus, transition zones become saturated by low-elevation species originating from outpost populations. The expanding populations may even

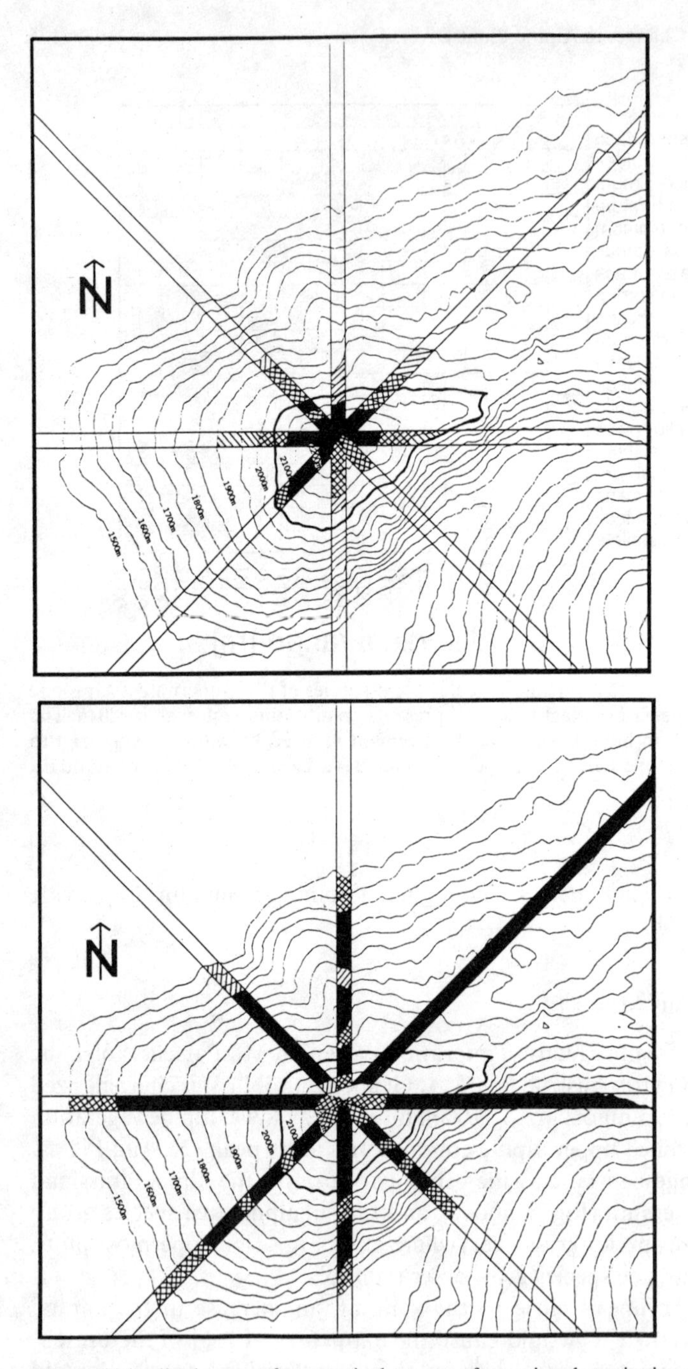

Fig. 5. Altitudinal range of two typical groups of associated species in relation to the compass directions at Mt Patscherkofel, Austria. *Above* Typical species of alpine grasslands (i.e. *Carex curvula, Tanacetum alpinum, Primula minima, Senecio incanus, Oreochloa disticha*); *below* the trees of the subalpine forest (i.e. *Larix decidua, Pinus cembra, Picea abies*), the latter growing as krummholz at the uppermost altitudes. Given are occurrences in the particular quadrats: *black* all species occur in the quadrat; *white* no species; *hatched* intermediate values. The *thick line* roughly indicates the altitudinal limit of closed subalpine forests.

grow downslope. This process should be envisaged as "filling" rather than as "moving".

Even now, such filling can be observed in many snow beds of the upper alpine zone in the Central Alps. Their mossy centers are frequently "infected" by patches or fairy rings of *Carex curvula*. The oldest ones may be 80–100 years old and date back to a germination event which was only possible under the more favourable climatic conditions after the last peak of the Little Ice Age in the mid-19th century.

12.4.3 Evidence for Actual Upward Movement

The example of *Carex curvula* populations encroaching snow beds indicates that movement is currently occurring and will continue. Since 1850, the average annual mean air temperature in Austria has increased by 1 °C, equivalent to a 200 m change in elevation (Auer et al. 1992). By comparing old records from Piz Linard (3411 m), Braun-Blanquet (1957) found increasing species richness. Hofer (1992) also found increased species richness on most summits investigated compared to the old records of species presence/absence in one geographically restricted area (Rübel 1912). However, a comprehensive study of the literature revealed many more old records from various parts of the Alps (Schibler 1898; Vaccari 1901, 1906, 1911; Braun-Blanquet 1913, 1955, 1957, 1958; Braun-Blanquet and Thellung 1921; Klebelsberg 1913; Gams 1936; Lüdi 1939; de Vilmorin and Guinet 1951; Reisigl and Pitschmann 1958). Here, we report the results of a 1992–1993 survey of 25 peaks of the Alps, most of them higher than 3000 m (M. Gottfried and H. Pauli, unpubl.). Species occurrences and abundances were documented using the same plot size. For 12 summits, Braun-Blanquet (1913, 1955, 1957) collected individual florulae in a meter-by-meter approach. These summits were sampled again in the same way.

Species richness has increased on most of the summits (Fig. 6). The average increase in weighted species richness since the begining (or middle) of this century is about 25% for the summits below 3200 m. As rare species are weighted, 25% means an increase of "at least" 25%.

At higher summits this change is smaller or absent. For example on Mt. Linard, the classical mountain for high alpine summit research, there was one species in 1835, three in 1864, four in 1895, eight in 1911, eleven in 1937, and ten in 1947. Our visit in 1992 revealed ten species, eight of which were also recorded by Braun-Blanquet (1957). The two new species were represented by only a few individuals. Thus, species composition at this summit has not changed over 45 years. However, compared to Braun-Blanquet's (1957) map of populations in 1947, our 1992 map indicates that most of the species have increased in abundance and that they grow at new sites. By contrast, on Mt. Hohe Wilde (3480 m), which was investigated in 1953 by Reisigl and Pitschmann (1958), the species number has changed from 10 to 19 (Fig. 6), an increase which cannot be explained by chance alone. This mountain has a rocky southern slope with uninterrupted vegetated corridors to the upper alpine grassland fragments. In fact, most of the new species can be considered characteristic for the alpine grassland zone.

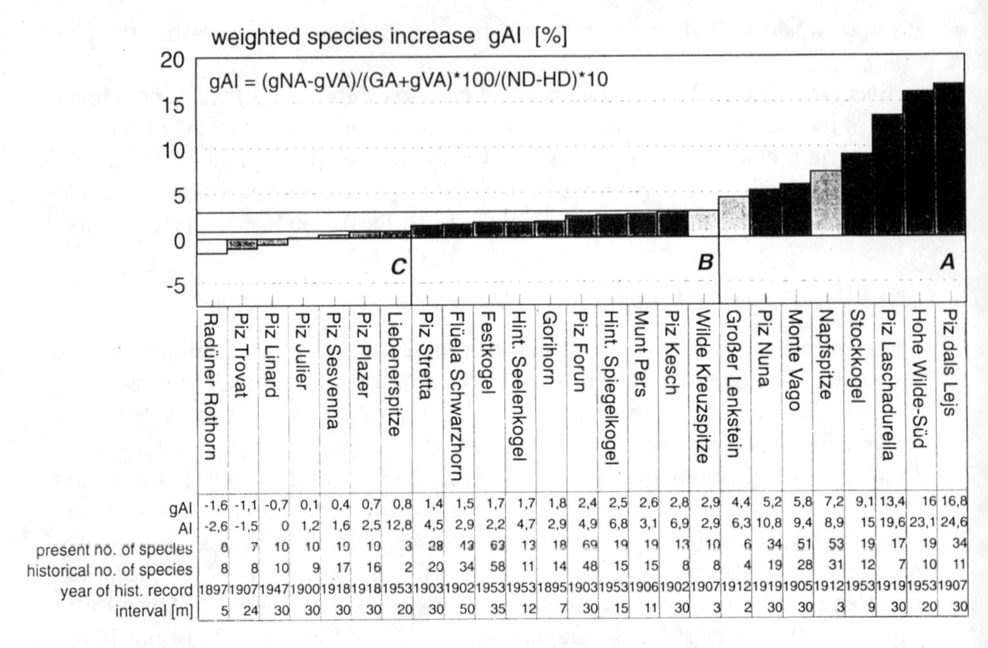

	Radüner Rothorn	Piz Trovat	Piz Linard	Piz Julier	Piz Sesvenna	Piz Plazer	Liebenerspitze	Piz Stretta	Flüela Schwarzhorn
gAI	-1,6	-1,1	-0,7	0,1	0,4	0,7	0,8	1,4	1,5
AI	-2,6	-1,5	0	1,2	1,6	2,5	12,8	4,5	2,9
present no. of species	0	7	10	10	10	10	3	28	43
historical no. of species	8	8	10	9	17	16	2	20	34
year of hist. record	1897	1907	1947	1900	1918	1918	1953	1903	1902
interval [m]	5	24	30	30	30	30	20	30	50

	Festkogel	Hint. Seelenkogel	Gorihorn	Piz Forun	Hint. Spiegelkogel	Munt Pers	Piz Kesch	Wilde Kreuzspitze
gAI	1,7	1,7	1,8	2,4	2,5	2,6	2,8	2,9
AI	2,2	4,7	2,9	4,9	6,8	3,1	6,9	2,9
present no. of species	63	13	18	69	19	19	13	10
historical no. of species	58	11	14	48	15	15	8	8
year of hist. record	1953	1953	1895	1903	1953	1906	1902	1907
interval [m]	35	12	7	30	15	11	30	3

	Großer Lenkstein	Piz Nuna	Monte Vago	Napfspitze	Stockkogel	Piz Laschadurella	Hohe Wilde-Süd	Piz dals Lejs
gAI	4,4	5,2	5,8	7,2	9,1	13,4	16	16,8
AI	6,3	10,8	9,4	8,9	15	19,6	23,1	24,6
present no. of species	6	34	51	53	19	17	19	34
historical no. of species	4	19	28	31	12	7	10	11
year of hist. record	1912	1919	1905	1912	1953	1919	1953	1907
interval [m]	2	30	30	3	9	30	20	30

Fig. 6. Increase in species richness at nival summits in th Alps during the last 100 years which can be related to the observed air temperature increase since the middle of the last century. The rate of migration depends on the altitude of the summit and its morphological character (rocky versus gravelly). Values are weighted for abundance and standardized by the time interval between the historical and recent record:
gAI weighted species increases (weight for common species = 1; for rare species = 0.25; for intermediate species = 0.5; this weighting takes into account that the probability that the first author overlooked one or the other rare species is comparatively high; errors in species identifications may also have occurred); $(gNA - gVA)$ difference between present number of species and historical number; $(GA + gVA)$ number of species present in historical record but missing today plus historical species also present today; $(ND - HD)$ time span between historical and recent record

The presence of migration corridors is certainly crucial. In Figure 6, three groups can be distinguished. Group A shows high rates of migration, indicating a clear increase in species number. The second (group B) shows a moderate increase, group C no increase or even a slight decrease. Besides altitudinal differences, a careful reconnaissance of summit size and shape shows that most group A summits are rocky with many crevices for plants to set roots; groups B and C consist mainly of "gravel mounts" (e.g. Mt. Linard) where plants find virtually no stable substrate.

12.4.4 Migration Rates

The precise information provided for 12 mountains by Braun-Blanquet (1913, 1957, 1958) enabled us to repeat his meter-by-meter approach. The observed upward movement for the eight most common species is maximally about 4 m per

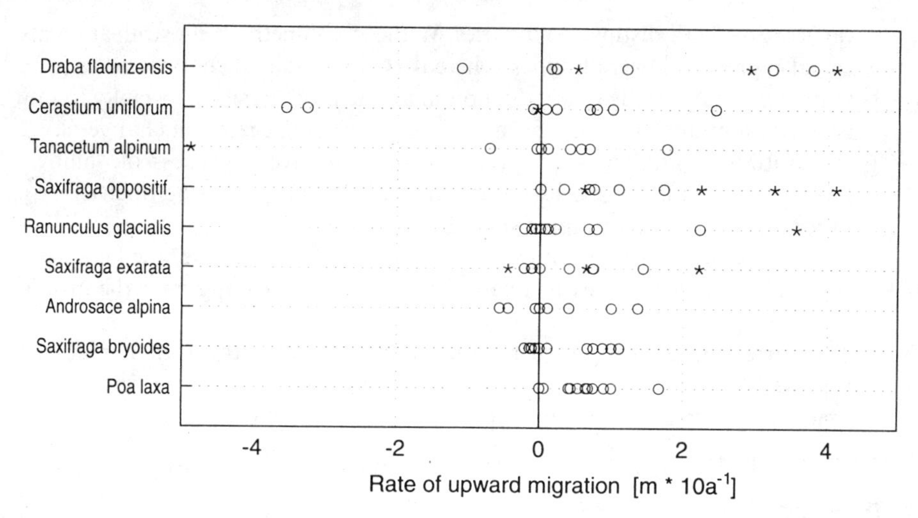

Fig. 7. Migration rates for the most common nival plant species based on precise historical observations. Values represent distances between the uppermost historical record and uppermost recent record at 12 nival summits in the Swiss Alps standardized over time (oldest record > 80 years). *Circles* Species noted in both records, historical and recent; *stars* species not noted in the historical record

10 years (Fig. 7). The zero increase recorded for *Ranunculus glacialis, Saxifraga bryoides* and some other species occurred because these species were already present at the uppermost altitude in the historical records. However, most species show an upward movement of at least 1.0 m per 10 years. Negative values can be interpreted as due to touristic impacts.

The annual mean air temperature in Austria has increased about 0.5–0.7 °C since the turn of the century (Auer et al. 1992), equivalent to a upward isothermic movement of at least 8–10 m per 10 years. The empirical rates presented for the eight nival species are far below this hypothetical value. As all these species possess a somewhat ruderal strategy, it can be assumed that no other species with faster moving rates exist.

12.5 Conclusions

1. Higher plants can survive at very high altitudes (uppermost record: 6400 m); in the (humid) tropics, most of the highest summits also have the potential to support higher plant life. Closed vegetation, however, is restricted to altitudes far below this limit (sometimes more than 1000 m lower).
2. Species richness of the different alpine/nival life zones is closely related to within-community diversity and available habitat types. Diversity does not change gradually with altitude. Between the different elevational zones true floristic discontinuities exist. Key species are distributed over a small altitudinal

range, some with distinct boundaries. Without competition most nival plants can also grow at lower altitudes, where they show ruderal characteristics.
3. Global warming leads to a reduction in alpine biodiversity, especially in low island mountains hosting a flora rich in endemics. Vegetation change along the altitudinal gradient may predominantly involve a process of "filling": foreposts of "lowland" species will expand and fill the upper zone. Such a filling process is described for alpine snow beds.
4. There is evidence for current upward migration in the alpine/nival environment of the Alps; this may be related to the climate warming since the turn of the century.
5. The current rates of upward migration are far below those that might be expected from the rate of temperature increase; this suggests a remarkable time lag between a change in climate and alpine vegetation.

References

Anchisi E (1986) Quatrieme contribution a l'etude de la flore valaisanne. Bull Murithienne 102: 115–126

Auer R, Böhm I, Dirmhirn H, Mohnl E, Putz E, Rudel E, Skoda G (1992) 3. Klimareihen: Analyse und Interpretation von Klimadaten. In: Bestandsaufnahme anthropogene Klimaänderungen: Mögliche Auswirkungen auf Österreich–mögliche Maßnahmen in Österreich. Österreichische Akademic der Wissenschaften, Wien

Beck E (1988) Plant life on top of Kilimanjaro (Tanzania). Flora 181: 379–381

Beck E, Scheibe R, Senser M (1983) The vegetation of the Shira Plateau and the western slopes of Kibo (Mt. Kilimanjaro, Tanzania). Phytocoenologia 11: 1–30

Boer MM, Koster EA, Lundberg H (1990) Greehouse impact in Fennoscandia–preliminary findings of European workshop on the effects of climate change. Ambio 19: 2–10

Bolin B, Döös B, Jäger J, Warrick A (1986) The greenhouse effect, climatic change and ecosystems. SCOPE 29

Braun J (1913) Die Vegetationsverhältnisse in der Schneestufe der Rätisch-Lepontischen Alpen. Neue Denkschr Schweiz Naturforsch Ges 48: 156–307

Braun-Blanquet J (1955) Die Vegetationsverhältnisse des Piz Languoirol, ein Maßslab für Klimaänderungen. Svensk Bot Vidsk Bd 49, H: 1–2

Braun-Blanquet J (1957) Ein Jahrhundert Florenwandel am Piz Linard (3414 m). Bull Jard Bot Bruxelles Vol Jubil W. Robyns: 221–232

Braun-Blanquet J (1958) Über die obersten Grenzen pflanzlichen Lebens im Gipfelbereich des Schweizerischen Nationalparks . Komm Schweiz Naturforsch Ges Wiss Erforsch Nationalparks 6: 119–142

Braun-Blanquet J, Thellung A (1921) Observations sur la vegetation et sur la flore des environs de Zermatt. Bull Murithienne F XLI Sion: 18–55

Costin AB, Gray M, Totterdell CJ, Wimbush DJ (1982) Kosciusco alpine flora. CSIRO, Melbourne

de Vilmorin R, Guinet C (1951) Haute Maurienne. Bull Soc Bot Fr 98: 97–103

Emanuel WR, Shugart HH, Stevenson MP (1985) Climate change and the broad scale distribution of terrestrial ecosystem complexes. Clim Change 7: 29–43

Gams H (1936) Beiträge zur pflanzengeographischen Karte Öslerreichs. Die Vegetation des Großglocknergebietes. Abh Zool Bot Ges Wien Vol 16, He/12

Grabherr G (1985) Numerische Klassifikation und Ordination in der alpinen Vegetationsökologie als Beitrag zur Verknüpfung moderner "Computermethoden" mit der pflanzensoziologischen Tradition. Tuexenia 5: 181–190

Grabherr G (1989) On commmunity structure in high alpine grasslands. Vegetatio 83: 223–227
Halloy S (1991) Islands of life at 6000 m altitude: the environment of the highest autotrophic communities on earth (Socompa vulcano, Andes). Arct Alp Res 23: 247–262
Heer O (1885) Über die nivale Flora der Schweiz. Neue Denkschr Allg Schweiz Ges Gesamt Naturwiss 24: 4–114
Hofer HR (1992) Veränderungen in der Vegetation von 14 Gipfeln des Berninagbietes zwischen 1905–1985. Ber Geobot Inst Eidg Tech Hochsch Stift Ruebelich Zuer 58: 39–54
Holten JI (1993) Potential effects of climate change on distribution of plant species, with emphasis on Norway. In: Holten J I, Paulsen G, Oechel WC (eds) Impacts of climate change on natural ecosystems. Norwegian Institute for Nature Research, Trondheim, pp 84–105
Klebelsberg R (1913) Das Vordringen der Hochgebirgsvegetation in den Tiroler Alpen. Österr Bot Z: 177–187, 241–254
Körner C (1992) Response of alpine vegetation to global climate change. Catena Suppl 22: 85–96
Lind EM, Morrison MES (1974) East African vegetation. Longman, London
Lüdi W (1939) Die Gipfelflora des Flüela-Schwarzhorns bei Davos. Ber Geobot Inst ETH, Zürich: 50–53
Markham S, Dudley N, Stolton S (1993) Some like it hot. WWF International, Gland
Matveyeva NV (1993) Climatic gradient and taxonomical structure of the flora of Taymyr peninsula. In: Gilmanov T, Holten JI, Maxwell B, Oechel WC, Sveinbjörnsson B (eds) Global change and arctic terrestrial ecosystems. Abst Int, Conf, Oppdal, Norwegian Institute for Nature Research Trondheim
Miehe G (1989) Vegetation patterns on Mount Everest as influenced by monsoon and föhn. Vegetatio 79: 21–32
Miehe G (1991) Der Himalaya, eine multizonale Gebirgsregion. In: Walter H, Breckle SW (eds) Ökologie der Erde 4: Spezielle Ökologie der Gemäßigten und Arktischen Zonen außerhalb Euro-Nordasiens. Fischer, Stuttgart, pp 181–230
Nilsson S, Pitt D (1991) Mountain world in danger. Earthscan Publ Ltd, London, 196 pp
Ozenda P, Borel JL (1991) Mögliche Auswirkungen von Klimaveränderungen in den Alpen. CIPRA-Kleine Schriften 8/91, 715
Pawlowski B (1970) Remarques sur le endemisme dans la flore des Alpes et des Carpates. Vegetatio 21: 181–243
Polunin O, Stainton A (1985) Flowers of the Himalaja. Oxford University Press, Oxford
Rehder H, Beck E, Kokwaro JO, Scheibe R (1981) Vegetation analysis of the upper Teleki valley (Mount Kenya) and adjacent areas. J East Afr Nat His Soc Natl Mus 171: 1–8
Rehder H, Beck E, Kokwaro JO (1988) The afroalpine plant communities of Mt. Kenya (Kenya). Phytocoenologia 16: 433–463
Reisigl H, Pitschmann H (1958) Obere Grenzen von Flora und Vegetation in der Nivalstufe der zentralen Ötztaler Alpen (Tirol). Vegetatio 8: 93–129
Rübel E (1912) Pflanzengeographische Monographie des Berninagebietes. Bot Jahrb 47/1–4
Schibler W (1898) Über die nivale Flora der Landschaft Davos. Jahrb Schweiz Alpenclubs 23: 262–291
Solomon AM, Shugart HH (1993) Vegetation dynamics and global change. Chapman & Hall, New York
Vaccari L (1901) Flora cacuminale Valle d'Aosta. Nuovo G Bot Ital 8: 416–439
Vaccari L (1906) La vegetatione della Grivola (3969 m). Club Alp Ital Riv Mensile 25
Vaccari L (1911) La flora nivale del Monte Rosa. Bull Soc Flore Valdotaine 7: 5–67

13 Anthropogenic Impacts on Biodiversity in the Arctic

O. R. YOUNG[1] and F. S. CHAPIN III[2]

The actions of humans are major determinants (Fig. 1) of the biological composition of the Earth's ecosystems at scales ranging from the genetic diversity of local populations to the level of landscapes. Although the anthropogenic impacts of modern industrial societies are obvious, early settlers using simple technologies also altered their environments dramatically through such practices as slash-and-burn clearing and the cultivation of crops in the rain forests of Central America; the diversion of riverwater for irrigation in the Near East; the intentional use of fire in the woodlands and prairies of North America's temperate zones, and alternations in the assemblages of harvested and unharvested species in the far north (Turner et al. 1990)

At the same time, some societies have proven far more destructive than others of the natural systems they inhabit; the impacts of individual societies on the environment have also varied greatly over time. The goals of this chapter are to (1) describe and account for this variance in the impacts of human action on the composition and diversity of ecosystems with particular reference to tundra systems located in the far north and (2) consider the responses of plant and animal communities to anthropogenic impacts under northern conditions. The chapter begins with the articulation of an analytic framework identifying the proximate causes of human action leading to changes in surrounding ecosystems as well as the social drivers underlying these forms of human action. It then proceeds to locate the principal anthropogenic impacts on arctic systems in terms of this framework and to comment on ecosystem responses to these impacts.

13.1 Proximate Causes

Although there is a sense in which the proximate causes of anthropogenic impacts form a continuum, it is helpful for purposes of analysis to group them into several broad categories.

[1] Dickey Center Institute of Arctic Studies, 6193 Murdough Center, Dartmouth College, Hanover, NH 03755, USA

[2] Department of Integrative Biology, University of California, Berkeley, CA 94720, USA

Ecological Studies, Vol. 113
Chapin/Körner (eds) Arctic and Alpine Biodiversity

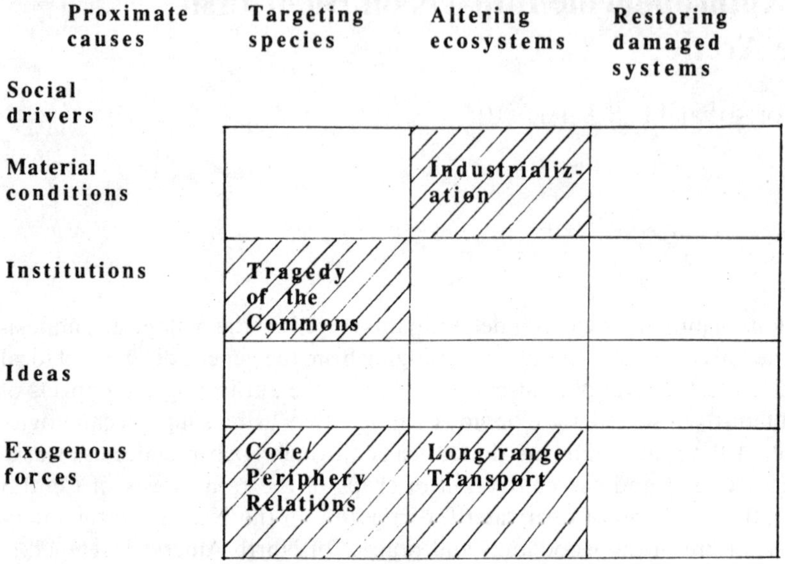

Social drivers / Proximate causes	Targeting species	Altering ecosystems	Restoring damaged systems
Material conditions		Industrialization	
Institutions	Tragedy of the Commons		
Ideas			
Exogenous forces	Core/ Periphery Relations	Long-range Transport	

Fig. 1. Anthropogenic determinants of biodiversity. *Shaded cells* indicate sources of the major anthropogenic impacts on tundra ecosystems in the Arctic

13.1.1 Targeting Species

The classic case of human action affecting the composition and diversity of ecosystems arises from the harvesting of targeted species for consumptive uses. In the Arctic, hunter/gatherers have harvested both terrestrial species (e.g. caribou, musk oxen) and marine species (e.g. bowhead whales, walrus, ringed seals) for subsistence purposes for several millennia. More recently, arctic species (e.g. sea otters, harp seals, reindeer, salmon) have become important to commercial and recreational harvesters. In some cases, humans have driven species to extinction (e.g. the great auk) or severely depleted major stocks (e.g. Atlantic walrus, bowhead whales). In other cases, humans have harvested species sustainably for long periods of time (e.g. ringed seals, arctic char, fox). Unsustainable practices are often attributed to ignorance of the population dynamics of relevant stocks or species, a problem that is particularly acute when the velocity and amplitude of natural swings in the relevant populations (e.g. North American caribou) are great. However, lack of knowledge is not a necessary condition for this type of human behavior. Coincidentally, the depletion or extinction of targeted species can have significant consequences for the status of non targeted species (e.g. the release of sea urchins and the decline in kelp resulting from harvesting sea otters).

In addition to harvesting targeted species, humans often endeavor to rearrange natural assemblages of species in ecosystems, either by enhancing the competitive advantage of preferred species or by eliminating undesired species. This may involve extreminating natural predators, like wolves, in order to increase stocks of moose or caribou for human consumption or eliminating indigenous species, like buffalo, in order to increase available forage for domesticated species, like reindeer or sheep. Alternatively, it may involve controlling a wide range of plant species – treated as weeds – in order to maximize production of crops like wheat, corn, or soybeans. In some cases, rearranging takes the form of introducing new species – both intentionally and inadvertently – that are not native to a region. Such impacts are not unknown in the Arctic, as illustrated by reindeer husbandry in the North American Arctic and fur farming in the Russian North. However, these actions are less prevalent in the far north than in areas of the mid-latitudes where numerous forms of agriculture, horticulture, and animal husbandry are widely practiced.

13.1.2 Altering Ecosystems

Sometimes human intervention goes beyond the rearrangement of assemblages of species to more drastic restructuring of whole ecosystems. The most dramatic examples probably involve irrigation, the clearing and plowing of large tracts of land for agriculture, and mineral and hydrocarbon extraction. In the far north, the restructuring of ecosystems often involves efforts to develop the region's oil and gas reserves, as in Prudhoe Bay on Alaska's North Slope and northwestern Siberia, and abundant hydroelectric power potential, as in James Bay and Churchill Falls in Canada as well as numerous areas in Fennoscandia and the Russian North. But other actions of this type have been proposed (e.g. plans to redirect water from the northward flowing rivers of Siberia, create a major harbor on Alaska's northwest coast using nuclear explosives, and establish a rail connection between Anchorage and Nome). Perhaps the most dramatic example involves the idea of damming the Bering Strait in order to control the flow of water moving between the Arctic Ocean and the Bering Sea/North Pacific.

Still other human actions affecting whole ecosystems involve side effects or unintended consequences of actions motivated by other concerns. Sometimes the link is relatively direct, as in the cases of the destruction of tundra and altered groundwater flows associated with oil and gas development in northwestern Siberia and northern Alaska and the disruption of marine habitat arising from commercial navigation and offshore hydrocarbon development in the north. Often cumulative, indirect impacts of such developments extend far beyond the area directly disturbed (Walker et al. 1987). The long-range transport of pollutants originating outside the Arctic has now become a major concern among those interested in northern ecosystems (Jaffe, et al. 1991); arctic haze as well as nuclear contamination resulting from the dumping of radioactive wastes into northward flowing rivers are prominent examples.

13.1.3 Restoring Damaged Ecosystems

Humans may also make conscious efforts to restore ecosystems damaged either by anthropogenic impacts or by natural forces. Such efforts sometimes entail relatively small-scale actions as in measures to clean up contaminants left at abandoned DEWLine sites or to reintroduce species like musk oxen to areas that were once part of their natural range. In other cases, restoration efforts are attempted on a considerably larger scale. A dramatic case in point, which illustrates the limits as well as the attractions of ecological restoration, is the clean up program for Prince William Sound carried out in the aftermath of the wreck of the *Exxon Valdez* in 1989. In all these cases, restoration involves conscious human action intended to revive damaged ecosystems. In this sense, restoration differs from regeneration or the natural evolution of ecosystems toward some prior state in the wake of human actions. The revival of depleted stocks (for example, sea otters in the North Pacific) following a cessation of human harvesting and the readjustment of surrounding ecosystems illustrates this latter case.

In terms of the prospects for biological diversity in the Arctic's tundra ecosystems, the issues of greatest concern today probably center on the harvesting of targeted species within the region and the disruption of whole ecosystems as an unintended by-product of industrial activities based outside the region. Because arctic ecosystems do not lend themselves to agriculture on a significant scale, conscious efforts on the part of humans to rearrange assemblages of tundra species have not been nearly as prevasive as in temperate systems. With a few striking exceptions, efforts to alter arctic ecosystems for industrial purposes are far less extensive than parallel actions in the mid-latitudes. The remoteness of the Arctic has generally made tundra ecosystems less prominent in the thinking of those concerned with restoration than severely damaged ecosystems located in the heavily populated mid-latitudes. On the other hand, there is a long tradition of harvesting targeted species in the Arctic on the part of both hunter/gatherers whose cultures have evolved under nothern conditions and commercial and recreational users whose activities are rooted in Western cultures. Recent decades have also brought a dramatic upsurge in the presence of various pollutants – often originating far beyond the southern boundaries of the Circumpolar North – that are already affecting arctic ecosystems and that could prove far more disruptive to these ecosystems in the future.

13.2 Social Drivers

The forces described in the preceding paragraphs are all proximate causes in the sense that they invovle human actions directly affecting impacted ecosystems and becoming, in the process, determinants of the biological composition and diversity of natural systems. But what drives these human actions, and why do social systems vary so much in terms of the extent to which they give rise to such anthropogenic impacts? To answer these questions requires delving into the driving forces underlying human actions. In this connection, the objective is not only to identify driving forces but also to gain some sense of how they interact with each other and what proportion of the variance in human behavior affecting ecosystems can be attributed to each. Here, too, a simple taxonomy will help to organize the discussion.

13.2.1 Material Conditions

Material conditions encompass demographic factors, including the size, distribution, and rate of change of human populations, technological factors, including means of harvesting or extracting natural resources, and economic factors, including the natue of the modes of production societies develop (Ehrlich and Holdren 1971). Focusing on material conditions leads to a concern for carrying capacity (Weeden 1992). Is there a maximum size or density of human population that is compatible with sustainable harvesting of wildlife in hunter/gatherer systems? Can participants in such systems continue to harvest wildlife sustainably once they have access to modern technologies in such forms as high-powered rifles, outboard motors, snowmobiles, all terrain vehicles (ATVs), and CB radios? Are the mixed subsistence/cash economies that have replaced traditional subsistence systems in many parts of the Arctic more or less likely to prove disruptive to surrounding ecosystems? Can advanced industrial societies succeed in controlling the dissemination of wastes or residuals or disrupting northern ecosystems far from their points of origin?

13.2.2 Social Institutions

Institutions are rules of the game or codes of conduct that serve to define social practices, assign roles to the participants in those practices, and guide interactions among the occupants of these roles (Ostrom 1990). The central questions to be considered in thinking about institutions as social drivers concern; the extent to which the effective institutions can control or channel human actions affecting the biological composition and diversity of ecosystems and the degree to which the variance among social systems with regard to their impact on ecosystems can be accounted for in terms of differences in the social institutions they develop. We want to know, for example, whether subsistence users can develop governance systems allowing them to harvest common property resources in a sustainable manner under the conditions prevailing in the Arctic and whether the overlay of Western systems of land tenure introduced in many arctic settings during recent decades has eroded or undermined the common property regimes characteristic of indigenous societies. Similarly, we want to ask whether modern societies can devise incentive-based systems capable of eliminating or controlling side effects of industrial activity – in such forms as arctic haze and nuclear contamination – likely to degrade habitats in remote areas like the far north.

13.2.3 Ideas

Although prevailing ideas are obviously affected by material conditions and institutions, most analysts of anthropogenic impacts on ecosystems assume that ideas also have an independent role to play as drivers of human action or collective behavior (White 1968). In this connection, the category of ideas includes scientific knowledge, traditional knowledge, belief systems, ideologies, and ethical or normative precepts. Thus, we want to ask whether sustainable subsistence practices require human users in remote regions like the Arctic to think of themselves as belonging to the land in contrast to owning the land themselves and whether indigenous belief systems concerning the behavior of animals and the role of the spirit world are critical to the operation of the mechanisms of social control characteristic of sustainable harvesting practices (Fienup-Riordan 1990). If so, does the dilution or erosion of such ideas resulting from the impact of Western world views pose a growing threat to the maintenance of biological diversity under northern conditions. Similarly, we ask whether the Judeo-Christian emphasis on human dominion over nature–in contrast to the Leopoldian vision of man as a biotic citizen (Leopold 1966)–is a significant force driving the transfomation of ecosystems and the destruction of habitat in areas like the Arctic affected directly and indirectly by the actions of Western, industrial societies.

13.2.4 Exogenous Forces

In some cases, societies that have developed sustainable relationships with the natural environments they occupy find themselves severely impacted by outside forces they are powerless to control (Jodha 1993). The result can and often will be profound changes in the impact of human actions on the biological composition and diversity of surrounding ecosystems. The European invasion of North America–including the continent's northern most reaches–constitutes a dramatic case in point. The death of up to 90% of the aboriginal human population as a result of the spread of non-indigenous diseases drastically reduced human harvesting of wild animals and human cultivation of the land during the interval between the destruction of indigenous populations and the spread of European settlements. On the other hand, many historians believe that the ravages of introduced diseases caused a collapse in indigenous belief systems and a concomitant dissolution of mechanisms of social control that had served to regulate human harvesting of wild animals in earlier times (Martin 1978). To take an Arctic-specific example, we want to inquire in this context about the impacts on northern ecosystems of the American and Canadian policies during the postwar era of gathering the Arctic's permanent residents into relatively large, year-round settlements to facilitate health, education, and welfare services.

13.3 Arctic Interactions

Which combinations of the proximate causes and social drivers identified in Fig. 1 most strongly affect the biological composition and diversity of northern ecosystems? We suggest that two of the most important processes governing the future of biological diversity in the Arctic are (1) interactions among institutions and exogenous forces that determine the sustainability of the harvest of targeted species in arctic settings and (2) impacts of industrial activities occurring both in the far north and, especially, outside the region on tundra ecosystems.

13.3.1 Determinants of Sustainable Harvesting

During the last two decades, a spirited debate has sprung up around the idea of the "tragedy of the commons" (Hardin 1968). This tragedy, the argument goes, occurs because harvesters of targeted species operating under common property arrangements will deplete available stocks (and even drive individual species to extinction) as a consequence of the absence of incentives to conserve renewable resources or the ecosystems of which they are a part. But this argument, which assumes that institutions in the form of structures of property rights are the principal social driver, has been challenged by anthropologists, cultural geographers, and others who have observed seemingly sustainable harvesting and land-use practices in a variety of small-scale traditional societies–including arctic societies–in which common property is widespread. There is a large and rapidly growing literature on "governing the commons," which details an array of mechanisms of social control that have proved more or less effective in regulating consumptive uses of renewable resources in the absence of some system of private or individual rights (McCay and Acheson 1987; Bromley 1992).

Accordingly, the debate over the impacts of human harvesting on the biological composition and diversity of affected ecosystems has turned increasingly toward an effort to identify conditions that are likely to determine the sustainability of traditional harvesting and land-use practices (McKean 1992). Sustainability depends, in part, on the biological margin for error, i.e., the extent to which affected ecosystems can withstand human misjudgments or delays in the ability of human users to adapt their social practices to biological changes. The conventional wisdom is that tundra ecosystems have a broad biological margin and that this attribute has played an important role in the ability of arctic hunter/gatherers to develop sustainable harvesting practices. Yet there is reason to question this view, at least in cases featuring large concentrations of organisms in a few locations or exhibiting dramatic swings in plant and animal populations (e.g. caribou) as a consequence of nonanthropogenic forces.

It is increasingly clear that exogenous forces–usually in the form of sustained encounters with Western societies–constitute another major factor affecting the sustainability of traditional harvesting practices. In the Arctic, these forces, which have become pervasive in modern times, include: the introduction of advanced harvesting technologies; the concentration of human populations in fixed settlements; the growth of mixed economies encompassing sizable cash sectors, and the emergence of Western systems of land tenure and corporate structures.

The introduction of advanced technologies has had a profound impact on human/nature relations in the Arctic during the last 30–50 years. Individual hunters now rely on high-powered rifles, outboard motors, snowmachines,

ATVs, and CB radios; reindeer herders in Fennoscandia and the Russian North use helicopters. High-endurance trawlers, along with the ability to flash freeze fish at sea, have revolutionized the capacity to harvest offshore fish stocks. These developments have given rise to new problems for those concerned with the protection of biological diversity in the far north. Social practices developed to regulate the human use of renewable resources in one technological setting commonly prove grossly inadequate once powerful new technologies are introduced; technological change often outstrips the capacity of social institutions to adapt to new situations. When the resultant lag times are coupled with ecosystems featuring a thin margin for biological error, threats to biological diversity rise sharply.

The advent of new technologies in the Arctic has been accompanied by the emergence of fixed human settlements and mixed economies. Driven in large part by the commitment of distant governments to provide health, education and welfare services to remote northern communities, these demographic/economic developments have led to new threats to biological diversity including: a rising demand for country food in communities with larger human populations; the capital intensification of subsistence practices; a growing need to make provisions for dealing with the wastes generated by people living in year-round settlements; the rise of opportunities to sell plant and animal products to outside buyers, and the emergence of jobs created by industries involved in the extraction of nonrenewable resources in the north. Although the consequences of these developments are not all negative, it would be foolish to allow our thinking to be guided by an outmoded conception of human settlements and social organizations in assessing the impact of anthropogenic forces on biological diversity in the Arctic.

To this we must add an appreciation for the dramatic social and cultural changes that have been occurring in the Circumpolar North for some time. This is not to say that traditional institutions are no longer relevant in the Arctic. But evidence of the growing significance of Western institutions abounds throughout the far north. This is so whether we focus on structures of property rights, corporate arrangements, or Western systems of government. The native peoples of Alaska have become the largest private landowners in the state, but their property rights in the land are now defined in terms of Western ideas of property. Native Greenlanders dominate the Home Rule government that has developed in Greenland since 1979, but this government rests on an overwhelmingly Western conception of governance. Throughout the Canadian north, Western-style corporations have been established under the terms of settlements of comprehensive land claims.

The significance of these sociocultural changes for biological diversity lies in the facts that two fundamentally different perspectives on the consumptive use of wildlife–the state system and the indigenous system are now in operation at the same time in the Arctic (Usher 1987; Huntington 1992). The rapid growth of interest in co-management (institutionalized cooperation between public officials and representatives of wildlife user groups) as a means of achieving a balance between the two is certainly promising in these terms (Osherenko 1990). However, it is too early to tell how effective co-management arrangements will prove in protecting harvested species in the far north over time.

13.3.2 Sources of Ecosystem Disruption

The case of the impact of industrial activities on whole ecosystems raises different questions. The primary threats to tundra ecosystems in arctic settings arise from the development of nonrenewable and flow resources on a large scale in the far north itself–the production of oil and gas on Alaska's North Slope and in northwestern Siberia and the production of hydroelectric power in the James Bay region of Canada are paradigmatic examples – and from airborne or waterborne industrial pollutants moving from the mid-latitudes, where they are produced, to the high latitudes, where they often accumulate in plant and animal communities.

Although there is some controversy about the scope of the environmental effects of resource-related industries in individual cases (e.g. the Prudhoe Bay development in Alaska), the overall impact of such resource extraction on ecosystems in both the North American Arctic and the Eurasian Arctic is not in doubt (Roginko 1993). Some of the most disturbing features of the development of these industries from the point of view of biological diversity involve their indirect effects (Walker et al. 1987). There is inexorable pressure to open transportation corridors (such as the haul road running alongside the Trans-Alaska Pipeline or the railroad under constuction on Russia's Yamal Peninsula), for example, to a variety of users outside the oil and gas industry, a development likely to produce ecological effects exceeding those associated with the exploitation of hydrocarbons per se. Similar indirect effects are easy to identify in other areas from the release of mercury in waters impounded in connection with the James Bay Project to the use of advaned technologis developed by the energy industry to open up other parts of the Arctic for new commercial ventures.

Even so, there is growing evidence to suggest that the most severe threats to Arctic ecosystems during the foreseeable future will stem from human activities taking place well beyond the boundaries of the Arctic itself. Many of these threats involve airborne and waterborne pollutants associated with industrial activities occurring in the mid-latitudes. Perhaps the most serious of these concerns today center on radioactive contamination, arctic haze, and a complex of heavy metals, organochlorines, and pesticides (Pfirman et al. 1993). Radioactive contamination, mostly originating in Russia and resulting from nuclear accidents, deliberate disposal of radioactive wastes in arctic waters, and the dumping of radioactive wastes into northward flowing rivers, constitutes a problem of massive proportions (Yablokov et al. 1993). Arctic haze, a form of air pollution identified originally in the 1950s, annually spreads sulfur dioxide and other airborne pollutants traceable largely to Eurasian industries across large areas of the Arctic Basin (Stonehouse 1986; Sturges 1991). Heavy metals and organochlorines, often transported to the Circumpolar North by ocean currents, are passed upward through the arctic food chain and are frequently ingested by humans (Jensen 1991). The emphasis on such long-range airborne and waterborne pollutants in the process of developing the Arctic Environmental Protection Strategy–an international agreement signed by all eight arctic states in 1991–is therefore a hopeful sign. Yet the capacity of this emerging international regime to mitigate these problems remains to be demonstrated.

Beyond this, the biological composition and diversity of arctic ecosystems will certainly be affected by global changes, like the depletion of stratospheric ozone and the buildup of greenhouse gases (Weller et al. 1991). Though ozone depletion in the Arctic does not match the much-publicized antarctic ozone 'hole', sharp seasonal fluctuations of stratospheric ozone have been observed for some years in the north polar region. Temperature increases in the high latitudes resulting from global warming are expected to exceed those occurring in the mid-latitudes by a factor of two or more (Etkin and Agnew 1991). Among the possible effects of these forcing functions on tundra ecosystems are large-scale changes in glaciers, the melting of layers of permafrost, and the movement of subarctic species northward. These global changes could well have profound consequences for biological diversity in the Arctic on a scale of decades to centuries (Edlund 1991).

A particularly disturbing feature of these sources of ecosystem disruption in the high latitudes is their association with patterns of core/periphery or metropole/hinterland relations (Young 1992). Natural conditions make the Arctic's raw materials expensive to extract and deliver to southern markets, but the geopolitics of northern peripheries make them secure sources of vital natural resources. Residents of these sparsely populated areas do not have the economic and political resources needed to acquire a strong voice in policymaking regarding the Arctic's raw materials or the long-range transport of airborne and waterborne pollutants. Because the Arctic is remote both physically and culturally from the concerns of residents of southern urban centers, it is difficult to engage mainstream institutions in the effort to protect arctic ecosystems. Although the environmental movement, at least in North America, has recently become conscious of these threats to arctic ecosystems, there are compelling reasons to be concerned about the growth of industrial activities both in the Arctic itself and beyond its southern boundaries as a source of change in the biological composition and diversity of tundra ecosystems during the foreseeable future.

13.4 Humans as Components of Arctic Ecosystems

Population sizes and biological diversity in ecosystems are controlled in part by positive and negative feedbacks. When population increases heighten the impact of a particular species, progressively larger changes in the species composition and functioning of the system often ensue. Increasing abundance of spruce trees during forest succession, for example, causes larger and larger changes in understory environment and leads to directional changes in all aspects of ecosystem structure and function (Van Cleve et al. 1991). When negative feedback predominates, by contrast, there are limits placed on the ecological impact a species can have. Thus, a predator that overexploits its prey reduces its own food supply, leading to a decline in the predator population and allowing the prey population to recover. Humans have been integral components of northern ecosystems since the Pleistocene. But have the forces described in the preceding sections altered the nature, timing, or magnitude of feedbacks between human populations and ecosystems to such an extent that heightened human impacts on arctic biological diversity are inevitable?

13.4.1 Harvesting of Targeted Species

Prior to the introduction of modern hunting technologies and mixed economies, there were strong negative feedbacks controlling the intensity of human harvesting. Any population crash of critical food species, whether caused by hunting or other natural factors, reduced the efficiency with which people could harvest additional animals and threatened to reduce local human populations. Social institutions governing the use of common food sources minimized the probability of episodes of overhunting. Modern hunting technologies, however, have enabled people to continue hunting efficiently even when prey populations are reduced to low levels. Mixed economies have made human population size less dependent on changes in local animal populations and, in many cases, have increased the demand for animal products. Thus, the negative feedbacks that previously limited the impacts of humans on animal populations have been seriously weakened by changes in material conditions and exogenous forces. At the end of the Pleistocene, an improvement in hunting technology in the form of microblades–stone chips used in arrows and spears–may have contributed to the extinction of mammoths, bison, and other components of the Pleistocene megafuna (Martin 1984). The impacts of recent changes in material conditions may prove even more disruptive to arctic ecosystems.

Protection of targeted species requires adjustments in negative feedback loops linking humans and their prey. Social institutions that recognize increased hunting efficiency must provide effective mechanisms for limiting harvests when prey populations are depleted. A number of options are available, based on both traditional and Western practices, including: fallowing (e.g. rotating trap lines), husbandry of targeted populations (e.g. reindeer), regulations limiting hunting of prey species (e.g, musk oxen), restrictions on exogenous demand for specific animal products (e.g. walrus ivory), and limitations on the use of advanced technologies in hunting targeted species (e.g. bowhead whales). The success of social institutions in adjusting harvesting practices to protect populations of many northern animals suggests that innovative solutions can limit the extent to which harvesting of targeted species will have negative effects on the biological diversity of arctic ecosystems.

13.4.2 Alterations of Ecosystems

The impact of pollutants originating outside the Arctic probably represents the single greatest threat to biological diversity and sustainability of arctic ecosystems today, because there are no natural feedback processes by which progressive alteration of arctic systems can lead to reductions in the magnitude of the impact. The pollutants are produced by advanced industrial societies, often in countries or regions remote from the point of impact. Traditional social institutions cannot reduce levels of these pollutants, except by interacting with Western institutions that have little or no incentive to show concem about the effects of pollution in remote areas.

Many pollutants have powerful effects on particular organisms and therefore affect biological diversity directly. Organochlorine pesticides, for example, are concentrated in food chains and disproportionately affect top carnivores, such as birds of prey or polar bears. Radionuclides produced by fallout are absorbed effectively by lichens and have strong effects on participants in lichen-based food chains, including reindeer, wolves, and humans. Lichens and mosses are particularly sensitive to heavy metals emanating from smelters (e.g. the nickel works on Russia's Kola Peninsula), a process that also contributes to a decline in lichen-based food chains. More indirectly, mosses are important to the thermal insulation of permafrost, and the impacts of pollutants on mosses may partially explain the soil warming and high rates of carbon dioxide flux observed in the Arctic (Oechel et al. 1993; Zimov et al. 1993). Consequent changes in the soil thermal regime may alter competitive relationships among plants and cause widespread changes in biological diversity and ecosystem functions.

13.5 Conclusions

Are anthropogenic impacts more or less threatening to biological diversity in tundra biomes than in other biomes? What are the prospects for the future in this regard?

1. So far, arctic systems have fared comparatively well in these terms: human populations are small and, with some exceptions, widely dispersed throughout the far north; tundra ecosystems do not lend themselves to most forms of agriculture; the high latitudes have not become centers of industrial activity other than the extraction of raw materials.
2. Even so, it would be a mistake to discount the threats to biological diversity arising from human activities affecting tundra ecosystems.
3. At present, the principal threats involve the harvesting of targeted species for subsistence, recreational, and commercial purposes; the extraction of raw materials, and, increasingly, the long-range transport of airborne and waterborne pollutants emanating from industrial centers located to the south.
4. Problems arising from excessive harvesting are undoubtedly serious in some areas, but, for the most part, these are local problems that do not threaten larger ecosystems, much less lanscapes.
5. Almost certainly, the disruption of ecosystems caused by airborne and waterborne pollutants entering the Arctic from the mid-latitudes coupled with the

peripheral status of this remote region will constitute the most serious threat to biological diversity in the high latitudes during the foreseeable future.

References

Bromley DW (ed) (1992) Making the commons work. ICS Press, San Francisco

Edlund S (1991) Climate change and its effects on Canadian arctic plant communities. In: Woo MJ, Gregor ID;)J (eds) Arcticenvironment: past, present and future. McMaster University, Hamilton, Ontario, pp 121–137

Ehrlich PR, Holdren JP (1971) Impact of population growth. Science 171: 1212–1217

Etkin D, Agnew T (1991) Arctic climate in the future. In: Woo MJ, Gregor, DJ (eds) Arctic environment: past, present and future. McMaster University, Hamilton, Ontario, pp 17–34

Fienup-Riordan A (1990) Eskimo essays: Yup'ik lives and how we see them. Rutgers University Press, New Brunswick

Hardin G (1968) The tragedy of the commons. Science 162: 1243–1248

Huntington HP (1992) Wildlife management and subsistence hunting in Alaska. University of Washington Press, Seattle

Jaffe DA, Honrath RE, Herring JA, Li SM, Kahl RD (1991) Measurements of nitrogen oxides at Barrow, Alaska during spring: evidence for regional and northern hemispheric sources of pollution. J Geophys Res 96: 7395–7405

Jensen J (1991) Report on organochlorines. In: The state of the arctic environment. University of Lapland, Rovaniemi, pp 335–384

Jodha NS (1993) Property rights and development. Beijer Institute, Stockholm

Leopold A (1966) Sand county almanac. Ballantine, New York

Martin C (1978) Keepers of the game: Indian-animal relationships and the fur trade. University of California Press, Berkeley

Martin PS (1984) Prehistoric overkill: the global model. In: Martin PS, Klein RG (eds) Quaternary extinctions. University of Arizona Press, Tucson, pp 354–403

McCay BJ, Acheson JM (eds) (1987) The question of the commons: the culture and ecology of communal resoures. University of Arizona Press, Tucson

McKean MA (1992) Success on the commons: a comparative examination of institutions for common property resource management. J TheoPol 4: 247–281

Oechel WC, Hastings SJ, Vourlitis G, Jenkins M, Riechers G, Grulke N (1993) Recent changes of arctic tundra ecosystems from a net carbon dioxide sink to a source. Nature 361: 520–523

Osherenko G (1990) Can comanagement save arctic wildlife? Environment 30: 6–13, 29–34

Ostrom E (1990) Governing the commons: the evolution of institutions for collective action. Cambridge University Press, Cambridge

Pfirman S. Crane K, Defur P (1993) Regional distribution of contaminants in the Arctic. Environmental Defense Fund, New York

Roginko A (1993) Arctic development, environment and northern natives in Russia. In: Kakonen J (ed) Politics and sustainable growth in the Arctic. Dartmouth Publishing Company Brookfield, VT, pp 25–35

Stonehouse B (ed) (1986) Arctic air pollution Cambridge University Press, Cambridge

Sturges WT (ed) (1991) Pollution of the arctic atmosphere. Elsevier, New York

Turner BL, Clark WC, Kates RW, Richards JF, Mathews JT, Meyer WB (eds) (1990) The earth as transformed by human action. Cambridge University Press, New York

Usher PJ (1987) Indigenous management systems and the conservation of wildlife in the Canadian north. Alternatives 14: 3–9

Van Cleve K, Chapin FS, Dryness CT, Viereck LA (1991) Element cycling in taiga forest: state-factor control. BioScience 41: 78–88

Walker DA, Webber PJ, Binnian EF, Everett KR, Lederer ND, Nordstrand EA, Walker MD (1987) Cumulative impacts of oil-fields on northern Alaskan landscapes. Science 238: 757–761

Weeden RB (1992) Messages from earth: nature and the human prospect in Alaska. University of Alaska Press, Fairbanks

Weller G, Wilson C, Severin BAB (eds) (1991) The role of the polar regions in global change. Geophysical Institute and Center for Global Change and Arctic System Research, Fairbanks

White L (1968) The historical roots of our ecologic crisis. Science 155: 1203–1207

Yablokov AV, Karasev VK, Rumyantsev VM, Kokeyev MYe, Petrov OI, Lystsov VN, Yemelyanenkov AF, Rubtsov PM (1993) Facts and problems related to radioactive waste disposal in seas adjacent to the territory of the Russian Federation. Russian Government Commission on Matters Related to Radioactive Waste Disposal at Sea, Moscow

Young OR (1992) Arctic politics: conflict and cooperation in the circumpolar north. University Press of New England, Hanover

Zimov SA, Zimova GM, Daviodov SP, Daviodova AI, Voropaev YV, Voropaeva ZV, Prosiannikov SF, Prosiannikova OV, Semiletova IV, Semiletov IP (1993) Winter biotic activity and production of CO_2 in Siberian soils: a factor in the greenhouse effect. J Geophys Res 98D: 5017–5023

Part III
Ecosystem Consequences of Diversity

14 Plant Functional Diversity and Resource Control of Primary Production in Alaskan Arctic Tundras

G.R. SHAVER

14.1 Introduction

The arctic tundra is an ecosystem type in which both vegetation composition and primary production are strongly limited by an extreme physical environment, including limitation by low temperatures, low light, a short growing season, and extremes of soil moisture (Billings and Mooney 1968; Bliss et al. 1981; Chapin and Shaver 1985; Chapin et al. 1992). In this physically stressful environment, both nutritional and biotic resources like soil-available nutrients or pollinators are also in short supply and thus contribute to its overall stressfulness. Finally, arctic tundras are frequently and extensively disturbed by freeze and thaw processes, and thus both the environment and the vegetation are continually changing on a fine scale even though on a coarse scale they may appear to be relatively stable (Sigafoos 1952; Churchill and Hanson 1958; Bliss and Peterson 1992).

In such a stressful and disturbed environment, it is not surprising that relatively few species can survive, grow, and reproduce. The flora of arctic tundra is quite small in comparison to the vast area that it covers, and some common plant forms such as trees, epiphytes, and annual plants are either missing entirely or very rare (Billings 1992). Yet, even .within this restricted flora there is significant variation in the functional characteristics of arctic tundra plants (e.g., in their phenology, leaf type, or rooting depth). The variation is expressed most dramatically as distinct differences in the growth form composition of tundra vegetation along local and regional environmental gradients (Billings 1973; Shaver et al. 1991). However, there is also a surprising amount of variation in plant functional characteristics *within* tundra communities, even when the community is strongly dominated by a single growth form (Shaver and Chapin 1991).

The aim of this chapter is to review what is known about within-community diversity in the functional characteristics of arctic tundra plants, and to discuss the implications of this diversity for ecosystem-level functions like primary production of the whole vegetation. The underlying hypotheses are: (1) Within-community diversity is at least in part a result of specialization in the spatial and temporal patterns of resource uptake among species within the tundra community. (2) This specialization at the species level results in greater total resource uptake by the community as a whole than would occur if there were only one

The Ecosystems Center, Marine Biological Laboratory, Woods Hole, MA 02543, USA

Ecological Studies, Vol. 113
Chapin/Körner (eds) Arctic and Alpine Biodiversity

species present or if there were no differences among species in their resource uptake patterns. Finally, (3) because community resource uptake is greater as a result of functional diversity, community productivity is also greater. Although a fair amount of data are available to apply to this review, the data are largely descriptive and comparative. Critical experiments have not been done, and thus the hypotheses cannot be tested directly.

The focus of this chapter is on soil resource *uptake*, especially soil nutrient uptake, because productivity of tundra ecosystems is strongly and consistently nutrient-limited (Berendse and Jonasson 1992; Shaver et al. 1992). Of course, once resources are taken up by the plants, diversity in the patterns of resource *use* in growth are also important to the regulation of community productivity. The ecosystem-level implications of interspecies differences in resource use are discussed in detail elsewhere (Berendse and Jonasson 1992; Shaver et al. 1994; Hobbie, this Vol.; Chapin et al. this Vol.).

14.2 Resource Uptake

14.2.1 Spatial Patterns

Arctic tundra vegetation typically includes about 10 to 20 common vascular plant species and a similar number of moss and lichen species, all densely intermixed when viewed from above. However, a vertical profile through this vegetation from the top of the permafrost to the top of the canopy often shows striking and consistent differences in species distribution. The vertical partitioning of space among plant species occurs even in sites with relatively uniform topography and vegetation that is dominated by a single growth form. For example, in coastal wet sedge tundra at Barrow, Alaska, three rhizomatous graminoid species (two sedges, *Carex aquatilis* and *Eriophorum angustifolium*, and one grass, *Dupontia fischeri*) account for 60–80%of the primary productivity of the landscape (Tieszen 1978; Brown et al. 1980). The relative abundances of these species vary considerably from site to site, but all three species are common at most sites. Above ground, the leaf distribution of these species is similar but below ground they differ strikingly in their root distributions: *E. angustifolium* roots extend all the way to the bottom of the active layer (25–30 cm), while *D. fischeri* roots are concentrated in the upper 5–15 cm and *C. aquatilis* roots have an intermediate distribution (Shaver and Billings 1975). A similar spatial partitioning of root distribution occurs in tundras with greater species and growth form diversity, such as Alaskan moist tussock tundra. In this case, the differences in rooting depth are apparent both within and between growth forms (Shaver and Cutler 1979; Miller et al. 1982), with graminoids and some forbs being more deeply rooted overall, evergreens often shallowly- rooted, and deciduous shrubs intermediate. In this tundra type, there is also considerable microtopographic heterogeneity assoiated with sedge tussocks (Chapin et al. 1979), and a distinct horizontal patterning to the vegetation and its root distributions in relation to the tussocks (Fetcher and Shaver 1982; Miller et al. 1982).

It is difficult to believe that this spatial partitioning in root distribution is not somehow related to a spatial partitioning in nutrient uptake among species at least where several species coexist. Yet, there have been few experiments to test whether such partitioning exists, or whether it has any impact on community nutrient uptake (e.g., Fetcher 1985; Jonasson 1992; discussed below). There also have been no systematic comparisons of root distribution patterns in sites where only one or two species dominate versus where several species dominate. If roots are distributed throughout the active layer in sites containing only one or two species, then one might conclude that the absence of some species would not result in a lesser community nutrient uptake because the remaining species could compensate by exploiting a larger portion of the soil in the absence of competition. However, Chapin and Shaver (1981) found that species growing on disturbed sites did not change their root distributions despite deeper soil thaw and loss of competing species.

Above ground, the species of tundra vegetation also differ in the distribution of their leaf area. In wet sedge tundra the differences are due mainly to differing leaf orientation with little difference in leaf height (Miller et al. 1980; Tieszen et al. 1980). In other tundras where a more diverse mix of growth forms occurs there are distinct differences in leaf height, both within and among growth forms (Tenhunen et al. 1993). For example, in moist tussock tundra, both creeping (*Empetrum nigrum, Vaccinium vitisidaea*) and semierect (*Ledum palustre*) evergreen shrubs occur together, and tall (0.5–2.0 m) deciduous shrub tundra, dominated by species such as *Salix* spp., often includes a discontinuous subcanopy of graminoids, forbs, and other deciduous and evergreen shrubs (Shaver and Chapin 1991). Except in the drier heath tundras there is usually a continuous ground cover of mosses and lichens. The implications of these differences in canopy distribution and orientation have been discussed elsewhere (Stoner et al. 1978; Miller et al. 1980, 1984; Tenhunen et al. 1992, 1993)

Mosses completely cover the surface of many arctic tundras, and may play a more significant role in regulating primary productivity than any other single plant form or species. In Alaskan boreal forests, moss growth alone adds a cation exchange capacity to the upper layers of soil that is equivalent to almost three times the annual cation deposition rates (Oechel and Van Cleve 1986); mosses in tundra should be expected to be equally capable of retaining cations in the ecosystem. Although nutrient uptake or immobilization by mosses might significantly reduce the movement of these nutrients deeper into the soil, where they could be taken up by vascular plant roots, the mosses' most important role may be to retain nutrients within the ecosystem that might otherwise be carried away in surface runoff and be lost.

14.2.2 Temporal Patterns

The species of tundra vegetation also differ in their temporal patterns of above- and belowground growth (Sørensen 1941; Webber 1978; Murray and Miller 1982; Shaver and Kummerow 1992. However, interspecific differences in growth phenology are not good indicators of differences in temporal patterns of resource uptake, because much of the current growth is supported from storage, not current uptake (Billings and Mooney 1968; Chapin et al. 1980, 1986; Chapin and Shaver 1985a, 1988). Storage is especially important for soil-available resources like N and P, because only a small portion of the soil is thawed at the start of the growing season when growth is most rapid. When most of the soil is frozen, uptake cannot keep up with the demands for growth, and the plants must use stored resources. These stores are replenished later in the growing season when the soil is thawed more deeply, often after aboveground growth has stopped. In addition to N and P storage, carbon storage is also important in support of both early-season growth and overwinter maintenance, when C fixation is limited either by environmental conditions or lack of leaves (Billings and Mooney 1968; Chapin and Shaver 1985a, 1988).

Because resource uptake and resource use in growth are not necessarily synchronous in tundra plants, the best way to document temporal patterns in resource uptake is through changes in overall carbon and nutrient budgets. For example, the total biomass and N and P mass of *Eriophorum vaginatum* plants actually *declines* early in the growing season, at the same time as new leaf and root growth is proceeding rapidly (Chapin et al. 1986; Shaver et al. 1986) These net losses occur mainly as a result of the death of storage tissues (stembases), which lose some of their initial N and P content in dying but also transfer much of their N and P into new leaves and roots. It is not until late in the.growing season that net N and P uptake results in recovery of total N and P stocks to about the same level as at the start of the season.

Unfortunately, relatively few data exist showing seasonal patterns of net, whole-plant element uptake from the soil by other tundra plant species, especially comparisons of more than one species growing within the same community. Although it is clear that tundra plant species differ widely in the chemical forms in which nutrients are stored for use in growth, and in the tissues that are used for storage (Chapin et al. 1980; Chapin and Shaver 1989), there is no clear evidence from past research that tundra plants differ greatly in their seasonal patterns of net uptake of soil-available resources. Interspecific differences in the seasonal pattern of uptake may in fact exist, but it is more likely that these differences are due at least as much to spatial differences in root distribution in relation to the timing of soil thaw (and thus nutrient availability) as to within-season differences in uptake requirements in support of growth. Furthermore, it is not clear that any such diversity in temporal uptake patterns should necessarily lead to greater total community uptake because the small number of past experiments have given inconsistent results (discussed below).

The seasonal patterns of C uptake, on the other hand, differ greatly and consistently among tundra plant species (Johnson and Tieszen 1976; Tieszen 1978; Tenhunen et al. 1993). These differences can be explained largely by seasonal changes in the abundance, photosynthetic competence, and spatial distribution of leaf area among the major plant functional types. Mosses and evergreen species, because they start and end the growing season with green leaves, have a particular advantage early in the growing season. Later, after the development of the deciduous canopy, mosses in particular may be overtopped and have lower C uptake. In the short term (within a growing season), this diversity in temporal patterns of C uptake clearly must lead to greater total community productivity because the species that are capable of the most rapid C uptake (deciduous shrubs) are not capable of C uptake early and late in the season, when the less productive species like mosses and evergreens still are. Here, though, it is not clear that this temporal diversity in C uptake necessarily leads to a greater community productivity in the long term, because long-term tundra productivity is so strongly limited by soil nutrients, not carbon supply (Shaver et al. 1992, 1994).

There is at least one mechanism by which diversity in the controls over temporal patterns of growth by different tundra species may be particularly important to long-term resource uptake and community productivity. That is, we know that tundra species differ in the effects of day length on senescence at the end of the growing season. Assuming that leaf or root senescence indicates decline or cessation of resource uptake, any interspecific differences in the controls by day length on end-of-season senescence may have particularly important implications for both species and whole-community resource uptake at the end of an arctic growing season. We know, for example, that roots of *Eriophorum vaginatum* are essentially ‘deciduous’ and begin to senesce when the day length is below about 21 h. *Carex aquatilis* roots, on the other hand, are perennial and decrease their growth rates in proportion to the decrease in day length at the end of a growing season, but remain fully functional (Shaver and Billings 1977). This difference between species suggests that end-of-season nutrient uptake, especially in a warmer climate with a longer growing season, will be reduced in communities comprised of species with strong photoperiodic constraints on senescence.

14.2.3 Uptake Sources and Mechanisms

Although C is taken up by plants in essentially only one form (CO_2) and from one source (the atmosphere, at least in terrestrial systems), soil-available resources may be taken up from a variety of sources and by several different mechanisms. For example, in arctic tundras, N may be taken up from the soil as NO_3 or as NH_4 and, at least in some tundra plants, as free amino acids (Kielland and Chapin 1992). These different forms of N, and also soil-available forms of P, may be taken up either directly by roots or indirectly through mycorrhizae. Furthermore, some tundra plant species are associated with a variety of N-fixing organisms and obtain much of their N through these organisms from the atmosphere (Chapin and Bledsoe 1992).

If all species were equally capable of taking up all available forms of these resources, then species diversity should have no effect on community resource uptake. On the other hand, if each species depended entirely on uptake of only one form, then the presence or absence of a given species might have considerable impact on community uptake. At present, differences in the relative importance of different forms of soil-available elements to species growing together in the same tundra vegetation are not well enough known to establish exactly where tundras fall between these two extremes. Most tundra plant species appear to be capable of taking up soil-available elements in more than one-form (Kielland 1990; Kielland and Chapin 1992), and most species show at least some mycorrhizal development (Miller and Laursen 1978; Miller 1982). On the other hand, some recent studies have shown large differences in nitrate reductase activity among species growing in the same sites (K. Nadelhoffer, unpubl. data), suggesting interspecific differences in the use of NO_3. Other recent studies have shown large and consistent interspecific differences in $\delta^{15}N$ in leaf tissues, suggesting that species obtain their N from different (but unknown) sources with different characteristic $\delta^{15}N$ values (K. Nadelhoffer, unpubl.).

The N-fixing plants are a special case in which the presence or absence of a single species or group of species can make a large difference to community resource availability and resource uptake. In tundras, these include legume species, alder (*Alnus*) species, and several species of lichens and mosses (Chapin and Bledsoe 1992). Although N-fixing plants are rarely dominant members of tundra vegetation, they control a very important, long-term input of N to the ecosystem that is a major determinant of future availability of N in various forms to other species.

14.2.4 Uptake Kinetics

The species of tundra plants differ widely in their relationships between available resource concentration and resource uptake (e.g., photosynthesis vs CO_2 concentration or phosphate uptake vs phosphate concentration; Johnson and Tieszen 1976; Tieszen 1978; Chapin 1978). This diversity in uptake kinetics at the species level may increase, decrease, or have no effect on total resource uptake by the whole community. For example, in moist tussock tundra, species of *Salix* have higher rates of phosphate absorption at all concentrations of available phosphate than any of the other common or dominant species of this community (Chapin and Tryon 1982; Kielland and Chapin 1992). However, this does not necessarily mean that a single-species community of *Salix* would always have a higher total phosphate uptake than a mixed-species community. Rather, if phosphate uptake occurs at same rate at which phosphate is supplied to the available pool, then the phosphate concentration in the available pool has little impact on the total amount taken up over time, and cumulative uptake would be the same for all species that are capable of uptake at whatever the rate of supply is. In this case, there would be no difference in community phosphate uptake between a single-species community and a mixed-species community. On the other hand, if the rate of phosphorus supply exceeds the uptake capacity of the vegetation, then a single-species community of Salix should have a higher total phosphate uptake than any other single- or multispecies community (of equal biomass or root mass).

14.3 Effects on Community Productivity

14.3.1 Diversity and Productivity in Relation to Climate Change

In an extreme climate like that of the Arctic tundra, one might expect that the productivity of all species should be strongly responsive to both annual climatic variation and longer-term climate change. However, all tundra plant species are not equally responsive to change in all climate variables. Rather, species respond individualistically to different combinations of changes in temperature, light, and nutrient availability (Chapin and Shaver 1985b). As a result, changes in community productivity from one year to the next are the sum of the changes in individual species, which are not always even in the same direction (i.e., increases in productivity of one species may compensate for decreases in productivity of another). In this way, the annual fluctuations in productivity of individual species may, in effect, serve to buffer community productivity against changes due to annual fluctuations in climate. For example, in a 5-year study in moist tussock tundra at Eagle Creek, Alaska, Chapin and Shaver (1985b) showed that community productivity had a coefficient of variation of only 11%, while variation in productivity of individual species was much greater.

Diversity in sources and spatial and temporal patterns of resource uptake may at least in part explain the individualistic responses of tundra plant species to climatic change, and thus the buffering effect these individualistic responses can have on changes in community productivity. The reason for this is that annual climate variation should change both the relative availability of different uptake sources (e.g., NO_3 vs NH_4) and the spatial and temporal patterns of their availability. These changes in availability should favor some species and constrain others. Unfortunately, though, no long-term data on year-to-year variation in soil resource availability in relation to productivity changes are yet available to test this idea.

Another, contributing explanation for the buffering of annual changes in community productivity by compensatory fluctuations in productivity of individual species is that productivity of virtually all tundra plants is at least as strongly dependent on storage from previous years as it is on resource uptake concurrent with new production. The biochemical forms and seasonal patterns of resource storage differ widely among tundra species (Chapin and Shaver 1988), so that current-year's productivity of a given species is affected not only by individualistic responses to current-year's conditions but also to previous year's resource availability.

Species diversity may also have important effects on the rate and pattern of change in community productivity in response to a consistent, multiyear trend in climate. For example, the decade of the 1980s was characterized by a series of warm summers in arctic Alaska, especially after 1985. In the moist tundra at Toolik Lake, community primary production decreased between 1983 and 1989 by about one-third. However, there were even more dramatic changes in the relative abundance and productivity of two of the dominant species, with *Eriophorum vaginatum* productivity decreasing by two-thirds, while productivity of *Betula nana* increased by 50% (Chapin et al. 1994). The mechanisms responsible for these compensatory changes in productivity of individual species are not entirely clear, but the two most likely mechanisms are (1) a change in the amount, form, and spatial or temporal pattern of resource availability such that resource availability to one species was increased and availability to the other was decreased, and (2) a change in the proportion of some shared, limiting resource that was taken up by one species versus the other. Of course, both mechanisms could also be operating simultaneously. However, it is clear that the importance of the first mechanism will be a function of (1) the extent to which these two tundra plant species do specialize in their patterns of resource uptake, and (2) the extent to which the form and pattern of resource availability is changed under sustained climatic change. Whatever the mechanism, changes in community productivity in response to sustained climate change appear to be effectively buffered by compensatory changes in resource uptake and productivity of individual species.

14.3.2 Diversity and Productivity in Disturbed or Manipulated Vegetation

Perhaps the most direct means of testing the importance of species diversity to community productivity is to add or remove species and then document any subsequent changes in productivity. Experiments that have been completed thus far tend to support the idea that diversity increases productivity because there is a strong displacement in sources of resource uptake among species in tundra communities. For example, Jonasson (1992) removed the dominant deciduous shrubs from three tundra communities near Abisko, Sweden, but was unable to detect any effect on abundance or diversity of the remaining species over three years. The lack of response by the other species suggests that they may not be capable of taking up the resources that are used by deciduous species, even in the absence of competition from deciduous species. In this case the presence or absence of a particular species or plant functional type seems clearly to lead to increased productivity through exploitation of a different nutrient source. Fetcher (1985) obtained similar results in a species removal experiment in Alaskan moist tussock tundra. Although he found that removal of shrubs from *Eriophorum vaginatum* tussocks had large effects on Eriophorum tiller size and tillering rates, Fetcher (1985) interpreted these changes as resulting from alterations in the light regime of the tussocks due to reduced shading by shrubs; the tussock nutrient regime was apparently little affected by the shrub removal.

The results of long-term fertilizer experiments have somewhat different implications for the effects of particular species on community resource uptake and productivity. In such experiments it is common for community productivity to be dramatically increased by fertilizer addition, with virtually all species responding initially (Shaver and Chapin 1980). The contribution of individual species to the increase is not equal, though, and it changes over time. For example, in moist tussock tundra the grass and sedge species are usually the quickest to respond

(Shaver and Chapin 1986), but over the longer term the community response is due to greater production by deciduous shrubs such as *Betula nana* (Chapin et al. 1994). This sequence of responses suggests that the different species do compete for at least some of the same nutrient resources but that they differ in their ability to take advantage of a sudden increase in this nutrient pool, either because they differ in their capacity to take the nutrients up or to use the nutrients in rapidly increased growth. In this case, the effect of having multiple species in the community is to increase the initial production response to fertilizer, assuming that a community without graminoids would be less responsive in the short term. A strong initial response may be important in retaining resources within the ecosystem that might otherwise be lost without early plant uptake.

In the longer term, however, the shift to dominance by deciduous species following fertilizer addition does not necessarily lead to greater fertilizer uptake or even greater productivity on fertilized plots relative to controls (Chapin et al. 1994). Rather, the long-term effect of fertilizer on both whole-community uptake and whole-community productivity is not quantitatively different from the short-term response, even though the species responsible for the long-term response (deciduous shrubs) are not the same as the species responsible for the short-term response (graminoids). This long-term change in species composition without any change in productivity or nutrient uptake on fertilized plots suggests that species compete strongly for the fertilizer nutrient pool. The 'winner' of this competition, however, does not obtain any more resources than the 'loser' but is simply more efficient at exploiting a limited fertilizer nutrient pool.

14.3.3 Interactions Between Resource Uptake, Resource Use, and Productivity

It is important to remember that resource uptake interacts strongly with controls on plant resource allocation and resource use efficiency in determining productivity of both tundra species and tundra communities (Berendse and Jonasson 1992). Thus, changes in resource uptake do not necessarily lead to strictly proportional changes in productivity. There are two principal ways in which changes in resource uptake can be decoupled from changes in productivity. First, resource uptake is not 'free', but requires the allocation of resources to the uptake mechanism itself that might otherwise be used to support growth. If the marginal costs of increased uptake are greater than the return in uptake of given resource, productivity may actually decrease (Bloom et al. 1985). Second, production per unit resource taken up (i.e., resource use efficiency) may vary greatly, as reflected in large changes in resource concentration within the plant but much lesser changes in production or biomass.

A detailed discussion of the effects of species differences in resource use on tundra community productivity is beyond the scope of this chapter and is available elsewhere (e.g., Shaver and Chapin 1991; Berendse and Jonasson 1992; Shaver et al. 1992; Hobbie, this vol.). However, at least in relatively stable tundra vegetation, variation in resource uptake appears to be a much more important regulator of productivity than variation in resource use (Shaver et al. 1992). For example, across a series of tundras that varied in productivity by an order of magnitude, Shaver and Chapin (1991) found little or no difference in several measures of overall community and whole-plant efficiency of N and P use, while the N and P requirements for productivity were closely correlated with production. They concluded that the effects on productivity of interspecies differences in resource use might be more important in tundras that were disturbed or rapidly changing than in relatively stable tundras.

14.4 Conclusions

1. The species of arctic tundra vegetation clearly do specialize in terms of their spatial patterns of soil resource uptake, and probably also in their temporal patterns of uptake. Although most species are capable of taking up resources in more than one chemical form, a growing body of evidence suggests further specialization with respect to the chemical form or perhaps the mechanism of uptake (e.g., mycorrhizal vs root uptake). This specialization is evident within as well as between the major plant life forms of arctic tundra.
2. Limited experimental evidence and observations on species-poor disturbed sites indicate that this specialization in resource uptake does not change greatly in the absence of competitors, suggesting that species diversity may be positively related with productivity, at least under some conditions.
3. On the other hand, because most tundra species are capable of taking up most forms of soil-available resources, when additional soil resources (e.g., fertilizer) are added most species respond initially with higher productivity. Later competition may reduce the abundance of some species below their initial abundance, or even eliminate some species, but both the short- and the long-term responses to fertilizer experiments contrast strongly with the results of species removal experiments and suggest at least some direct, interspecific competition for the same resources.
4. The more species-rich tundras are more likely to include species that exploit more fully the range of spatial, temporal, and chemical variation in soil resource supply. Thus, different but relatively diverse tundra communities should have greater and more similar productivity than less diverse communities growing on similar sites. However, in the Arctic where element cycles are relatively closed and uptake is close to replenishment rate, diversity should have less of an impact on community productivity than in sites with more open element cycles and more rapid element turnover.
5. In a fluctuating environment (stable on average, no directional change), diversity in resource uptake patterns may serve as a buffer of tundra community productivity due to individualistic responses to current weather and individualistic patterns of storage versus uptake in support of current growth.
6. Following disturbance, different but relatively species-rich communities should be more similar in the pattern of changes in productivity than less

species-rich communities because chances are greater that the species-rich communities will share species that respond similarly. Also, diverse communities are more likely to contain species that are capable of relatively high nutrient uptake following disturbance, so diverse communities should respond more rapidly than less diverse communities following disturbance.

7. At least in comparisons of relatively stable tundras, productivity is closely correlated with resource uptake and variation in resource use efficiencies is relatively small and unimportant in its effect on tundra productivity. This suggests that species specialization in the spatial and temporal patterns and chemical forms of soil resource uptake should be more important in regulating community productivity in relatively stable than in disturbed tundras.

References

Berendse F, Jonasson S (1992) Nutrient use and nutrient cycling in northern ecosystems In: Chapin FS III, Jefferies R, Reynolds J, Shaver G, Svoboda J (eds) Arctic ecosystems in a changing climate: an ecophysiological perspective. Academic Press, New York, pp 337–356

Billings WD (1973) Arctic and alpine vegetations: similarities, differences, and susceptibility to disturbance. BioScience 23: 697–704

Billings WD (1992) Phytogeographic and evolutionary potential of the arctic flora and vegetation in a changing climate. In: Chapin FS III, Jefferies R, Reynolds J, Shaver G, Svoboda J (eds) Arctic ecosystems in a changing climate: an ecophysiological perspective. Academic Press, New York, pp 91–110

Billings WD, Mooney HA (1968) The ecology of arctic and alpine plants. Biol Rev. 43: 481–530

Bliss LC, Peterson KM (1992) Plant succession, competition, and physiological constraints of species in the Arctic. In: Chapin FS III, Jefferies R, Reynolds J, Shaver G, Svoboda J (eds) Arctic ecosystems in a changing climate: an ecophysiological perspective. Academic Press, New York, pp 111–138

Bliss LC, Heal OW, Moore JJ (eds) (1981) Tundra ecosystems: a comparative analysis. Cambridge University Press, Cambridge

Bloom AJ, Chapin FS III, Mooney HA (1985) Resource limitation in plants – an economic analogy. Annu Rev Ecol Syst 16: 363–392

Brown J, Miller PC, Tieszen LL, Bunnell FL (eds) (1980) An arctic ecosystem: the coastal tundra at Barrow, Alaska. Dowden, Hutchinson and Ross, Stroudsburg

Chapin DM, Bledsoe CS (1992) Nitrogen fixation in arctic plant communities. In: Chapin FS III, Jefferies R, Reynolds J, Shaver G, Svoboda J (eds) Arctic ecosystems in a changing climate: an ecophysiological perspective. Academic Press, New York, pp 111–138

Chapin FS III (1978) Phosphate uptake and nutrient utilization by Barrow tundra vegetation. In: Tieszen LL (ed) Vegetation and production ecology of an Alaskan arctic tundra. Springer, Berlin, Heidelberg, New York, pp 37–112

Chapin FS III, Shaver GR (1981) Changes in soil properties and vegetation following disturbance of Alaskan arctic tundra. J Appl Ecol 18: 605–617

Chapin FS III, Shaver GR (1985a) Arctic. In: Chabot B, Mooney HA (eds) Physiological ecology of North American plant communities. Chapman and Hall, London, pp 16–40

Chapin FS III, Shaver GR (1985b) Individualistic growth response of tundra plant species to manipulation of light, temperature, and nutrients in a field experiment. Ecology 66: 564–576

Chapin FS III, Shaver GR (1988) Differences in carbon and nutrient fractions among arctic growth forms. Oecologia 77: 506–514

Chapin FS III, Shaver GR (1989) Differences in growth and nutrient use among arctic plant growth forms. Funct. Ecol. 3: 73–80

Chapin FS III, Tryon PR (1982) Phosphate absorption and root respiration of different plant growth forms from northem Alaska. Holarct Ecol 5: 164–171

Chapin FS III, Van Cleve K, Chapin MC (1979) Soil temperature and nutrient cycling in the tussock growth form of *Eriophorum vaginatum* L. J Ecol 67: 169–189
Chapin FS III, Johnson DA, McKendrick JD (1980) Seasonal nutrient movements in various plant growth forms in an Alaskan tundra: implications for herbivory. Ecology 68: 189–210
Chapin FS III, Shaver GR, Kedrowski RA (1986) Environmental controls over carbon, nitrogen, and phosphorus chemical fractions in *Eriophorum vaginatum* L. in Alaskan tussock tundra. J Ecol 74: 167–196
Chapin FS III, Jefferies R, Reynolds J, Shaver G, Svoboda J (eds) (1992) Arctic ecosystems in a changing climate: an ecophysiological perspective. Academic Press, New York
Chapin FS III, Hobbie SE, Shaver GR (1994) Impacts of global change on composition of arctic communities: implications for ecosystem functioning. Proc Int Conf on Global change and arctic terrestrial ecosystems, Oppdal,Norway, Aug 1993 (submitted)
Churchill EC, Hanson HC (1958) The concept of climax in arctic and alpine vegetation. Biol Rev 24: 127–191
Fetcher N (1985) Effects of removal of neighboring species on growth, nutrients, and microclimate of *Eriophorum vaginatum* Arct Alp Res 17: 7–17
Fetcher N, Shaver GR (1982) Growth and tillering patterns within tussocks of *Eriophorum vaginatum*. Holarct Ecol 5: 180–186
Johnson DA, Tieszen LL (1976) Aboveground biomass, leaf growth, and photosynthesis patterns in tundra plant forms in arctic Alaska. Oecologia 24: 159–173
Jonasson S (1992) Plant responses to fertilization and species removal in tundra related to community structure and clonality. Oikos 63: 420–429
Kielland K (1990) Processes controlling nitrogen release and turnover in arctic tundra. PhD Thesis, University of Alaska, Fairbanks
Kielland K, Chapin FS III (1992) Nutrient absorption and accumulation in arctic plants. In: Chapin FS III, Jefferies R, Reynolds J, Shaver G, Svoboda J (eds) Arctic ecosystems in a changing climate: an ecophysiological perspective. Academic Press, New York, pp 321–336
Miller OK (1982) Mycorrhizae, mycorrhizal fungi, and fungal biomass in subalpine tundra at Eagle Summit, Alaska. Holarct Ecol 5: 125–134
Miller OK, Laursen GA (1978) Ecto- and endomycorrhizae of arctic plants at Barrow, Alaska. In: Tieszen LL (ed) Vegetation and production ecology of an Alaskan arctic tundra. Springer, Berlin, Heidelberg, New York, pp 3: 229–238
Miller PC, Webber PJ, Oechel WC, Tieszen LL (1980) Biophysical processes and primary production. In: Brown J, Miller PC, Tieszen LL, Bunnell FL (eds) An arctic ecosystem: the coastal tundra at Barrow, Alaska. Dowden, Hutchinson and Ross, Stroudsburg, pp 66–101
Miller PC, Mangan R, Kummerow J (1982) Vertical distribution of organic matter in eight vegetation types near Eagle Summit, Alaska. Holarct Ecol 5: 117–124
Miller PC, Miller PM, Blake-Jacobsen M, Chapin FS III, Everett KR, Hilbert DW, Kummerow J, Linkins AE, Marion GM, Oechel WC, Roberts SW, Stuart L (1984) Plant-soil processes in *Eriophorum vaginatum* tussock tundra in Alaska: a systems modeling approach. Ecol Monogr 54: 361–405
Murray C, Miller PC (1982) Phenological observations of major plant growth forms and species in montane and *Eriophorum vaginatum* tussock tundra in central Alaska. Holarct Ecol 5: 109–116
Oechel WC, VanCleve K (1986) The role of bryophytes in nutrient cycling in the taiga. In: VanCleve K, Chapin FS III, Flanagan PW, Vierek LA, Dyrness CT (eds) Forest ecosystems in the Alaskan taiga: a synthesis of structure and function. Ecological Studies 57. Springer, Berlin, Heidelberg, New York, pp 121–137
Shaver GR, Billings WD (1975) Root production and root turnover in a wet tundra ecosystem, Barrow, Alaska. Ecology 56(2): 401–410
Shaver GR, Billings WD (1977) Effects of daylength and temperature on root elongation in tundra graminoids. Oecologia 28: 57–65
Shaver GR, Chapin FS III (1980) Response to fertilization by various plant growth forms in an Alaskan tundra: nutrient accumulation and growth. Ecology 61(3): 662–675
Shaver GR, Chapin FS III (1986) Effect of fertilizer on production and biomass of tussock tundra, Alaska, USA . Arct Alp Res 18: 261–268

Shaver GR, Chapin FS III (1991) Production/biomass relationships and element cycling in contrasting arctic vegetation types. Ecol Monogr 61: 1–31
Shaver GR, Cutler JC (1979) The vertical distribution of phytomass in cottongrass tussock tundra. Arct Alp Res 1(3): 335–342
Shaver GR, Kummerow J (1992) Phenology, resource allocation, and growth of arctic vascular plants. In: Chapin FS III, Jefferies R, Reynolds J, Shaver G, Svoboda J (eds) Arctic ecosystems in a changing climate: an ecophysiological perspective. Academic Press, New York, pp 193–212
Shaver GR, Chapin FS III, Gartner BL (1986) Factors limiting growth and biomass accumulation in *Eriophorum vaginatum L.* in Alaskan tussock tundra. J Ecol 74: 257–278
Shaver GR, Nadelhoffer KJ, Giblin AE (1991) Biogeochemical diversity and element transport in a heterogeneous landscape, the North Slope of Alaska. In: Turner MG, Gardner RH (eds) Quantitative methods in landscape ecology. Springer, Berlin, Heidelberg, New York, pp 105–126
Shaver GR, Billings WD, Chapin FS III, Giblin AE, Nadelhoffer KJ, Oechel WC, Rastetter EB (1992) Global change and the carbon balance of arctic ecosystems. BioScience 42: 433–441
Shaver GR, Giblin AE, Nadelhoffer KJ, Rastetter EB (1994) Plant functional types and ecosystem change in arctic tundras. In: Smith T, Shugart H, Woodward I (eds) (in press) Plant functional types, Cambridge University Press, Cambridge
Sigafoos RS (1952) Frost action as a primary physical factor in tundra plant communities. Ecology 33: 480–487
Sørensen T (1941) Temperature relations and phenology of northeast Greenland flowering plants. Medd Gronl 125: 1–305
Stoner WA, Miller PC, Oechel WC (1978) Simulation of the effects of the tundra vascular canopy on the productivity of four moss species. In: Tieszen LL (ed) Vegetation and production ecology of an Alaskan arctic tundra. Springer, Berlin, Heidelberg, New York pp 371–388
Tenhunen JD, Lange OL, Hahn S, Siegwolf R, Oberbauer SF (1992) The ecosystem role of poikilohydric tundra plants. In: Chapin FS III, Jefferies R, Reynolds J, Shaver G, Svoboda J (eds) Arctic ecosystems in a changing climate: an ecophysiological perspective. Academic Press, New York, pp 213–237
Tenhunen JD, Siegwolf RA, Oberbauer SF (1993) Effiects of phenology, physiology, and gradients in community composition, structure, and microclimate on tundra ecosystem CO_2 exchange. In: Schulze E-D, Caldwell MM (eds) Ecophysiology of photosynthesis. Ecological Studies 100. Springer, Berlin, Heidelberg, New York, pp 431–460
Tieszen LL (ed) (1978) Vegetation and production ecology of an Alaskan arctic tundra. Springer, Berlin, Heidelberg, New York
Tieszen LL, Miller PC, Oechel WC (1980) Photosynthesis. In: Brown J, Miller PC, Tieszen LL, Bunnell FL (eds) An arctic ecosystem: the coastal tundra at Barrow, Alaska. Dowden, Hutchinson and Ross, Strudsburg, pp 102–139
Webber PJ (1978) Spatial and temporal variation of the vegetation and its productivity, Barrow, Alaska. In: Tieszen LL (ed,Vetation and production ecology of an Alaskan arctic tundra. Springer, Berlin, Heidelberg, New York, pp 37–112

15 Direct and Indirect Effects of Plant Species on Biogeochemical Processes in Arctic Ecosystems

S.E. Hobbie

15.1 Introduction

Predicting how biogeochemical cycles in arctic systems will respond to global environmental change requires knowledge of how environmental change will directly affect biological activities controlling biogeochemical cycles, and whether changes in species' distributions will further influence such cycles. For example, environmental factors like temperature directly influence processes such as microbial respiration and nutrient mineralization (Marion and Black 1986; Nadelhoffer et al. 1991), photosynthesis (Limbach et al. 1982), plant nutrient uptake (Chapin and Bloom 1976; Chapin and Tryon 1982), and plant growth (Kummerow and Ellis 1984). As such processes respond to environmental change, species' distributions will shift because of species' differences in physiological tolerance, indirect effects of environmental change (Chapin 1983), and subsequent changes in competitive hierarchies and/or trophic interactions. Indeed, species' distributions in arctic ecosystems are quite sensitive to the environmental parameters expected to change in northern ecosystems such as temperature (Chapin and Shaver 1985b; Havström et al. 1993), light (op cit.), nutrient availability (Chapin and Shaver 1985b; Jonasson 1992; Havstrom et al. 1993;), and soil moisture (Chapin et al. 1988; Hastings et al. 1989; Shaver and Chapin 1991).

In this chapter, I will outline a framework for predicting which plant species will strongly affect biogeochemistry. Making such predictions requires generating hypotheses based on our knowledge of interspecific variation in plant traits involved in biogeochemical processes. Once we have generated hypotheses about the importance of particular species, we can test those hypotheses using experiments and observations along species' gradients.

Plant species can potentially influence biogeochemistry in two major ways. First, plants directly control certain biogeochemical processes such as carbon (C) acquisition and loss, nutrient uptake and loss, and transpiration. Second, plants indirectly control biogeochemical cycling by influencing how microbes and herbivores cycle elements. For example, plants influence how microbes cycle C and nutrients by controlling abiotic factors that affect microbial activity and substrate quality. Plants influence how herbivores cycle nutrients because of variation in palatability and in tolerance to herbivory.

Department of Integrative Biology, University of California, Berkeley, CA 94720, USA

Ecological Studies, Vol. 113
Chapin/Körner (eds) Arctic and Alpine Biodiversity

Changes in plant species' relative abundance will affect biogeochemistry directly if (1) species vary in traits that affect biogeochemical processes or (2) species vary in the response of those traits to environmental change (Figs. 1, 2). For example, if species A exhibits higher rates of some process than species B, and both species respond to environmental perturbation similarly, replacement of species B by A will lead to higher rates of that process (Fig. 1a, b). In addition, if the magnitude of the interspecific difference is large relative to the change in the process induced by environmental change, we must consider species' effects when predicting the process rates in the new environment (Fig. 1, cf. a and b). Alternatively, if species A and B respond differently to environmental perturbation, replacement of B by A could lead to either higher or lower rates, depending on the

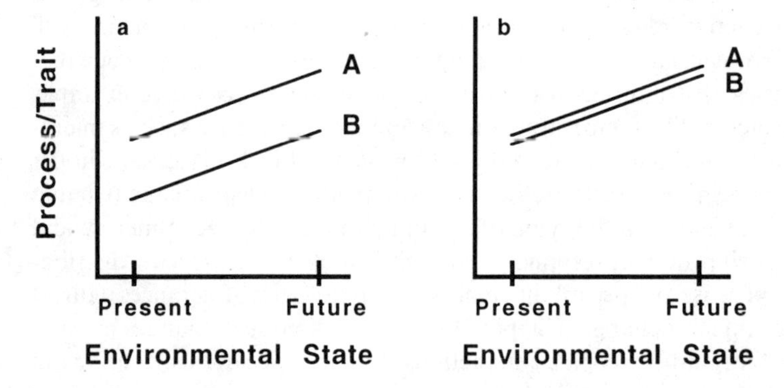

Fig. 1 a, b. The response of some plant trait to environmental change in two hypothetical species (*A* and *B*) for which no environment by species interaction exists. **a** Interspecific variation in the trait is large relative to response of the trait to environmental change. **b** Interspecific variation in the trait is small relative to the response of the trait to environmental change

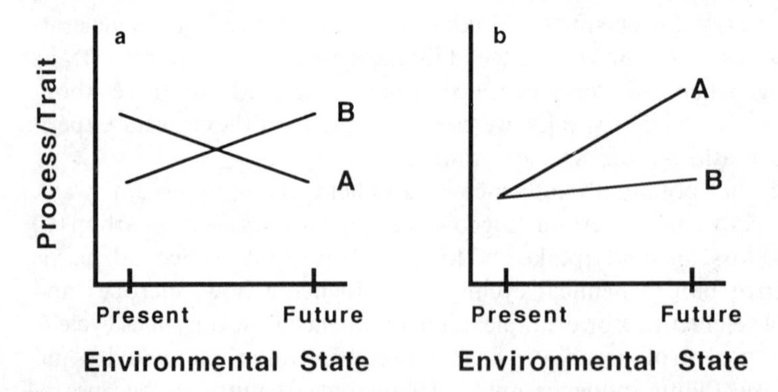

Fig. 2 a, b. The response of some plant trait to environmental change in two hypothetical species (*A* and *B*) for which an environment by species interaction exists. **a** Species respond oppositely to environmental change. **b** Species respond /similarly in sign, but differently in magnitude to environmental change

nature of the species-by-environment interaction (Fig. 2a, b). Thus, predicting species' effects on biogeochemistry requires knowledge of both species' differences in traits that influence biogeochemistry under ambient conditions and under conditions accompanying predicted environmental change.

To predict if and how changes in species' distributions will affect biogeochemistry, we can take the following approach. First, we must identify which plant traits most strongly influence or control the biogeochemical process of interest. Second, we must determine the magnitude of interspecific variation both in those traits and in the response of those traits to environmental change. We can then formulate hypotheses about which species will most affect biogeochemical cycles, and, finally, use experiments and observations to test our hypotheses.

15.2 Direct Influence over Biogeochemistry

A number of plant traits directly influence biogeochemical cycling, including growth rate, nutrient uptake rate, and litter production (Table 1). Growth rate controls the rate of C acquisition by plants (Lambers and Porter 1992), thereby influencing C cycling through ecosystems. Faster growing plants, or plants with higher production:biomass ratios, take up more CO_2 than slow-growing species. Thus, if fast-growing plants increase in abundance relative to slow-growing species, they will change rates of C cycling, potentially influencing the sign of ecosystem CO_2 flux in nonsteady state systems. Growth forms in tundra differ strongly in both absolute and relative growth rates, with deciduous shrubs, graminoids, and forbs having higher growth rates than evergreen shrubs (Table 2; Bliss 1956; Shaver and Chapin 1991). Similarly, graminoids and forbs have high, deciduous shrubs have intermediate, and evergreen shrubs and mosses have low production:biomass ratios (Table 2; Murray et al. 1989; Shaver and Chapin 1991). Interspecific differences in photosynthetic rates correlate positively with those in growth rates with forbs, graminoids, and deciduous shrubs having the highest and evergreen shrubs and mosses and

Table 1. Examples of plant traits that directly control biogeochemical processes

Process	Mode of influence	Traits involved
C cycling	C acquisition and loss	Growth Photosynthesis Respiration
Nutrient cycling	Nutrient acquisition and loss	Growth Nutrient uptake Tissue turnover Retranslocation efficiency
H_2O cycling	Evapotranspiration	Stomatal conductance Leaf area Rooting depth

lichens having the lowest photo synthetic rates (Table 2; Chapin and Shaver 1985a; Oberbauer and Oechel 1989).

Tundra species also differ in their growth response to environmental perturbations. In general, vascular plants respond to.increased temperature and nutrients with increased growth, although the response is not always consistent across sites (Shaver and Chapin 1980; Chapin and Shaver, 1985b; Shaver et al. 1986; Havstrom et al. 1993). Mosses, in contrast, grow less with increased nutrients and temperature (Chapin and Shaver 1985b; F.S. Chapin and G.R. Shaver unpubl; S.E. Hobbie and F.S. Chapin, unpubl). Canopy species, such as *Carex bigelowii*, *Eriophorum vaginatum*, *Betula nana*, and *Ledum palustre* grow less when shaded, as do mosses, while shading does not affect growth of understory species like *Vaccinium vitis-idaea* and *Rubus chamamorus* (Chapin and Shaver 1985b). Mosses and some shrubs grow more when irrigated (Hastings et al. 1989; Murray et al. 1989).

Greater nutrient uptake accompanies faster growth in tundra plants. Fast-growing species thus promote more rapid rates of nutrient cycling than slow-growing species. Faster growing species have a higher nutrient uptake capacity (Chapin and Tryon 1982) and relative nutrient accumulation rates (Shaver and Chapin 1991), although the importance of new uptake vs movement of nutrients into and out of belowground reserves has not been determined. Those community types dominated by faster growing shrubs and graminoids cycle nutrients most tightly, suggesting that plant growth rate controls nutrient uptake in tundra (Chapin et al. 1992). Most tundra species exhibit increased nutrient uptake rates with increased temperature and nutrient availability, although their response to increased temperature is less than that of temperate species (Kielland and Chapin 1992).

The rate at which C and nutrients are lost from plants to soils also differs among tundra species. Assuming steady-state condition (biomass production = litter production), graminoids and forbs have the highest litter production, and the fastest biomass turnover rates, with deciduous shrubs exhibiting inter-

Table 2. Production: biomass ratios, relative growth rates, and photosynthetic rates for the major arctic tundra growth forms

Growth form	Relative growth rate [(log(g g^{-1})]	Production: biomass (g yr^{-1} g^{-1})	Photosynthetic rate (nmol g $leaf^{-1}s^{-1}$)
Graminoid	0.48[a]	0.65[a]	120.3[c]
Deciduous shrub	0.38[a]	0.33[a]	133.3[c]
Forb	1.25[a]	0.94[a]	200.0[c]
Evergreen shrub	−0.09[a]	0.27[a]	45.5[c]
Moss	–	0.18[b]	15.3[d]
Lichen	–	–	3.3[d]

[a]Shaver and Chapin (1991).
[b]Calculated from Shaver and Chapin (1991) and Murray et al. (1989) for *Sphagnum* spp.
[c]Oberbauer and Oechel (1989).
[d]Tenhunen et al. (1992).

mediate rates, and evergreen shrubs and mosses exhibiting slow rates (Oechel and Sveinbjörnsson 1978; Shaver and Chapin 1991).

15.3 Indirect Influence over Biogeochemistry

15.3.1 Effects on Soil Microbes

In addition to directly controlling biogeochemical processes, plants indirectly control biogeochemistry by influencing how microbes and herbivores cycle elements. Plant species potentially affect microbial cycling of C and nutrients because of interspecific variation in substrate quality, and in how they affect physical controllers of microbial activity such as temperature, moisture, pH, and the availability of electron acceptors, such as O_2 (Table 3).

Plant species affect microbial substrate quality, and thus potentially decomposition rates and C and nutrient cycling, because of differences in litter quality. Decomposition rates of tundra species' litter correlate best with initial lignin content and C/N ratio (Van Cleve 1974). Deciduous leaves and all graminoid organs have low lignin concentrations relative to shrub stems and roots, while mosses have relatively high concentrations of recalcitrant compounds (Chapin et al. 1986). In general, deciduous leaf litter decomposes faster than evergreen leaf litter, and mosses, lichens, roots and woody stems decompose more slowly still (Heal and French 1974; Clymo and Hayward 1982; Nadelhoffer et al. 1992). Interspecific differences in leaf C/N ratios are due mainly to initial differences in leaf chemistry, rather than differences in retranslocation efficiency, which shows no clear pattern among species (Chapin and Shaver 1989; Shaver and Chapin 1991).

Intraspecific differences in litter quality of different plant organs (roots, stems, and leaves) may be large relative to interspecific differences in the quality of any

Table 3. Examples of plant traits that indirectly control biogeochemical processes by influencing soil microbial activity and herbivory

Process	Mode of influence	Traits involved
Microbial C and nutrient cycling	Substrate quality	Litter production
		Litter chemistry
		Allocation
	Oxidant availability	Aerenchyma
		Root respiration
		NO_3^- uptake
	Temperature	Thermal conductivity
	Moisture	Rooting depth
		Leaf area
	pH	Ion uptake
Herbivore nutrient cycling	Palatability	Tissue chemistry
		Allocation
	Regrowth potential	Meristem location

given plant part. Thus, plant species may differ in their effects on decomposition because of interspecific variation in allocation, as well as tissue chemistry. Graminoids invest less production into stems than into leaves and roots (Fig. 3; Shaver and Billings 1975; Shaver and Chapin 1991; Shaver et al. 1992). Furthermore, because their stems are not woody (Shaver 1986), graminoid stems contain relatively little lignin compared to shrubs (Chapin et al. 1986). Both deciduous and evergreen shrubs invest relatively more into stems and leaves than into roots (Fig. 3; Shaver et al. 1992), and their stems are highly lignified (Chapin et al. 1986). Thus, based on allocation and quality differences, graminoid litter should decompose more quickly than litter of deciduous or evergreen shrubs.

The response of litter quality to environmental change depends on the nature of the perturbation. Fertilization increases the nutrient concentration of leaf litter in a variety of species and growth forms, such that differences between control and fertilized leaves are greater than differences among species (Chapin and Shaver 1989). Shade similarly increases leaf nutrient concentrations of most species' tissue (F.S. Chapin and G.R. Shaver unpubl). Increased temperature does not change shoot C/N ratios in a variety of species (S.E. Hobbie and F.S. Chapin unpubl).

Plant species can also influence microbial activity by altering temperature, moisture, pH, and redox. Mosses promote cold soil temperatures and permafrost development by conducting heat under cool, moist conditions and insulating soils under warm, dry conditions (Oechel and Van Cleve 1986). Species influence evapotranspiration and thus soil moisture because of interspecific variation in rooting depth and leaf area. *Sphagnum* spp. maintain acid conditions (Clymo and Hayward 1982) and vascular tundra plants also likely decrease pH through NH_4^+ uptake and associated H^+ excretion. Plants can also affect microbial activity by altering the availability of electron acceptors, especially O_2. In wet tundra soils, roots may compete with heterotrophic microbes and nitrifiers for O_2. In saturated soils, however, aerenchymatous sedges may transport O_2 to anaerobic soils, such that their net effect is to increase O_2 availability to soil microbes. Plants may also compete with microbes for NO_3^-.

15.3.2 Effects on Herbivory

Plant species can potentially influence biogeochemistry by affecting rates of herbivory (Table 3). Interspecific variation in palatability affects browsing and grazing rates, since various herbivores prefer different plant foods. In general, browsers and grazers prefer deciduous and graminoid over evergreen species, although herbivores differ greatly in their preferred foods (Jefferies et al. 1992). Herbivores may increase rates of nutrient cycling directly and because of increased nutrient concentrations in plant regrowth as long as plants can sustain herbivory (Bryant et al. 1991; Bryant and Reichardt 1992). If plants cannot tolerate herbivory, because of low nutrient availability or low plant regrowth potential, sustained herbivory results in a community shift toward less palatable species that also reduce nutrient cycling rates (Pastor et al. 1988).

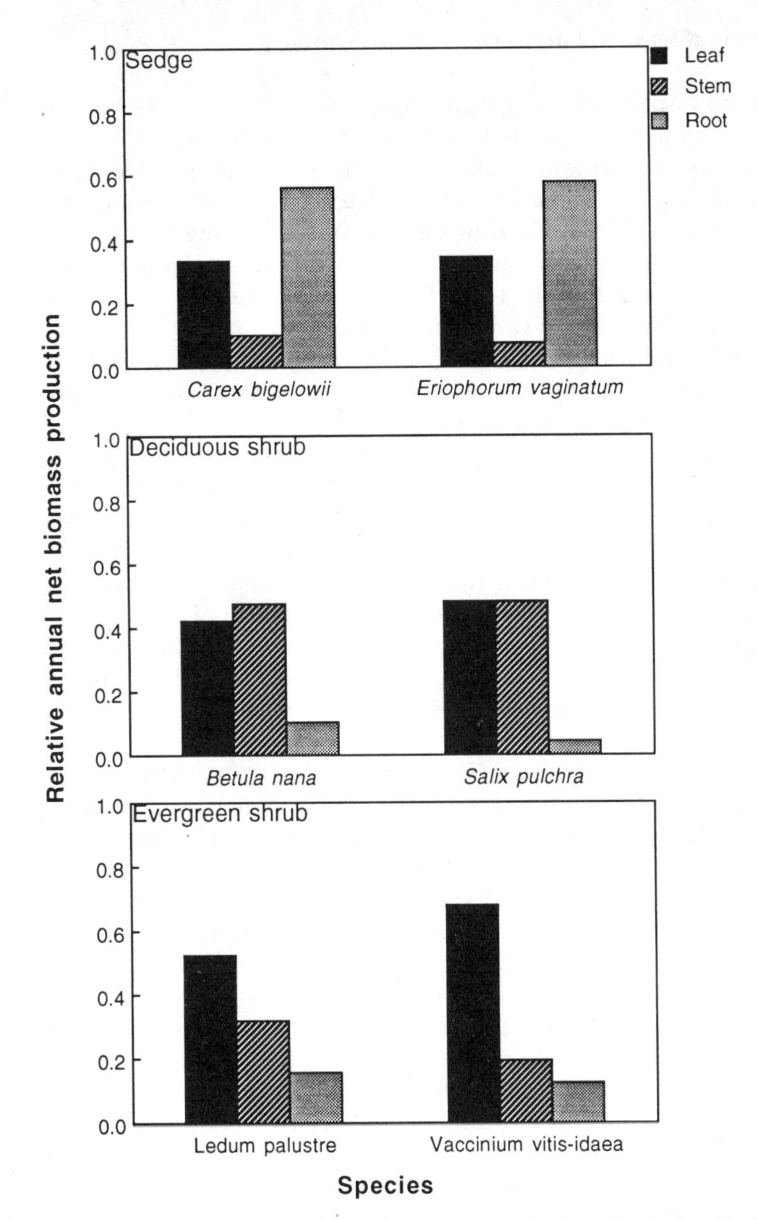

Fig. 3. The proportion of annual net primary production allocated to leaf, stem, and root biomass in three tundra growth forms. Leaf and stem biomass production are from Shaver and Chapin (1991, Appendix II). Root production was estimated using leaf biomass from Shaver and Chapin (1991) and leaf:new root biomass ratios from Shaver and Kummerow (1992) and J. Kummerow (pers. comm.)

15.4 Predictions of Species' Effects with Climate Warming

Knowledge of how species differ in their direct and indirect effects on biogeochemistry allows us to formulate hypotheses about how species will affect biogeochemical processes in the future. Because of the rising atmospheric concentration of greenhouse gases, temperatures in the Arctic will increase significantly in the next century (Maxwell 1992). How soil moisture will change is less clear, since general circulation models disagree about how cloudiness and precipitation will change. Thus, we can imagine two climate change scenarios: one that is warmer and drier and one that is warmer and wetter.

A warmer and drier climate will likely favor an increased abundance of deciduous shrubs and forbs and reduced abundance of graminoids, evergreen shrubs, and mosses, since deciduous shrubs exhibit the strongest positive response to manipulated warming (Chapin and Shaver 1985b) and increase in abundance just north of the forest-tundra border (Bliss and Matveyeva 1992), and mosses in particular are sensitive to drying (Tenhunen et al. 1992). The direct effect of such a change in plant community composition will be to increase rates of C cycling due to the higher growth rates and production:biomass ratios of deciduous shrubs compared with evergreen shrubs and mosses. Faster rates of nutrient cycling will accompany faster C cycling, and rates of H_2O cycling will also increase due to higher leaf area and overall canopy photosynthetic rates.

The indirect effects of changes in species' relative abundance will be less pronounced. Increased deciduous shrub abundance will have little effect on litter quality since higher quality due to increased deciduous leaf litter production will be offset by greater stem production (Fig 3). However, reduced moss abundance should increase rates of C and nutrient cycling because both their poor litter quality (Chapin et al. 1986) and their tendency to create cold, waterlogged, acidic soils (Clymo and Hayward 1982; Oechel and Van Cleve 1986) inhibit decomposition. Replacement of graminoids and mosses by deciduous shrubs may initially favor herbivory, thereby increasing rates of nutrient cycling via herbivores. However, heavy browsing may ultimately depress rates of nutrient cycling by favoring a plant community dominated by unpalatable evergreen species (Pastor et al. 1988; Bryant et al. 1991). If warmer and drier conditions favor wildfire in such communities, a cycle could begin in which deciduous communities favoring rapid rates of nutrient cycling and herbivory are replaced by evergreen communities that do not support herbivores but promote wildfires, again creating opportunities for establishment of deciduous species (Bryant, this Vol.).

A warmer and wetter climate will favor increased relative abundance of deciduous shrubs, aerenchymatous graminoids, and mosses (Hastings et al. 1989; Tenhunen et al. 1992). Such a change in community composition will have little direct or indirect effect on C and nutrient cycling. Although graminoids have relatively high growth rates, production:biomass ratios, and litter quality, and tend to aerate wet soils, thereby increasing rates of C and nutrient cycling, mosses offset these effects with their low production:biomass ratios, poor litter quality, and insulating effect.

15.5 Approaches to Studying the Role of Species in Biogeochemistry

Assessing interspecific variation in plant traits and trait responses to environmental change allows formation of hypotheses regarding the importance of shifts in species' distributions for biogeochemistry. To assess the importance of species' traits and the relative importance of species' vs abiotic effects in controlling biogeochemistry, researchers have used both field and modeling experiments, as well as existing gradients in climate and/or species' distributions. Field experimental approaches to studying species' effects include planting monocultures, removing species, adding species, or studying existing removals or additions (i.e. extinctions or invasions). Monocultures have revealed significant effects of grass species on N mineralization (Wedin and Tilman 1990). Species removals have revealed significant effects on ecosystem CO_2 flux in tussock tundra (S.E. Hobbie and F.S. Chapin, unpubl.) and CH_4 flux from wet meadow tundra (Torn and Chapin 1993). The invasion of an alien N-fixer into primary successional seres in Hawaii has altered N cycling there (Vitousek and Walker 1989).

Another way of assessing species' effects is to examine the effect of one particular plant trait, such as litter quality, using reciprocal transplants or controlled laboratory experiments. Often, these studies also manipulate abiotic factors, allowing comparison between species' and abiotic effects on biogeochemical cycling. Reciprocal litter transplants among sites that differ in aspect (and thus temperature) in the Alaskan taiga revealed that substrate quality better explained variation in decomposition rate than site characteristics (Flanagan and Van Cleve 1983). Similarly, soils from an Alaskan tundra toposequence differing mainly in species composition exhibited greater variation in C and N mineralization when incubated at the same temperature than did the same soils incubated at different temperatures (Nadelhoffer et al. 1991). In contrast, a variety of tundra litters incubated across a range of sites varied little in decomposition rate among species, but greatly across sites (Heal et al. 1982), perhaps reflecting the importance of climate at large, and substrate at small, spatial scales.

Another useful approach for discerning species' and abiotic effects takes advantage of existing gradients in abiotic factors where species composition remains fairly constant and vice versa. For example, in Hawai'i, an elevation gradient that represents mainly a temperature gradient, but contains virtually the same species and soils, allows examination of the effects of temperature on ecosystem processes (Vitousek et al. 1992). Similarly, comparing different community types within constant macroclimates may facilitate understanding species' roles in ecosystem processes.

Modeling experiments that parallel those described in the field are rare. To do such experiments, models must include species-level information and information about the controls over biogeochemical cycles. Most ecosystem models do not incorporate species, but rather include average species' characteristics like lignin content, allocation, and tissue element ratios (Parton et al. 1987; Rastetter et al. 1991). Such models may be most appropriate for ecosystems dominated by single growth forms or species, such as grasslands and some forests. Models that do include species as well as decomposition and nutrient cycling allow litter-quality feedbacks resulting from plant population changes to constrain subsequent population and ecosystem responses. For example, in a simulation of the effects of climate warming on boreal and temperate forests, warming decreased water availability, limiting the growth of deciduous species, but favoring the growth of evergreen species on sandy compared with silty loam soils. The high lignin content of evergreen species reduced N mineralization rates, further reducing productivity on sands compared to loams (Pastor and Post 1988).

The most powerful models for examining species effects on biogeochemistry would be models that similarly link population dynamics and biogeochemistry, but in which modelers could manipulate both species and climate. Thus, modelers could factorially combine species 'removals' and 'additions' with changes in climate to understand net effects on biogeochemical processes.

15.6 Conclusions

1. Plant species potentially affect biogeochemical cycles both directly, because they acquire and lose C, nutrients, and water, and indirectly, because they influence microbial activity and herbivory.
2. Predicting the response of biogeochemical cycles to environmental change will require knowledge of plant species' responses to such change if interspecific variation in plant traits that control or influence biogeochemistry is large.
3. The largest species' effects on biogeochemical cycling will occur if drying accompanies global warming, since replacement of mosses by deciduous species will result in more rapid rates of C and nutrient cycling.

References

Bliss LC (1956) A comparison of plant development in microenvironments of arctic and alpine plants. Ecol Monogr 26:303–337

Bliss LC, Matveyeva NV (1992) Circumpolar arctic vegetation. In: Chapin FS III, Jefferies RL, Reynolds JF, Shaver GR, Svoboda J (eds) Arctic ecosystems in a changing climate: an ecophysiological perspective. Academic Press, San Diego, pp 59–89

Bryant JP, Reichardt PB (1992) Controls over secondary metabolite production in arctic woody plants. In: Chapin FS III, Jefferies RL, Reynolds JF, Shaver GR, Svoboda J (eds) Arctic ecosystems in a changing climate: an ecophysiological perspective. Academic Press, San Diego, pp 377–390

Bryant JP, Provenza FP, Pastor J, Reichardt PB, Clausen TP, du Toit JT (1991) Interactions between woody plants and browsing mammals mediated by secondary metabolites. Annu Rev Ecol Syst 22:431–446

Chapin FS III, (1983) Direct and indirect effects of temperature on arctic plants. Polar Biol 2:47–52

Chapin FS III, Bloom A (1976) Phosphate absorption: adaptation of tundra graminoids to low temperature, low-phosphorus environment. Polar Biol 2:37–52

Chapin FS III, Shaver GR (1985a) Arctic. In: Chabot BF, Mooney HA (eds) Physiological ecology of North America. Chapman and Hall, New York, pp 16–40

Chapin FS III, Shaver GR (1985b) Individualistic growth response of tundra plant species to environmental manipulations in the field. Ecology 66:564–576
Chapin FS III, Shaver GR (1989) Differences in growth and nutrient use among arctic plant growth forms. Funct Ecol 3:73–80
Chapin FS III, Tryon PR (1982) Phosphate absorption and root respiration of different plant growth forms from northern Alaska. Holarct Ecol 5:164–171
Chapin FS III, McKendrick JD, Johnson D (1986) Seasonal changes in carbon fractions in Aloaskan tundra plants of differing growth form: implications for herbivores. J Ecol 74:707–731
Chapin FS III, Fetcher N, Kielland K, Everett KR, Linkins AE (1988) Productivity and nutrient cycling of Alaskan tundra: enhancement by flowing soil water. Ecology 69:693–702
Chapin FS III, Vance ED, Zhong H (1992) Plant-microbial competition for nitrogen does not regulate productivity of arctic tundra. Bull Ecol Soc Am 73:136
Clymo RS, Hayward PM (1982) The ecology of *Sphagnum*. In: Smith AJE (ed) Bryophyte ecology. Chapman and Hall, London, pp 229–289
Flanagan PW, Van Cleve K (1983) Nutrient cycling in relation to decomposition and organic matter quality in taiga ecosystems. Can J For Res 13:795–817
Hastings SJ, Luchessa SA, Oechel WC, Tenhunen JD (1989) Standing biomass and production in water drainages of the foothills of the Philip Smith Mountains, Alaska, USA. Holarct Ecol 12:304–311
Havström, Callaghan TV, Jonasson S (1993) Differential growth responses of *Cassiope tetragona*, an arctic dwarf-shrub, to environmental perturbations among three contrasting high- and sub-arctic sites. Oikos 66:389–402
Heal OW, French DD (1974) Decomposition of organic matter in trundra. In: Holding AJ, Heal OW, Maclean SF, Flanagan PW (eds) Soil organisms and decomposition in tundra. Tundra Biome Steering Committee, Stockholm, pp 279–310
Heal OW, Flanagan PW, French DD, Maclean SF Jr (1982) Decomposition and accumulation of organic matter in tundra. In: Bliss LC, Heal OW, Moore JJ (eds) Tundra ecosystems: a comparitive analysis. Cambridge University Press, Cambridge, pp 587–633
Jefferies RL, Svoboda J, Henry G, Raillard M, Ruess R (1992) Tundra grazing systems and climatic change. In: Chapin FS III, Jefferies RL, Reynolds JF, Shaver GR, Svoboda J (eds) Arctic ecosystems in a changing climate: an ecophysiological perspective. Academic Press, San Diego, pp 391–412
Jonasson S (1992) Plant responses to fertilization and species removal in tundra related to community structure and clonality. Oikos 63:420–429
Kielland K, Chapin FS III, (1992) Nutrient absorption and accumulation in arctic plants. In: Chapin FS III, Jefferies RL, Reynolds JF, Shaver GR, Svoboda J (eds) Arctic ecosystems in a changing climate: an ecophysiological perspective. Academic Press, San Diego, pp 321–335
Kummerow J,Ellis B (1984) Temperature effect on biomass production and root/shoot biomass ratios in two arctic sedges under controlled environmental conditions. Can J Bot 62:2150–2153
Lambers H, Poorter H (1992) Physiology and ecology of growth rate variation. Adv Ecol Res 23:187–261
Limbach WE, Oechel WC, Lowell W (1982) Photosynthetic and respiratory responses to temperature and light of three Alaskan tundra growth forms. Holarct Ecol 5:150–157
Marion GM, Black CH (1986) The effect of time and temperature on nitrogen mineralization in arctic tundra soils. Soil Sci Soc Am J 51:1501–1508
Maxwell B (1992) Arctic climate: potential for change under global warming. In: Chapin FS III, Jefferies RL, Reynolds JF, Shaver GR, Svoboda J (eds) Arctic ecosystems in a changing climate: an ecophysiological perspective. Academic Press, San Diego, pp 11–34
Murray KJ, Tenhunen JD, Kummerow J (1989) Limitations on moss growth and net primary production in tussock tundra areas of the foothills of the Philip Smith Mountians, Alaska. Oecologia 80:256–262
Nadelhoffer KJ, Giblin AE, Shaver GR, Laundre JA (1991) Effects of temperature and substrate quality on element mineralization in six arctic soils. Ecology 72:242–253

Nadelhoffer KJ, Giblin AE, Shaver GR, Linkins AE (1992) Microbial processes and plant nutrient availability in arctic soils. In: Chapin FS III, Jefferies RL, Reynolds JF, Shaver GR, Svoboda J (eds) Arctic ecosystems in a changing climate: an ecophysiological perspective. Academic Press, San Diego, pp 281–300

Oberbauer SF, Oechel WC (1989) Maximum CO_2-assimilation rates of vascular plants on an Alaskan arctic tundra slope. Holarct Ecol 12:312–316

Oechel WC, Sveinbjörnsson B (1978) Primary production processes in arctic bryophytes at Barrow, Alaska. In: Tieszen LL (ed) Vegetation and production ecology of an Alaskan arctic tundra. Springer, Berlin Heidelberg New York, pp 269–298

Oechel WC, Van Cleve K (1986) The role of bryophytes in nutrient cycling in the taiga. In: Van Cleve K, Chapin FS III, Flanagan PW, Viereck LA, Dyrness CT (eds) Forest ecosystems in the Alaskan taiga. Springer, Berlin Heidelberg New York, pp 121–137

Parton WJ, Schimel DS, Cole CV, Ojima DS (1987) Analysis of factors controlling soil organic matter levels in Great Plains grasslands. Soil Sci Soc Am J 51:1173–1179

Pastor J, Post WM (1988) Responses of northern forests to CO_2-induced climate change. Nature 334:55–58

Pastor J, Naimen RJ, Dewey B, McInnes P (1988) Moose, microbes, and the boreal forest. BioScience 38:770–777

Rastetter EB, Ryan MG, Shaver GR, Melillo J M, Nadelhoffer KJ, Hobbie JE, Aber JD (1991) A general biogeochemical model describing the responses of the C and N cycles in terrestrial ecosystems to changes in CO_2, climate, and N deposition. Tree physiol 9:101–126

Shaver GR, (1986) Woody stem production in Alaskan tundra shrubs. Ecology 56:401–410

Shaver GR, Billings WD (1975) Root production and root turnover in a wet tundra ecosystem, Barrow, Alaska. Ecology 56:401–410

Shaver GR, Chapin FS III (1980) Response to fertilization by various plant growth forms in an Alaskan tundra: nutrient accumulation and growth. Ecology 61:662–675

Shaver GR, Chapin FS III (1991) Prodcution: biomass relationships and element cycling in contrasting arctic vegetation types. Ecol Monogr 61:1–31

Shaver GR, Kummerow J (1992) Phenology, resource allocation, and growth of arctic vascular plants. In: Chapin FS III, Jefferies RL, Reynolds JF, Shaver GR, Svoboda J (eds) Arctic Ecosystems in a changing climate. Academic Press, San Diego, pp 193–211

Shaver GR, Chapin FS III, Gartner BL (1986) Factors limiting seasonal growth and peak biomass accumulation in *Eriophorum vaginatum* in Alaskan tussock tundra. J Ecol 74:257–278

Shaver GR, Billings WD, Chapin FS III, Giblin AE, Nadelhoffer KJ, Oechel WC, Rastetter EB (1992) Global change and the carbon balance of arctic ecosystems. BioScience 61:415–435

Tenhunen JD, Lange OL, Hahn S, Siegwolf R, Oberbauer SF (1992) The ecosystem role of poikilohydric tundra plants. In: Chapin FS III, Jefferies RL, Reynolds JF, Shaver GR, Svoboda J (eds) Arctic ecosystems in a changing climate. Academic Press, San Diego, pp 213–237

Torn MS, Chapin FS III (1993) Environmental and biotic controls over methane flux from arctic tundra. Chemosphere 26:357–368

Van Cleve K (1974) Organic matter quality in relation to decomposition. In: Holding AJ, Heal OW, Maclean SF, Flanagan PW (eds) Soil organisms and decomposition in tundra. Tundra Biome Steering Committee, Stockholm, pp 311–324

Vitousek PM, Walker LR (1989) Biological invasion by *Myrica faya* in Hawai'i: plant demography, nitrogen fixation, ecosystem effects. Ecol Monogr 59:247–265

Vitousek PM, Aplet G, Turner D, Lockwood JJ (1992) The Mauna Loa environmental matrix: foliar and soil nutrients. Oecologia 89:372–382

Wedin DA, Tilman D (1990) Species effects on nitrogen cycling: a test with perennial grasses. Oecologia 84:433–441

16 Causes and Consequences of Plant Functional Diversity in Arctic Ecosystems

F. S. CHAPIN III, S. E. HOBBIE, M. S. BRET-HARTE, and G. BONAN

16.1 Introduction

Broad-scale predictions about the response of terrestrial ecosystems to global changes in climate and land use and feedbacks to the atmosphere require the simplest possible framework to be useful. This framework must contain sufficient detail to incorporate the major controlling mechanisms and be broadly applicable but omit details that are site-specific or unnecessarily complicate predictive models (O'Neill et al. 1989). What is the simplest possible framework for understanding the role of arctic ecosystems in global change? More specifically, how much do we need to know about the species composition and diversity of arctic ecosystems to predict how these ecosystems will respond to global change? At one extreme, we might consider vegetation to be a single component, as it is currently treated in most models of biogeochemistry (Agren et al. 1991), hydrometeorology (Jarvis and McNaughton 1986; Field 1991), and land-surface interfaces to general circulation models (Running and Coughlan 1988). At the opposite extreme, we might include each species in a multi-species biogeochemical model (Miller et al. 1984). In this chapter, we argue that, by grouping plant speices into functional groups, we achieve an intermediate level of detail that is useful in predicting response to and effects on ecosystem processes. More specifically, we present a framework for classifying species into functional groups in arctic ecosystems and evaluate its utility in addressing two questions:

1. What kinds of species will respond most strongly to global changes in climate and land use?
2. What are the consequences of changes in species composition for ecosystem function?

16.2 Predicting Species Response to Global Change

16.2.1 Functional Group Classification

The most useful classification of species into functional groups would incorporate both species *response* to environmental change and the *effect* of species on

Department of Integrative Biology, University of California, Berkeley, CA 94720, USA

Ecological Studies, Vol. 113
Chapin/Körner (eds) Arctic and Alpine Biodiversity

ecosystem processes. However, if different plant traits control response to each environmental factor and the effect of vegetation on each ecosystem process, we would need a separate functional-group classification for each factor or process. Thus, considering all processes, we might end up with nearly as many functional groups as species, in which case the functional-group concept loses its utility. Alternatively, if the rate of acquisition of one resource is correlated with acquisition rates of other resources (e.g. nutrients, carbon, water), we may be able to define a limited number of functional groups that are broadly applicable (Chapin 1993; Korner 1993). In this case, the most useful functional-group classification would be one that groups species according to their effect on the availability of resources that directly limit plant growth (e.g. nitrogen in most tundra ecosystems; Chapin 1987).

Plant size and relative growth rate (RGR) are traits useful in distinguishing broad functional groups because (1) together these traits determine the resource demand of an individual and (2) these traits have several important ecosystem consequences. Furthermore, commonly recognized growth forms of tundra plants generally differ in size and RGR (Shaver and Kummerow 1992; Shaver et al. 1994; Table 1) so that we can use these growth forms as our functional-group classification. Species within each functional group share similar responses to resource supply, as in other ecosystems (Grime 1977; Chapin 1980; Tilman 1990), and have similar effects on resource availability and water and energy balance, as described below.

16.2.2 Response to Resource Supply

Species belonging to different functional groups exhibit a characteristic pattern of distribution along resource gradients. Deciduous vascular plants (deciduous shrubs, forbs, and graminoids) dominate sites that are more nutrient-rich than

Table 1. Major arctic plant functional groups, examples of species from tussock tundra, and physiological attributes. Data from Chapin and Shaver (unpubl. observ.) and Shaver et al. (1994)

Functional group	Species	Maximum Height (cm)	Growth rate	Leaf longevity	Litter quality
Forb	*Polygonum bistorta*	20	High	12 weeks	Very high
Graminoid	*Eriophorum vaginatum*	30	High	8 weeks	Moderate
	Carex bigelowii	20	High	12 weeks	Moderate
Deciduous shrub	*Betula nana*	50	Moderate	10 weeks	Moderate
	Salix pulchra	100	Moderate	10 weeks	Moderate
Evergreen shrub	*Ledum palustre*	40	Low	1.5 yr	Low
	Vaccinium vitis-idaea	10	Low	2 yr	Low
Moss	*Hylocomium splendens*	10	Low	3 yr	Low
	Aulacomnium turgidum	5	Low	3 yr	Very low
	Sphagnum spp.	15	Low	8 yr	Very low
Lichen	*Cetraria* spp.	5	Very low	5 yr?	Very high
	Peligeria aphthosa	5	Very low	3 yr?	Unknown

those occupied by evergreen shrubs (Fig. 1). For example, Scandinavian calcareous substrates support many species of forbs and deciduous shrubs, whereas acidic soils or thin coarse-textured soils overlying deglaciated bedrock have a greater abundance of evergreen shrubs (Kalela 1961; Sjors et al. 1965; Ebeling 1978). In the Alaskan Arctic, young glacial till ($<60\,000$ yr) has many forbs and deciduous shrubs and few evergreen shrubs, whereas older glacial till has fewer forbs and more evergreen shrubs (Walker and Everett 1991). Ground-squirrel dens and other sites of regular nutrient enrichment by animals have more deciduous species and fewer evergreen shrubs than surrounding tundra (Batzli and Sobaski 1980; Walker and Everett 1991). The woody support structure of deciduous shrubs gives them a competitive advantage over shorter growth forms where nutrient supply is adequate to support their high nutrient demand,

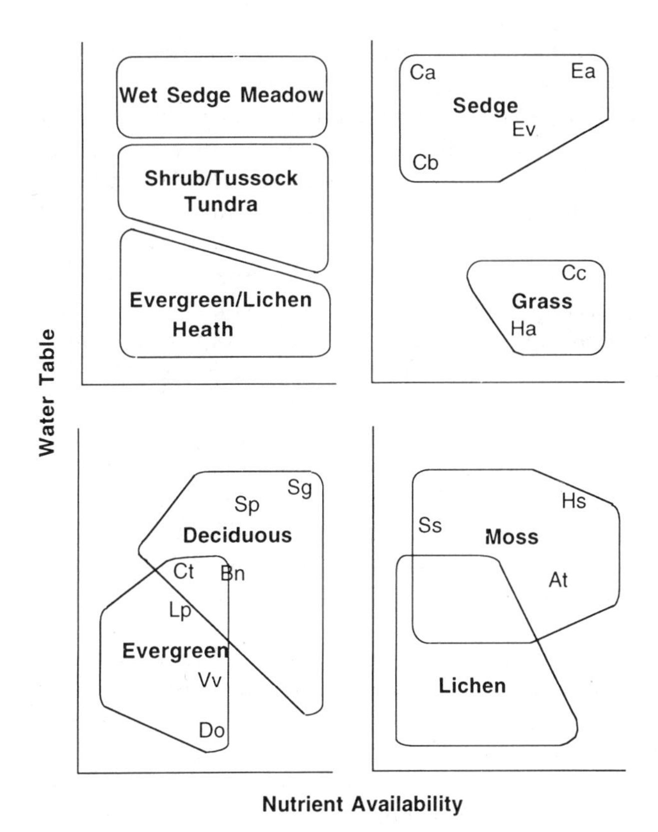

Fig. 1. Distribution of arctic communities, functional groups, and species along gradients of nutrient availability and water-table depth. Sedges are *Carex aquatilis* (*Ca*), *C. bigelowii* (*Cb*), *Eriophorum angustifolium* (*Ea*) and *E. vaginatum* (*Ev*). Grasses are *Calamagrostis canadensis* (*Cc*) and *Hierochloe alpina* (*Ha*). Deciduous shrubs are *Salix pulchra* (*Sp*), *S. glauca* (*Sg*), and *Betula nana* (*Bn*). Evergreen shrubs are *Cassiope tetragona* (*Ct*), *Ledum palustre* (*Lp*), *Vaccinium vitis-idaea* (*Vv*), and *Dryas octapetala* (*Do*). Mosses are *Sphagnum* spp. (*Ss*), *Hylocomium splendens* (*Hs*), and *Aulacomnium turgidum* (*At*)

as in zones of flowing groundwater (Shaver and Chapin 1991). Aerenchyma for oxygen transport gives sedges an advantage in areas where the water table is frequently close to or above the soil surface. Thus, arctic functional groups are arrayed predictably along gradients of nutrient and water supply (Fig. 1; Webber 1978).

Nonvascular plants are shorter in stature than most vascular plants and lack roots, so their growth is promoted where inadequate nutrient supply prevents a dense litterfall from broad-leaved vascular plants (Oechel and Van Cleve 1986) and where water supply is adequate. Mosses dominate dependably wet sites such as bogs and fens, whereas lichens dominate drier sites because of greater tolerance to frequent to occasional droughts. Lichens grow most rapidly in sites of frequent rain and fog (e.g. Scandinavian coastal mountains and western Alaskan hills), where dry periods prevent moss dominance and inadequate nutrient supply prevents dominance by broad-leaved vascular plants.

These patterns of functional-group distribution are consistent with their responses to manipulation of resource supply. Nutrient enhancement through fertilizer addition (Shaver and Chapin 1980, 1986), animal remains (McKendrick et al. 1980), or increased nitrogen mineralization associated with fire or human disturbance (Challinor and Gersper 1975; Chapin and Shaver 1981) consistently increases the abundance of deciduous species and reduces the abundance of evergreen shrubs. Graminoids initially dominate these high-fertility sites and maintain their dominance in sites of animal grazing and disturbance. However, in the absence of disturbance, deciduous shrubs dominate sites with regular nutrient enrichment (Chapin et al. 1994), just as observed along topographic gradients.

The responses of nonvascular plants to manipulation of resource supply are less well documented but are generally consistent with predictions based on habitat distribution. Although mosses initially show little response (Skre and Oechel 1979) or a small positive response to nutrient addition (Chapin et al. 1994), those nutrient additions that greatly enhance the development of a vascular-plant canopy cause a long-term decline in mosses and lichens due to the impact of vascular litterfall on nonvascular plants (Chapin et al. 1994). However, complete removal of the vascular-plant canopy causes moss desiccation and a decline in moss growth (Tenhunen et al. 1992).

The patterns of plant functional-group abundance along gradients of resource availability and the response of functional groups to resource manipulation are predictable throughout the circumpolar Arctic and in temperate bogs and fens and thus form a useful basis for predicting plant response to future changes in environment. However, predictions of species responses *within* a functional group are often site-specific and thus their prediction requires new studies in each region. For example, in arctic Alaska *Salix pulchra* grows more rapidly and tends to occur on more fertile sites than *Betula nana*, whereas in interior Alaska other willows (*S. glauca* and *S. arbusculoides*) grow more slowly and occur on less fertile sites than do birches (*B. glandulosa* and *B. papyrifera*). In other cases, generalizations about species responses within a functional group are broadly applicable. For example, *Sphagnum* mosses characteristically occupy less fertile sites than other moss species and are displaced by other mosses when nutrient supply increases in Alaska (Walker and Everett 1991), Canada (Johnson and Damman 1991), and England (Clymo and Hayward 1982).

16.2.3 Physiological Basis of Response to Soil Resources

The predictable response to soil resources of species belonging to the same functional group follows logically from the interdependence of physiological processes that link resource capture to patterns of size and growth rate, as described in detail by Hobbie (this vol.) and Chapin (1993). In brief, plant relative growth rate (RGR) is a central indicator of plant physiological strategies. To support a high RGR, a plant must have high rates of resource capture, i.e., high photosynthetic rates per gram tissue, requiring high tissue-nitrogen concentrations or a large leaf allocation, also requiring substantial nitrogen investment. The photosynthetic potential and stomatal conductance of a plants are physiologically linked at the leaf and whole-plant levels such that plants with high photosynthetic capacity also have high a transpiration rate. Rates of photosynthesis and nutrient absorption tend to decline as tissues age, so continued high rates of resource capture require high rates of root and leaf turnover. The associated nutrient loss can be replaced at relatively low cost in a high-resource environment. Maintenance of a high RGR also requires a large allocation to production of resource-acquiring tissues. Under conditions of high nutrient and water availability or low light, a large allocation of resources to leaves maximizes RGR, whereas under low nutrient and water availability or high light availability relatively large investment in roots maximizes RGR.

By contrast, in low-resource environments slow growth permits slow tissue turnover, conserving the carbon and nutrients contained therein. Slow growth also minimizes the growth respiration associated with production of new tissues. By reducing resource loss and the rate of resource capture, slow growth reduces nutrient demand from the environment. The low rate of photosynthesis at the whole-plant level which is associated with slow growth also reduces water loss, which reduces plant water requirements (Chapin et al. 1993a).

16.2.4 Response to Climate

Major physiognomic groups of plants are predictably distributed with respect to climate. For example, succulents dominate in deserts, trees are absent in extremely cold environments, ring-porous hardwoods do not grow in areas that experience −40 °C (due to intracellular freezing), and broad-leaved evergreen trees do not grow in areas that experience −15 °C (Woodward 1987). However, more detailed responses to climate are difficult to predict by functional group. For example, there were no predictable responses of species or functional groups to increased air temperature in Alaskan tussock tundra, whereas functional groups responded predictably to changes in nutrient supply and light (Chapin and Shaver 1985, 1994). Similarly, in the paleoecological record, species have migrated individualistically in response to changes in climate in a fashion that would be difficult to predict by functional group (Davis 1981), but the distribution of functional groups on different soil types is quite predictable (Pastor and Post 1988).

In summary, we suggest that there is a stronger physiological basis for predicting functional-group responses to changes in resource supply than to changes in climate (Chapin et al. 1993c). Thus, any "simple" framework for understanding the impact of global changes in climate on vegetation composition must include the effects of climatic change on resource supply (Fig. 2).

16.2.5 Responses of "Missing" Functional Groups to Environment

Introduction of new functional groups such as nitrogen fixers (Vitousek et al. 1987) or fire-prone grasses (D'Antonio and Vitousek 1992) often cause large changes in ecosystem structure and function, providing "surprises" that could not be predicted from responses of existing vegetation to changes in environment. Important new functional groups will probably invade the Arctic in response to global changes in climate and land use. Evergreen and deciduous trees have invaded arctic tundra in response to past climatic warming events and appear to be advancing in areas of current climatic warming (D'Arrigo et al. 1987). Decidous trees occur in areas within the current Arctic where soils are warmed by thermal springs or steep south-facing slopes (Murray 1980), suggesting that soil temperature and/or soil resources are critical factors restricting the advance of

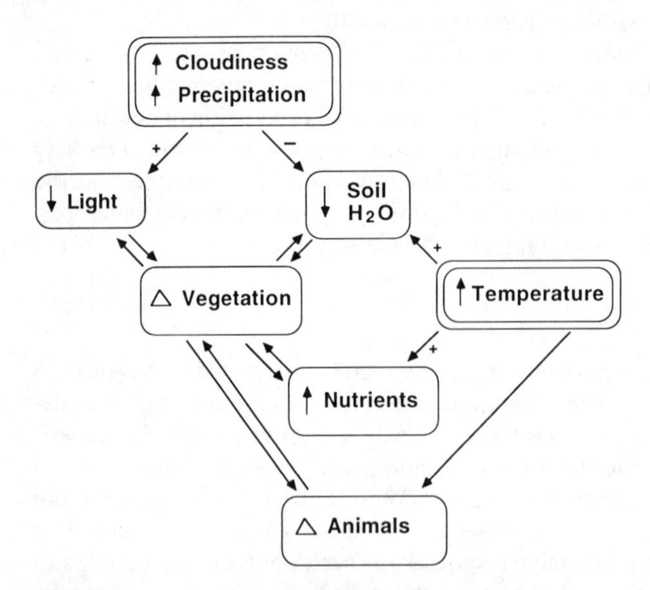

Fig. 2. The major effects of climatic changes on vegetation operate through changes in resources (light, water, and nutrients) required by plants and changes in interactions with animals

trees into the Arctic. Tall deciduous shrubs (*Salix* and *Alnus*) occur in well-drained river and stream valleys and other well-drained sites. Large areas dominated by tall shrubs are currently restricted to areas of forest tundra, particularly in the low arctic of Alaska and eastern Russia (Bliss and Matveyeva 1992). These patterns suggest that warmer conditions or perhaps warmer, drier soils might cause invasion of tall shrubs and trees into areas where vegetation is now largely snow-covered in winter. Because some of these tall shrubs (*Alnus*) support nitrogen-fixing symbionts, there could be substantial changes in nitrogen cycling as well.

16.3 Effects of Species on Ecosystem Processes

Size and RGR are important indicators of species' effects on energy flux, trace-gas flux, and biogeochemical cycling (Chapin 1993). Therefore, the functional-group system developed above provides a useful framework for predicting species effects on ecosystem processes.

16.3.1 Energy Budget

Through the partitioning of net radiation into sensible and latent heat and through radiative exchange, vegetation has an important effect on regional and global climates. For example, differences in winter albedo of tundra and boreal forest vegetation have a profound effect on the energy balance of northern ecosystems. Model simulations in which boreal forest is replaced by tundra vegetation, which remains snow covered in winter, suggests that the loss of emergent vegetation would cause a substantial decrease in absorbed solar radiation and a large, permanent regional cooling (Bonan et al. 1992). Conversely, we expect northward advance of the tree line or an increased abundance of tall shrubs in tundra regions as a result of the current warming trend (Chapman and Walsh 1993). This could substantially increase absorbed radiation and reinforce the warming trend.

The effects of vegetation on atmospheric processes have been included in global climate models via "land-surface process models" such as BATS (Dickenson et al. 1993) or SiB (Sellers et al. 1986). These models are implemented globally by recognizing major vegetation types that differ in important physiological and biophysical properties (e.g. stomatal resistance, leaf reflectance and transmittance, rooting depth). These models include only the variability in physiology thought to be most important for atmospheric processes. Consequently, the number of vegetation types is minimal: BATS has 15 vegetation types; SiB has 11.

Two approaches are used when implementing these models. The first is to apply the model to functional types, similar to those discussed in this chapter, that constitute the vegetation types. Both BATS and SiB distinguish broadleaved evergreen trees, broad-leaved deciduous trees, needle-leaf evergreen trees, needle-leaved deciduous trees, evergreen shrubs, deciduous shrubs, and grasses. An alternative approach is to apply the model to the major vegetation types directly.

BATS and SiB both have a mixed needle-leaved evergreen tree and broad-leaved deciduous tree vegetation type, with physiological and biophysical properties for the "average" tree. SiB represents savanna through separate tree and grass lifeforms, but BATS represents savanna using tall grass as a "typical" plant. The SECHIBA land surface model (Ducoudre et al. 1993) is typical of this latter approach, with required vegetation parameters estimated for tundra, grassland, grassland and shrub, grassland and tree, deciduous forest, evergreen forest, and rain forest vegetation types.

Regardless of whether these models are applied to functional types that comprise vegetation types or to vegetation types directly, atmospheric sciences have clearly adopted an approach in which species-level variability is ignored for variability at a higher level.

16.3.2 Biogeochemistry and Trace-Gas Flux

Plant functional groups have predictable effects on biogeochemical cycling and trace-gas flux, as described by Hobbie (this Vol.). To summarize her conclusions, functional groups with large size and/or rapid growth rates (e.g. deciduous shrubs, graminoids, and forbs) stimulate nutrient cycling through both high rates of resource capture (photosynthesis and nutrient absorption) and high litter quality which promotes carbon and nitrogen mineralization. Sedges also promote soil microbial activity in waterlogged soils by transporting oxygen through aerenchymatous roots. By contrast, functional groups with slow growth (e.g. evergreen shrubs and mosses) retard nutrient cycling through slow rates of resource uptake and low litter quality.

Shaver et al. (1994) argue that these differences among functional groups in biogeochemistry are more important at times of change in resource supply than under "equilibrium" conditions. They point out that, when one considers the whole plant, rather than just current-year's tissues, there are relatively minor differences among functional groups in the field in terms of nutrient uptake, tissue turnover, and litter quality (Jonasson 1983; Shaver et al. 1994). However, following disturbance functional groups with more rapid growth rates can exploit more effectively any pulse of nutrient supply.

Regardless of the magnitude of differences among functional groups and the conditions under which these are important, differences among functional groups are probably greater than species differences within functional groups (Tab. 1). However, the magnitude of these species effects has not received careful study in the field. Species-specific effects could include differences in phenology of nutrient uptake or loss, litter placement (reflecting root: shoot ratio), leaf: stem ratio, and the effects of specific secondary metabolites on decomposition (Hobbie, this Vol.)

16.3.3 Disturbance Regime

Arctic tundra is a highly disturbed ecosystem, with many sources of disturbance (e.g. frost action of various types, landslides, mud and slush flows, and fire).

However, functional groups of plants strongly influence disturbance regime. For example, northward movement of coniferous trees will increase fire frequency (due to greater fuel load and high flammability) in areas that are currently treeless (Viereck 1973). Species that provide effective thermal insulation (e.g. mosses) or which dry soils through high transpiration rates (e.g. deciduous species) may reduce the probability of frost features. Species with strong woody roots (e.g. deciduous shrubs) may reduce the probability of landsides. In general, these species effects on disturbance regime should be predictable at the level of functional group, although these effects have received little study.

16.3.4 Species Interactions

Plant functional groups provide a strong basis for predicting diet preference of herbivores. For example, mammals, birds, and most insects prefer deciduous species over evergreens and generally avoid mosses, probably because the same secondary metabolities that retard decomposition inhibit gut microbes (Batzli and Jung 1980; Bryant and Kuropat 1980; MacLean and Jensen 1986). At the functional-group level, some mammals (e.g. *Microtus*) specialize on deciduous shrubs and forbs, others (e.g. *Lemmus*) concentrate on monocots (Batzli and Jung 1980), and others (e.g. reindeer in winter) eat primarily lichens. Many insects also specialize at the level of functional group, although others specialize on single species or plant parts within a species (MacLean and Jensen 1986) and will be particularly sensitive to any changes in species diversity in reponse to climatic change. Even generalist vertebrate herbivores could be quite sensitive to changes in species diversity because they require a mixed diet to minimize intake of the secondary metabolites associated with each plant species (Jefferies and Bryant this Vol.). Loss of species diversity reduces the extent to which mixed feeding can occur. Conversely, animals with a mixed feeding strategy reduce species diversity because they preferentially feed on rare species and avoid common species to reduce intake of each secondary metabolite below the toxicity threshold.

Plant-pollinator interactions range from being highly specific to generalized. Many arctic species are either wind-pollinated (all graminoids and many common deciduous shrubs) or are pollinated by generalist pollinators such as bumble bees. Many of the less widespread arctic plant species are insect pollinated and vary in the degree of specificity of pollinator interactions (Inouye and McGuire 1991; McGuire 1993). The greater the degree of specificity, the more each member of the mutualism will depend on abundance of the complementary species. Even in the case of generalized pollinator systems, insects may depend on a sequence of flowering species to provide pollen throughout their foraging period. In this case, loss of a plant species that occupies a key phenological window could eliminate the insect and, therefore, perhaps other plant species that depend on the same pollinator at other times of year. These cascading indirect effects where loss of certain key species are critical to other species have recived little study.

Plant-microbial interactions also differ predictably among plant functional groups but are sensitive to species differences within functional groups. For example, ericoid mycorrhizae of heath species (generally evergreens) effectively exploit organic nitrogen, whereas VA mycorrhizae of graminoids are often restricted to inorganic forms of nitrogen (Read 1991; Schimel, (this Vol.), but see Chapin et al. 1993b). Ectomycorrhizae associated with deciduous shrubs are generally intermediate. Some mycorrhizal associations, especially among ericoids, are species-specific for both plant and fungus, but VA mycorrhizal associations are generally less species-specific (Allen 1991). Woody species often have lower rates of root exudation than herbaceous species (Lynch and Whipps 1991) but we currently know nothing about how these rates vary among individual species. These plant-microbial interactions should strongly influence rates of carbon and nutrient cycling and are clearly important at the level of functional group. However, we can only speculate as to how strongly individual plant species differ in their interaction with soil microbes.

16.4 Species Diversity as Insurance Against Loss of Function

Diversity within functional groups may have long-term implications for ecosystem function by providing insurance against loss of the functional groups. Because each species shows a unique response to climate, any change in climate or climatic extremes that is severe enough to cause extinction is unlikely to eliminate all members of a functional groups (Fig. 3). The more species there are

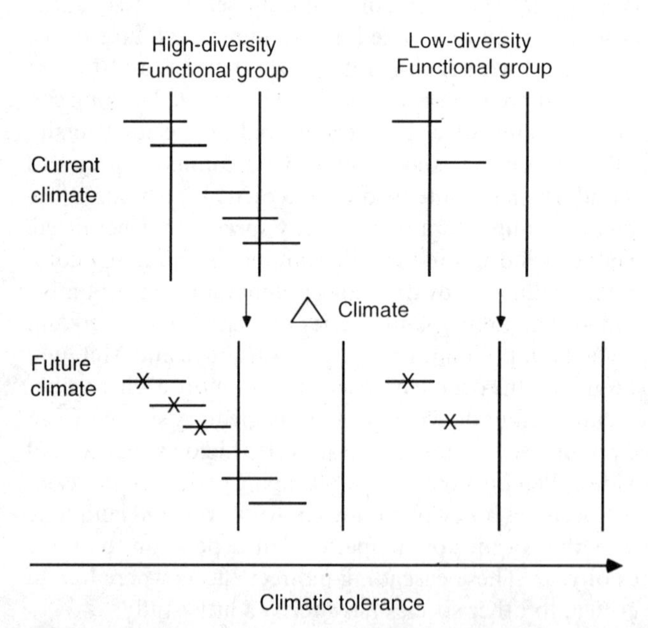

Fig. 3. Effect of climatic change on extinction of species (X) in high-diversity and low- diversity functional groups

in a functional group, the less likely that any extinction event or series of such events will have serious ecosystem consequences. By the same argument, high genetic diversity buffers a species from extinction. For these reasons, genetic and species diversity per se are important to maintain ecosystem function.

16.5 Conclusions

1. Plant functional groups are a simple framework to describe the physiognomic and physiological diversity present in natural ecosystems. Functional groups defined by plant size and RGR allow predictions of the effects of climate, soil resources, and animals on major functional groups of plants.
2. Conversely, functional groups of plants which differ in size and RGR have predictably different effects on a wide range of ecosystem processes, including biogeochemical cycling, trace-gas fluxes, herbivory, pollination, plant-microbial interactions, and disturbance.
3. Biogeochemical cycling, trace-gas flux, and disturbance rate are less sensitive to the identity of species within a functional group than are species interactions such as herbivory, pollination, and plant-microbial interactions.
4. Genetic and species diversity are critical to providing insurance against major changes in ecosystem functioning, if global change causes extinction of populations and species.

Acknowledgments. The research leading to these generalizations was supported by the National Science Foundation (DEB82-05344, BSR91-22791, DPP92-14906), the Department of Energy (92-SC-DOE-1003), and the National Atmospheric and Space Administration (2482-EDBECO-0007).

References

Agren GI, McMurtrie RE, Parton WJ, Pastor J, Shugart HH (1991) State of the art models of production-decomposition linkages in conifer and grassland ecosystems. Ecol Appl 1:118–138
Allen MF (1991 The ecology of mycorrhizae. Cambridge University Press, Cambridge
Batzli GO, Jung HG (1980) Nutritional ecology of microtine rodents: resource utilization near Atkasook, Alaska. Arct Alp Res 12: 483–499
Batzli GO, Sobaski S (1980) Distribution, abundance, and foraging patterns of ground squirrels near Atkasook, Alaska. Arct Alp Res 12: 501–510
Bliss LC, Matveyeva NV (1992) Circumpolar arctic vegetation. In: Chapin FS III, Jefferies RL, Reynolds JF, Shaver GR, Svoboda J (eds) Arctic ecosystems in a changing climate: an ecophysiological perspective. Academic Press, San Diego, pp 59–89
Bonan GB, Pollard D, Thompson SL (1992) Effects of boreal forest vegetation on global climate. Nature 359: 716–718
Bryant JP, Kuropat PJ (1980) Selection of winter forage by subarctic browsing vertebrates: the role of plant chemistry. Annu Rev Ecol Syst 11: 261–285
Challinor JL, Gersper PL (1975) Vehicle perturbation effects upon a tundra soil-plant system. II. Effects on the chemical regime. Soil Sci Soc Am Proc 39: 689–695
Chapin FS III (1980) The mineral nutrition of wild plants. Annu Rev Ecol Syst 11: 233–260

Chapin FS III (1987) Environmental controls over growth of tundra plants. Ecol Bull 38: 69–76
Chapin FS III (1993) Functional role of growth forms in ecosystem and global processes. In: Ehleringer JR, Field CB (eds) Scaling physiological processes: leaf to globe. Academic Press, San Diego, pp 287–312
Chapin FS III, Shaver GR (1981) Changes in soil properties and vegetation following disturbance in Alaskan arctic tundra. J Appl Ecol 18: 605–617
Chapin FS III, Shaver GR (1985) Individualistic growth response of tundra plant species to environmental manipulations in the field. Ecology 66: 564–576
Chapin FS III, Autumn K, Pugnaire F (1993a) Evolution of suites of traits in response to environmental stress. Am Nat 142: S78–S92
Chapin FS III, Moilanen L, Kielland K (1993b) Preferential use of organic nitrogen for growth by a non-mycorrhizal arctic sedge. Nature 361: 150–153
Chapin FS III, Rincon E, Huante P (1993c) Environmental responses of plants and ecosystems as predictors of the impact of global change. J Biosci 18: 515–524
Chapin FS III, Shaver GR, Giblin AE, Nadelhoffer KG, Laundre JA (1994) Response of arctic tundra to experimental and observed changes in climate. Ecology (in press)
Chapman WL, Walsh JE (1993) Recent variations of sea ice and air temperature in high latitudes. Bull Am Meteorol Soc 74: 33–47
Clymo RS, Hayward PM (1982) The ecology of *Sphagnum*. In: Smith AJE (eds) Bryophyte ecology. Chapman and Hall, London, pp 229–289
D'Antonio CM, Vitousek PM (1992) Biological invasions by exotic grasses, the grass-fire cycle, and global change. Annu Rev Ecol Syst 23: 63–87
D'Arrigo R, Jocoby GC, Fung IY (1987) Boreal forests and atmosphere-biosphere exchange of carbon dioxide. Nature 329: 321–323
Davis MB (1981) Quaternary history and the stability of forest communities. In: West DC, Shugart HH, Botkin DB (eds) Forest succession: concepts and applications. Springer-Berlin Heidelberg, New York, pp 132–153
Dickenson RE, Henderson-Sellers A, Kennedy PJ (1993) Biosphere-Atmosphere Transfer Scheme (BATS) version le as coupled to the NCAR community climate model. NCAR Tech Note NCAR/TN-397+STR: 1–72
Ducoudre NI, Laval K, Perrier A (1993) SECHIBA, a new set of parameterizations of the hydrologic exchanges at the land-atmosphere interface within the LMD atmospheric general circulation model. J Climatol 6: 248–273
Ebeling F (1978) Nordsvenska skogstyper. Sver Skogsvardsforb Tidskr 76:340–381
Field CB (1991) Ecological scaling of carbon gain to stress and resource availability. In: Mooney HA, Winner WE, Pell EJ (eds) Integrated responses of plants to stress. Academic Press, San Diego, pp 35–65
Grime JP (1977) Evidence for the existence of three primary strategies in plants and its relevance to ecological and evolutionary theory. Am Nat 111: 1169–1194
Inouye DW, McGuire AD (1991) Effects of snowpack on timing and abundance of flowering in *Delphinium nelsonii*: implications for climate change. Am J Bot 78: 997–1001
Jarvis PG, McNaughton KG (1986) Stomatal control of transpiration: scaling up from leaf to region. Adv Ecol Res 15: 1–49
Johnson LC, Damman AWH (1991) Species controlled *Sphagnum* decay on a South Swedish raised bog. Oikos 61: 234–242
Jonasson S (1983) Nutrient content and dynamics in north Swedish shrub tundra. Holarct Ecol 6: 295–304
Kalela A (1961) Waldvegetationszonen Finnlands und ihre klimatischen paralleltypen. Arch Soc Fauna Flora Vanamo 16C: 65–83
Korner C (1993) Scaling from species to vegetation: the usefulness of functional groups. In: Schulze E-D, Mooney HA (eds) Biodiversity and ecosystem function. Springer-Berlin Heidelberg, New York pp 117–140
Lynch JM, Whipps JM (1991) Substrate flow in the rhizosphere. In: Keister DL, Cregan PB (eds) The rhizosphere and plant growth. Kluwer, The Hague, pp 15–24
MacLean SF, Jensen TS (1986) Food plant selection by insect herbivores in Alaskan arctic tundra: the role of plant life form. Oikos 44: 211–221
McGuire AD (1993) Interactions for pollination between two synchronously blooming *Hedysarum* species (Fabaceae) in Alaska. Am J Bot 80: 147–152

McKendrick JD, Batzli GO, Everett KR, Swanson JC (1980) Some effects of mammalian herbivores and fertilization on tundra soils and vegetation. Arct Alp Res 12: 565–578
Miller PC, Miller PM, Blake-Jacobson M, Chapin FS III, Everett KR, Hilbert DW, Kummerow J, Linkins AE, Marion GM, Oechel WC, Roberts SW, Stuart L (1984) Plant-soil processes in *Eriophorum vaginatum* tussock tundra in Alaska: a systems modeling approach. Ecol Monogr 54: 361–405
Murray DF (1980) Balsam poplar in arctic Alaska. Can J Anthr 1: 29–32
Oechel WC, Van Cleve K (1986) The role of bryophytes in nutrient cycling in the taiga. In: Van Cleve K, Chapin FS III, Flanagan PW, Viereck LA, Dyrness CT (eds) Forest ecosystems in the Alaskan taiga. Springer-Berlin Heidelberg New York, pp 121–137
O'Neill RV, Johnson AR, King AW (1989) A hierarchical framework for the analysis of scale. Landsc Ecol 3: 193–205
Pastor J, Post WM (1988) Responses of northern forests to CO_2-induced climate change. Nature 334: 55–58
Read DJ (1991) Mycorrhizas in ecosystems. Experientia 47: 376–391
Running SW, Coughlan JC (1988) A general model of forest ecosystem processes for regional applications. I. Hydrologic balance, canopy gas exchange and primary production processes. Ecol Model 42: 125–154
Sellers PJ, Mintz Y, Sud YC, Dalcher A (1986) A simple biosphere model (SiB) for use within general circulation models. J Atmos Sci 43: 505–531
Shaver GR, Chapin FS III, (1980) Response to fertilization by various plant growth forms in an Alaskan tundra: nutrient accumulation and growth. Ecology 61: 662–675
Shaver GR, Chapin FS III, (1986) Effect of fertilizer on production and biomass of tussock tundra, Alaska, U.S.A. Arct Alp Res 18: 261–268
Shaver GR, Chapin FS III, (1991) Production: biomass relationships and element cycling in contrasting arctic vegetation types. Ecol Monogr 61: 1–31
Shaver GR, Kummerow J (1992) Phenology, resource allocation, and growth of arctic vascular plants. In: Chapin FS III, Jefferies RL, Reynolds JF, Shaver GR, Svoboda J (eds) Arctic ecosystems in a changing climate. Academic Press, San Diego, pp 193–211
Shaver GR, Giblin AE, Nadelhoffer KJ, Rastetter EB (1994) Plant functional types and ecosystem change in arctic tundras. In: Smith T, Shugart HH, Woodward FI (eds) Plant functional types. Cambridge University Press, Cambridge (in press)
Sjors H et al. (1965) The plant cover of Sweden. Acta Phytogeogr Suec 50: 1–314
Skre O, Oechel WC (1979) Moss production in a black spruce *Picea mariana* forest with permafrost near Fairbanks, Alaska, as compared with two permafrost-free stands. Holarct Ecol 2: 249–254
Tenhunen JD, Lange OL, Hahn S, Siegwolf R, Oberbauer SF (1992) The ecosystem role of poikilohydric tundra plants. In: Chapin FS III, Jefferies RL, Reynolds JF, Shaver GR, Svoboda J (eds) Arctic ecosystems in a changing climate. Academic Press, San Diego, pp 213–237
Tilman D (1990) Constraints and tradeoffs: toward a predictive theory of competition and succession. Oikos 58: 3–15
Viereck LA (1973) Wildfire in the taiga of Alaska. Quat Res 3: 465–495
Vitousek PM, Walker LR, Whiteacre LD, Mueller-Dombois D, Matson PA (1987) Biological invasion by *Myrica faya* alters ecosystem development in Hawaii. Science 238: 802–804
Walker DA, Everett KR (1991) Loess ecosystems of northern Alaska: regional gradient and toposequence at Prudhoe Bay. Ecol Monogr 61: 437–464
Webber PJ (1978) Spatial and temporal variation of the vegetation and its productivity, Barrow, Alaska. In: Tieszen LL (eds) Vegetation and production ecology of an Alaskan arctic tundra. Springer-Berlin Heidelberg, New York, pp 37–112
Woodward FI (1987) Climate and plant distribution. Cambridge University Press, Cambridge

17 Ecosystem Consequences of Microbial Diversity and Community Structure

J. SCHIMEL

17.1 Introduction

Biodiversity has become a major theme in ecological research and environmental policy (Schulze and Mooney 1993). This concern has arisen because people value diversity both for its own sake and because diversity may control important ecosystem services (food, fiber, animal production, tourism). While the first rationale for concern over biodiversity should apply to microbes, they lack charisma. I therefore doubt that arguments about microbial biodiversity for its own sake will carry much weight for most people, and our concerns with the issue will rest primarily on the implications of their diversity for ecosystem function. While several papers have discussed the effect of functional diversity on ecosystem processes (Meyer 1993; Beare et al. 1994), they basically conclude that microbes carry out many processes that are important to ecosystem function and that their interactions are complex. Formulating meaningful conclusions about the importance of diversity within functional groups, however, has been difficult.

The challenge in relating microbial diversity to ecosystem function is fundamentally one of relating the scales of microbial life to the scale of the ecosystem. Studies whose primary focus is at the ecosystem level may range from the landscape scale (to put the system into a larger perspective), to the process scale (to explain why the system behaves the way it does). When we discuss "microbes", however, we are considering organisms whose cells (for bacteria) or individual hyphae (for fungi) are microns in diameter, though individual fungal mycelia can extend over many square meters. At a small enough scale, microbial community structure *must* be a dominant control on ecological processes, but as we move up in scale toward the ecosystem and integrate across many individual communities, the influence of individual community structures decreases. The central question for this chapter can therefore be usefully phrased and diagrammed (Fig. 1):

Is there some minimal scale necessary to adequately explain ecosystem processes at which microbial community structure still has a measurable influence on the nature and rates of those processes?

This approach forces us to first determine whether we can adequately explain the dynamics of ecological processes without specifically considering the

Institute of Arctic Biology, POB 757000, University of Alaska, Fairbanks, AK 99775, USA

Ecological Studies, Vol. 113
Chapin/Körner (eds) Arctic and Alpine Biodiversity

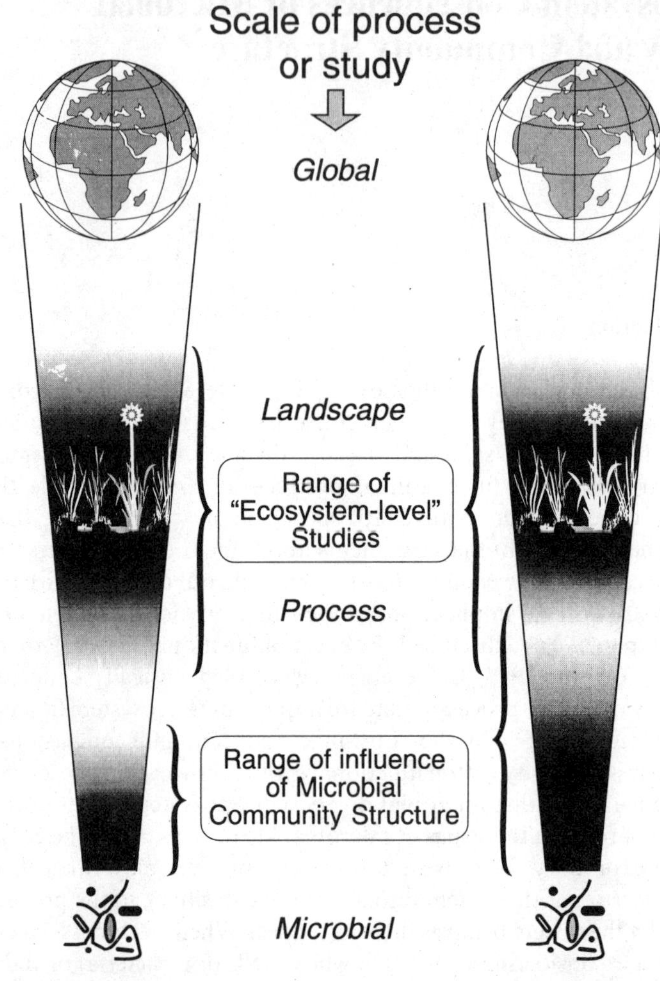

Fig. 1. Framework for deciding which kinds of ecological processes require understanding microbial community structure in order to understand ecosystem-level dynamics

microbial community. Only if this fails, do we need to examine microbial community structure. Thus, initially at least, we avoid difficult issues such as the vast diversity of unculturable organisms in soil. While such issues may be important in *microbial* ecology, their significance to *ecosystem* ecology remains unclear. My objective is to provide a framework for research in this area by identifying the kinds of processes for which microbial diversity and community structure may play a role at the ecosystem level. I will focus on tundra systems, but many examples will come from other systems because there has been very little appropriate study in the tundra.

17.2 "Broad" Processes

The initial challenge in addressing the importance of microbial diversity comes from ecosystem modeling. Few, if any, ecosystem-level models incorporate the biomass or diversity of microbial populations to drive C and nutrient flows (e.g. Rastetter et al. 1992). Even process-oriented models of decomposition and organic matter dynamics do not include microbial community structure (e.g. Grant et al. 1993), yet, they generally predict carbon, nutrient, and microbial biomass dynamics reasonably well. This poses a challenge to microbial ecologists: Does microbial community structure have *any* system-level consequences in terrestrial ecosystems?

That ecosystem models "work" suggests that large-scale biogeochemical cycling is insensitive to microbial community structure. The next step in determining whether microbial community structure can affect ecosystem level dynamics therefore should be to shift the perspective to the next smaller scale and to examine the more specific processing of substrates during carbon and nutrient turnover in soil. Several studies provide information on this issue.

Sugai and Schimel (1993) tested the hypothesis that as the inputs of carbonhydrates and phenolics in litter varied through succession in the taiga, adaptation of the microbial community to a changing substrate would cause changes in its ability to use these compounds. The primary polymeric material of plant and soil organic matter must be broken down to sugars and simple phenolics by extracellular enzymes before they can be used by microbes, so we therefore tested the ability of different communities to use sugar and phenolic monomers. We collected soils from six sites dominated by tree species ranging from alder to white spruce and assayed their ability to metabolize ^{14}C-labeled model compounds. We measured the conversion of glucose, p-hydroxybenzoic acid, and salicylic acid (o-hydroxybenzoic acid) into CO_2 and microbial biomass over a 48-h period. While each compound was metabolized quite differently from the others, the differences with successional state or season were minimal (Fig. 2). Even the time courses over the 48-h assays were remarkably similar. While the microbial communities presumably varied between the sites, the metabolic pathways and even C-use efficiencies for each substrate were consistent across all of them. At the functional level examined by this experiment, the communities all behaved the same, suggesting limited importance of microbial diversity in controlling metabolism of simple C-compounds. This is not to say that a different process, such as lignin breakdown, would not exhibit functional differences related to microbial population structure.

A second study examined microbial uptake kinetics across a range of tundra soils. Using techniques similar to those of Schimel and Firestone (1989), I measured short-term (6-h) $^{15}NH_4^+$ uptake in soil slurries with different NH_4^+ to determine the microbial N-uptake kinetics. There was a strong relationship between %N in the soil material and the maximum NH_4^+ uptake rate (Fig. 3). Tussock and intertussock soils showed very similar N-contents and N-assimilation kinetics despite very different chemical and physical composition. Tussock soils consist of dead *Eriophorum vaginatum* roots in a densely tangled mass, while intertussock soil is primarily composed of decomposed mosses. Intertussock soils are wetter and thaw much later in the season than tussocks. Thus, it is likely that the microbial communities would be substantially different, yet their N-assimilation kinetics were remarkably similar.

These studies suggest that for broad-scale processes and for broad physiologies, there is little apparent influence of microbial community structure. The next level to consider is therefore "narrower" processes or physiologies, those carried

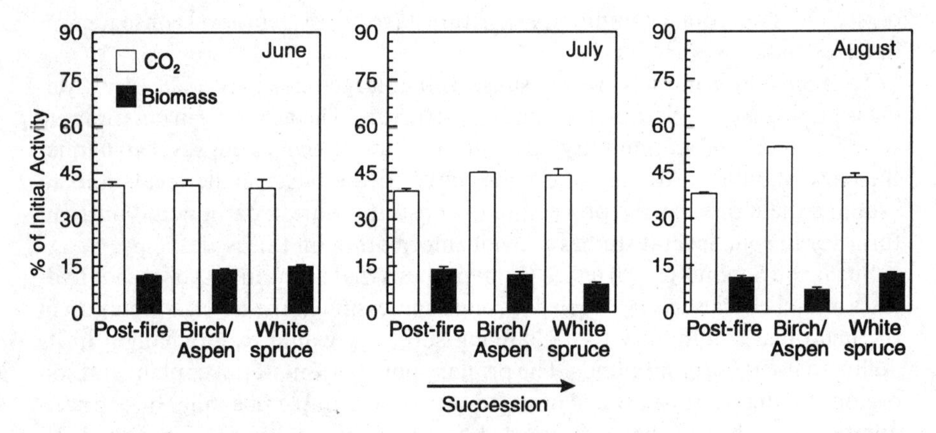

Fig. 2. Metabolism of p-hydroxybenzoic acid in soils from a post-fire secondary succession in the Alaskan taiga. Data represent the % of initial ^{14}C activity found in either CO_2 or K_2SO_4 extractable biomass. (Sugai and Schimel 1993 with permission of Pergamon Press)

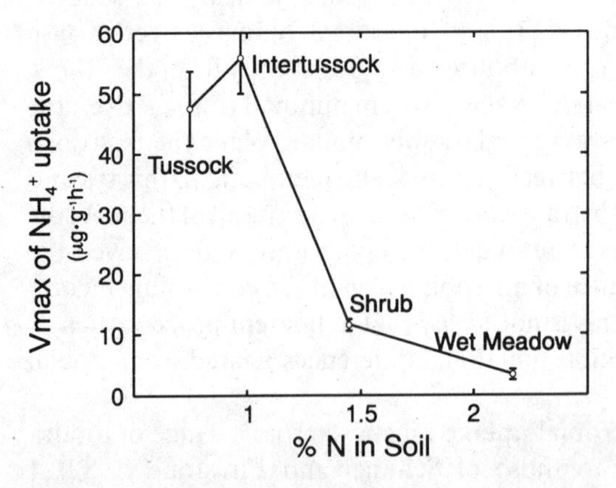

Fig. 3. V_{max} of NH_4^+ uptake by tundra soil in slurry assays

out by restricted groups of microorganisms. I will consider three groups of processes that have broad ecosystem importance, yet may be considered "narrow" and should therefore test whether at this level microbial population structure may have ecosystem-level significance. These are plant-microbe interactions, litter decomposition, and trace gas production.

17.3 "Narrow" Processes

17.3.1 Plant-Microbe Interactions

Plant-microbe interactions occur at different levels ranging from nutrient competition to highly specific symbiotic and pathogenic relationships, which may affect ecosystem structure and function at very large scales.

Mycorrhizae have diverse effects on plants and these effects are sensitive to the specific fungi that form the mycorrhizae (Read 1993). Many of these effects are reviewed in depth in Harley and Smith (1983) and Allen (1992) and so will only be mentioned briefly here. Direct effects that are sensitive to the fungiforming mycorrhizae include: inorganic N uptake (Finlay et al. 1992), organic N use (Abuzinadah and Read 1986), P metabolism (Ho and Zak 1979), pathogen resistance (Azcón-Aguilar and Barea 1992), and drought tolerance (Read 1992). Indirect effects include: the rate of soil organic matter decomposition (Dighton et al. 1987; Haselwandter et al. 1990), altered rhizosphere populations (Azcón Aguilar and Barea 1992), altered soil fauna populations (Fitter and Saunders 1992), and altered mammal diets–many animals eat mycorrhizal fruit bodies (Trappe and Maser 1977). These effects may not only be sensitive to the species of mycorrhizal fungus present but even in some cases to the specific fungal strain (Ho 1987). The fungi infecting the roots of ectomycorrhizal plants also change with both age of the plant and successional time (Dighton and Mason 1985), thus changing the nature of mycorrhizal communities and their effects on system behavior.

Not surprisingly, mycorrhizae affect the dynamics of plant communities. Ectomycorrhizae can enhance the reestablishment of forests after disturbance (Perry et al. 1989). Mycorrhizae also affect plant competition after colonization, though these effects are complex, sometimes favoring one species, while other times alleviating negative interactions between species (Miller and Allen 1992). For example, VA mycorrhizal infection enhances the competitive ability of late-successional grasses over nonmycorrhizal early invaders (Allen and Allen 1984; Miller and Allen 1992). Ectomycorrhizal infection of Douglas fir and Ponderosa pine seedlings, however, removed the mutual growth inhibition between the species that occurred with uninfected plants (Perry et al. 1989); this effect was dependent on the specific infecting fungi.

Many ectomycorrhizae form extensive mycelial mats, which may cover square meters with a single mycelium. Mycelial mats substantially alter the rates of decomposition and nutrient cycling within them. Studies by Griffiths et al. (1990) and Entry et al. (1991) have shown that N and P loss from decomposing litter is greater in mat than nonmat soil and that nitrate concentrations are significantly lower in the mats as well (Fig. 4). This may result partially from increased nutrient uptake and transfer into the plants but it may also result from specific changes in decomposition as suggested previously (Gadgil and Gadgil 1975). Some saprophytic fungi also produce mycelial mats. The presence or absence of a single species of fungus can therefore significantly alter important nutrient cycling processes.

N fixers probably show less dramatic but similar effects on community dynamics. Different strains of N fixers vary considerably in their abilities to infect roots and to fix N; some strains are highly infectious but are poor N fixers, while other strains are effective N fixers but less efficient at infection. This is true in both the legume-rhizobium (Alexander 1985) and the actinorhizal-Frankia system (Benson and Silvester 1993). The establishment and survival of legumes and actinorhizal plants depend, at least in part, on the availability of appropriate symbionts and on the effective functioning of those symbionts. Thus, establishment of plant communities in primary succession, where N fixing plants are often critical, may be sensitive to the dynamics of N fixer populations (Halvorson et al. 1991).

Microbial pathogens can also spread over wide ranges and dramatically alter ecosystem dynamics. For example, the potato crop of Ireland was nearly wiped out by potato blight during the 19th century and the forests of the eastern USA are still recovering from the introduction of chestnut blight (Stephenson 1986). Under stable, natural conditions without long-range transport of pathogens, plants often coevolve with their pathogens to limit the damage done by them (Burdon 1993). Human transport of novel pathogens into new territory, however, can cause large-scale damage to plant populations sensitive to the new organisms. Additionally, factors such as changing climate, pollution, and UV radiation

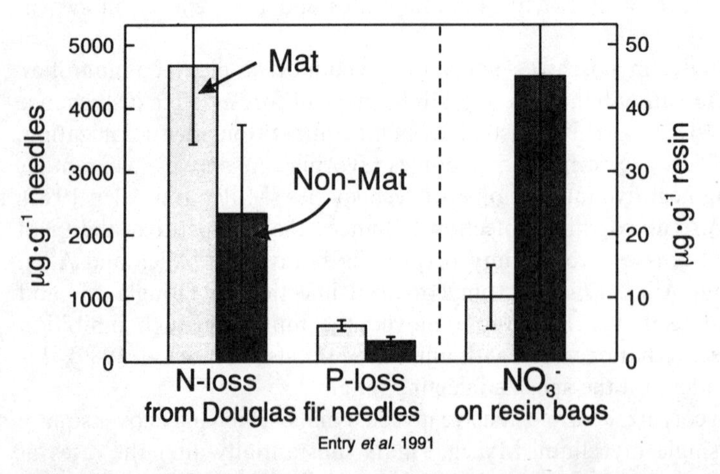

Fig. 4. Effect of *Hysterangium* mats on nutrient loss from decomposing Douglas fir needles and on NO_3^- accumulation on resin bags. Data from Entry et al. (1991)

may stress plant populations sufficiently to allow minor endemic pathogens to become major problems (Nihlgård 1985). Human factors may therefore enhance the importance of microbial pathogens as controls on ecosystem structure at a large scale.

The ecosystem-level impact of specific plant-microbe interactions, both mutualistic and pathogenic, appears to be greatest under conditions of change: plant migration, organism introduction, and environmental stress changing the susceptibility of hosts. Tree-line movement into tundra systems may be the process most sensitive to specific plant-microbe interactions. Tree establishment may rely on the availability of appropriate mycobionts, yet such fungi are unlikely to be found in tundra soils. As many ectomycorrhizal fungi fruit underground, their migration may also be slow and at least partially dependent on the activity of small mammals (Molina et al. 1992). Shifting the balance of woody shrubs in tundra may also be influenced by mycorrhizal interactions, but most of the common tundra shrubs are widely distributed, and so, probably, are their mycobionts. The introduction of novel plant pathogens, I suspect is unlikely in the Arctic. The dominant arctic flora is circumpolar (see chapters by Walker and others, this Vol.) and many species have migrated extensively in post-glacial time (see Murray, this Vol.). Therefore, it seems unlikely a plant would encounter a pathogen that it had not been previously exposed to.

17.3.2 Litter Decomposition

While litter decomposition may not seem to be a "narrow" physiology, litter is relatively chemically and physically homogeneous, composed largely of cellulose, hemicellulose, and lignin in a well-defined physical structure. Leaf litter also falls onto the soil surface where fluctuations of moisture and temperature are the most extreme, and episodic stress may limit microbial populations to those capable of surviving harsh conditions.

The labile components of litter (protein, lipids, etc.) are probably broken down by a wide range of different organisms and processes. The structural materials (cellulose and lignin), however, are broken down by exoenzymes, endo- and exocellulase for cellulose (Ljungdahl and Eriksson 1985) and a small variety of peroxidases for lignin (Gold and Alic 1993). Each of these groups of enzymes is produced by a relatively narrow group of microorganisms, which can therefore act as keystone species in litter decomposition. Cellulase is produced by both bacteria and fungi; in forest systems it is generally assumed that fungi are the more important group (Ljungdahl and Eriksson 1985), but in wet tundra bacteria may be important as well. Lignin is decomposed largely by white rot fungi, though some actinomycetes also have lignolytic activity (Gold and Alic 1993). Because only limited groups of microbes are responsible for litter decomposition, it is possible for large-scale effects to result from differences between litter communities. Several studies illustrate such effects.

First, it has commonly been reported that white rot fungi only produce substantial amounts of lignin-degrading enzymes under N-limiting conditions

(Gold and Alic 1993), suggesting important linkages between wood decomposition and nutrient availability. However, this is based largely on lab studies using *Phanaerochaete chrysosporium*. Other fungi, such as *Dichomitus squalens* and *Bjerkandera* sp., show lignolytic activity at high N concentrations (Périé and Gold 1991; Kaal et al. 1993). Thus, the response of lignin degradation to N availability in the field may depend on the white rot fungi present.

Hunt et al. (1988) examined decomposition of grass and pine litters in prairie, meadow, and forest sites. Over 1 year in the prairie and meadow, grass decomposed rapidly while pine litter showed essentially no mass loss; in the forest site, however, pine litter decomposed as rapidly as the grasses. Since the different decomposition rates of grass in the two sites should have accounted for microclimate effects, the relatively greater decomposition of pine litter in the forest resulted from the presence of a soil community adapted to processing pine litter. The microbial biomass in the pine site was dominated by fungi; the ratio of fungal to bacterial biomass was 8, while in the meadow the ratio was 0.1 (Ingham et al. 1989). Additionally, the fungi were different in the sites (2.5-μm-diameter hyphae in the meadow vs 4 μm in the pine site) and the bacterial communities appear different as well (J. C. Moore, pers. comm.). McClaugherty et al. (1985) found similar results in a litter transplant study: forest litters decomposed better in their native sites than could be predicted by assuming no interaction between litter and system, indicating microbial adaptation to the native litter. These studies illustrate that the specific composition of the microflora can affect litter decomposition in different sites.

The last study is a litter bag experiment at the Bonanza Creek Long Term Ecological Research Site near Fairbanks, Alaska. We collected litter bags monthly over 2 years and subsampled each bag to measure respiration potential (short-term respiration rate after adjusting samples to 15 °C and 50% of water holding capacity). This assay integrates substrate quality and the ability of the microbial community to use the available substrate. The respiration potential dropped from the first year to the second (Fig. 5), reflecting changes in substrate quality and the resulting changes in microbial activity. The response to moisture, however, changed as well. During the first year, there was a strong positive correlation between respiration potential and litter moisture at the time of sampling (before moisture adjustment), but in the second year the correlation became negative. These data therefore indicate that the litter microbial community changed during decomposition, and its response to moisture also changed. Thus, the response of litter decomposition to moisture is sensitive to the composition of the microbial community.

17.3.3 Trace Gases

A third major group of processes that may be sensitive to microbial community structure is the production and consumption of trace gases, particularly CH_4, N_2O, and NO. In tundra, CH_4 is of greatest concern but globally N_2O and NO are also important. Metabolism of these gases is generally carried out by very

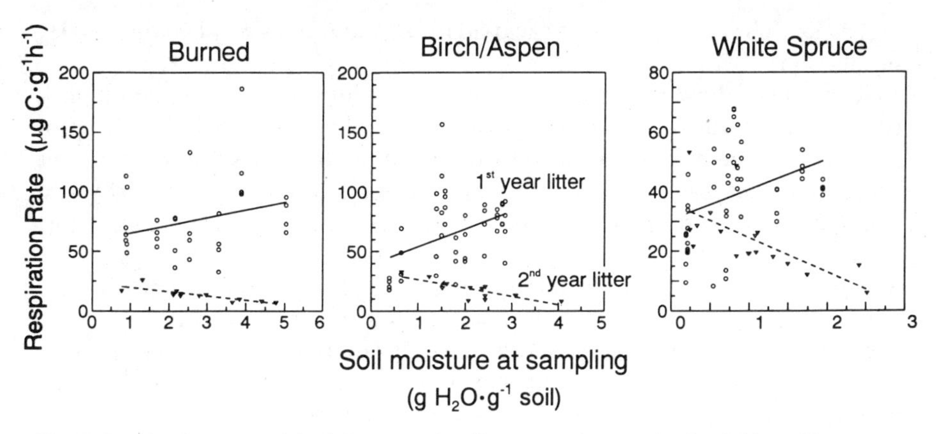

Fig. 5. Respiration potential of decomposing litter over 2 years in the field. o, First- year litter; ▼ Second-year litter

taxonomically restricted groups of bacteria. Methane is produced exclusively by a group of archaebacteria (Jones 1991) while it is oxidized primarily by methanotrophs, a small, distinctive group of bacteria (Lidstrom 1992). Nitrous and nitric oxides can be produced via both nitrification and denitrification (Firestone and Davidson 1989). While denitrification is carried out by a taxonomically diverse group of organisms (Tiedje 1988), it is physiologically narrow with only one or two basic enzyme systems for each step in the pathway (Zumft 1992). Autotrophic NH_4^+ oxidizers, on the other hand, are both taxonomically and physiologically narrow (Bock et al. 1992). We might therefore expect that the specific dynamics of most of these gases should be sensitive to the dynamics of the responsible populations and that these populations may be sensitive to environmental factors.

Even so, it is hard to find examples where the structure of the active community has effects on gas production dynamics, mostly because few appropriate studies have been done. An exception to this is the differential control of N_2O and NO production by nitrifiers and denitrifiers. The active group of organisms in the field strongly affects the rate and controls on gas production (Firestone and Davidson 1989; Schuster and Conrad 1992). Nitrifiers are aerobic autotrophs (Bock et al. 1992), while essentially all soil denitrifiers are heterotrophs that denitrify anaerobically (Zumft 1992). Since the product of nitrification is the substrate for denitrification, however, almost all production of these gases is controlled by nitrification. Thus, changes in the ammonium oxidizer community could affect NO and N_2O production from both nitrification and denitrification.

Few studies have allowed determination of any noticeable effects of microbial community structure on methane dynamics. One study on CH_4 production in peat soils, however, showed that the pH sensitivity of CH_4 production differed among communities (Valentine et al. 1994). In the neutral peats of most of their sites, methanogenesis was acid-sensitive, but in the acid, nutrient-poor "Black

Hole" sites, CH_4 production was acid-adapted, and increasing pH reduced CH_4 production.

So, from the above discussion of plant-microbe interactions, litter decomposition, and trace gas dynamics, I suggest that for physiologically "narrow" processes carried out by a restricted group of organisms, there exists the potential for substantial ecosystem-level effects from differences in microbial diversity and community structure. Even for "broad" processes, however, if stress of some sort reduces the diversity of the active population, it may become effectively "narrow". The characteristics of the organisms that survive the stress may therefore have noticeable effects at a larger level than they would otherwise (Salonius 1981). Further, the nature of the stress may be unrelated to the process of interest and so the effects of stress on a process may be unpredictable; for example, it would be hard to predict the effects of freezing on cellulose degraders.

17.4 Stress

Stress, both chemical and physical, can reduce both microbial biomass and community diversity (Atlas 1984; Domsch 1984). Chemical stress may result naturally from changing inputs of specific plant compounds to the soil or from human pollution. While many plant compounds are microbial substrates, others such as phenolics and terpenoids, may be toxic (Rice 1984; White 1986). The effects of different phenolic acids on soil populations vary with compound, concentration, type of organism, and soil horizon, often stimulating but sometimes inhibiting microbial activity (Blum and Shafer 1988). Of specific relevance to the Arctic, many lichens produce complex phenolics that are strongly antimicrobial (Lawrey 1989). Anthropogenic chemical stresses that have been shown to reduce biomass or alter communities include acidification (Visser and Parkinson 1989), heavy metals (Domsch 1984; Brookes et al. 1986), and organic compounds such as pesticides, oil, or industrial chemicals (Domsch 1984; Leahy and Colwell 1990). Chemical stress can have complex effects on microbial communities. For example, additions of jet fuel to soil increased total populations of bacteria, fungi, and hydrocarbon degraders, but reduced total microbial activity, suggesting that breakdown of native soil organic matter was inhibited (Song and Bartha 1990).

Physical stresses may be more common than chemical stresses and include aggregate disruption by tillage, roots, or microfauna, freezing, drought, and rapid variation in these factors such as drying/rewetting and freeze/thaw events. Drought is a major stress on microbial populations and has different effects on different groups of bacteria and fungi (Harris 1981). The stresses from drought and reduced water potential include direct water stress on the organisms (Harris 1981), reduced substrate diffusion in dry soils (Skopp et al. 1990), and increased microbial demand for C and N (Harris 1981; Schimel et al. 1989).

Despite the range of stress imposed by drought, rewetting may be worse. Drying and rewetting can kill a large part of the microbial biomass and substantially alter the composition of the surviving community (Kieft et al. 1987;

Van Gestel et al. 1991). While drying and rewetting soils and litters often increase the rate of carbon mineralization (Taylor and Parkinson 1988; Van Gestel et al. 1993), this is not always the case. Clein and Schimel (1994) showed that birch litter decomposition was reduced in a lab experiment by up to 25% over 2 months by even a single 1-day drying and rewetting (Fig. 6). It was suggested that this effect may have been due to the reduction in some key group of enzymes or organisms. In the field it is likely that these organisms would have been able to recolonize from the forest floor below, but their ability to do so should therefore control the recovery of decomposition activity. Other physical stresses may be less damaging than drying and rewetting. For example, freezing and thawing also kill off microorganisms, but the effects seem to be generally less than those of drying and rewetting (Skogland et al. 1988; DeLuca et al. 1992).

17.5 Implications for Tundra

Several properties of tundra systems may be particularly sensitive to microbial community structure: plant nitrogen uptake, vegetation response to changing climate, and the effects of arctic pollution. Nutrient supply to plants may be more sensitive to microbial community structure in tundra than in most other systems. Tundra plants often appear to acquire nutrients released by freeze/thaw events or other stresses that kill microbes (Gersper et al. 1980; Schimel et al. 1994). Thus, the specific ability of organisms to survive the stress and the ability of the survivors to grow and compete for the nutrients released may affect nutrient availability to plants. Second, amino acids appear to be an important source of N for many tundra plants (Chapin et al. 1993). Amino acids are excellent microbial substrates as well as breakdown products of polymer metabolism. The release and uptake of different amino acids involve a complex suite of processes,

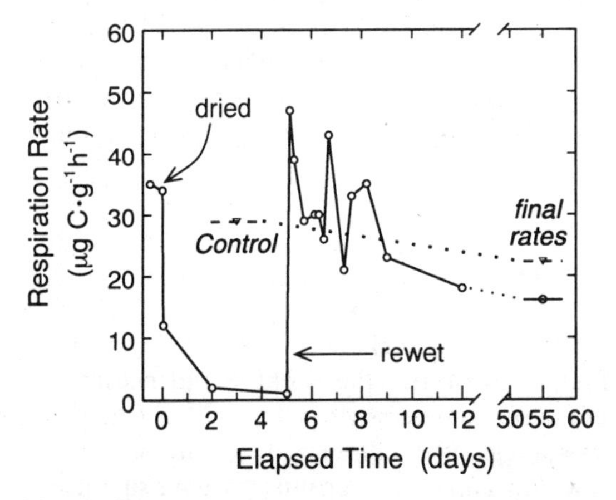

Fig. 6. Effects of drying and rewetting on microbial activity in birch litter

intimately linked to fundamental microbial biochemistry; it therefore may be sensitive to the specific structure of the microbial community. The second area that may be sensitive to the composition of the microbial community is the invasion of shrubs and trees into tundra. As previously discussed, mycorrhizal and rhizosphere populations may control plant community development. Finally, increased pollution in the Arctic could reduce critical microbial populations and thereby alter the dynamics of ecological processes in significant, though possibly unpredictable ways.

17.6 Overall Considerations

It appears that for "narrow" processes or where diversity is reduced by stress, the specific composition of the microbial community may have large-scale effects on ecological processes. This suggests that even for broad processes, the structure of the community may influence process dynamics, but without being able to link a process to different populations under different conditions, we are unable to demonstrate this. For these "broad" processes that involve many organisms, it is easy to see the diversity as redundancy, but each organism may only carry out the process under a specific set of conditions, rather than being truly redundant. Thus, in some cases, redundancy may be more apparent than real. Our inclination to see microbial diversity as redundancy may result in part from our very broad definitions of many microbial functional groups, e.g. denitrifiers. If we used equally broad categories for plants, we might only be able to divide plants into trees, shrubs, forbs, graminoids, and bryophytes.

There are ways in which microbial community structure and diversity have impacts at the ecosystem level, yet we know very little about many of those impacts. Because of this, I have included much speculation in this chapter. Only recently have techniques become available for studying microbial community ecology and linking community studies with process studies. I therefore expect that as such studies develop, we will see progressively more of the variation and surprises in microbial processes explained as differences in microbial communities from site to site and system to system. The Arctic may provide a valuable testing ground for some of these ideas because of its sensitivity to belowground processes and environmental perturbations.

17.7 Conclusions

1. For "broad" processes and physiologies, such as overall C flows in an ecosystem and metabolism of simple substrates, there is little evidence for any ecosystem-scale significance of microbial diversity or community structure.
2. For a range of "narrow" physiologies, the presence of specific groups of microorganisms or differences within a functional group may have significant impacts on process dynamics. Specific plant-microbe interactions are highly

sensitive to the presence of specific microbial species (e.g. pathogens and mycorrhizae), but processes such as litter decomposition and trace gas dynamics are also sensitive to changes in microbial communities.
3. Chemical and physical stresses (pollution, drying-rewetting, etc.) can reduce microbial numbers and diversity to such an extent that the size of specific populations, rather than substrate availability, may become the limiting factor in a range of processes.

Acknowledgments. I am grateful to Mike Beare, Jay Gulledge, Joy Clein and Jon Lindstrom for valuable discussions and contributions to my thinking for this paper. The work described was supported by grants from the National Science Foundation Long Term Ecological Research program, the Dept. Energy R4D program, and the Athens Environmental Research Laborartory of the US Environmental Protection Agency.

References

Abuzinadah RA, Read DJ (1986) The role of proteins in the nitrogen nutrition of ectomycorrhizal plants. I. Utilization of peptides and proteins by ectomycorrhizal fungi. New Phytol 103: 481–493

Alexander M (1985) Ecological constraints on nitrogen fixation in agricultural ecosystems. Adv Microb Ecol 8: 163–183

Allen MF (1992) Mycorrhizal functioning. Chapman and Hall, New York

Allen EB, Allen MF (1984) Competition between plants of different successional stages: mycorrhizae as regulators. Can J Bot 62: 2625–2629

Atlas RM (1984) Use of microbial diversity measurements to assess environmental stress. In: Klug MJ, Reddy CA (eds) Current perspectives in microbial ecology. American Society for Microbiology, Washington DC, pp 540–545

Azcón-Aguilar C, Barea JM (1992) Interactions between mycorrhizal fungi and other rhizosphere microorganisms. In: Allen MF (ed) Mycorrhizal functioning. Chapman and Hall, New York, pp 163–198

Beare MH, Coleman DC, Crossley DA Jr, Hendrix PF, Odum EP (1994) A hierarchical approach to evaluating the significance of soil biodiversity to biogeochemical cycling. Plant and Soil (in press)

Benson DR, Silvester WB (1993) Biology of Frankia strains, actinomycete symbionts of actinorhizal plants. Microbiol Rev 57: 293–319

Blum U, Shafer SR (1988) Microbial populations and phenolic acids in soil. Soil Biol Biochem 20: 793–800

Bock E, Koops H-P, Ahlers B, Harms H (1992) Oxidation of inorganic nitrogen compounds as energy source. In: Ballows A, Trüper HG, Dworkin M, Harder W, Schleifer K-H (eds) The prokaryotes 2nd edn. Springer, Berlin, Heidelberg New York, pp 414–430

Brookes PC, Heijnen CE, Vance ED (1986) Soil microbial biomass estimates in soils contaminated with metals. Soil Biol Biochem 18: 383–388

Burdon JJ (1993) The role of parasites in plant populations and communities. In: Schultze E-D, Mooney HA (eds) Biodiversity and ecosystem function; Ecological Studies 99, Springer, Berlin, Heidelberg New York, pp 165–179

Chapin FS III, Moilanen L, Kielland K (1993) Preferential use of organic nitrogen for growth by a non-mycorrhizal arctic sedge. Nature 361: 150–153

Clein JS, Schimel JP (1994) Reduction in microbial activity in birch litter due to drying and rewetting events. Soil Biol Biochem (in press) 26: 403–406

DeLuca TH, Keeney DR, McCarty GW (1992) Effect of freeze-thaw events on mineralization of soil nitrogen. Biol Fertil Soils 14: 116–120

Dighton J, Mason PA (1985) Mycorrhizal dynamics during forest tree development. In: Moore D, Casselton LA, Wood DA, Frankland JC (eds) Developmentalbiology of higher fungi. Cambridge University Press, Cambridge, pp 117–139

Dighton J, Thomas ED, Latter PM (1987) Interactions between tree roots, mycorrhizas, a saprotrophic fungus and the decomposition of organic substrates in a microcosm. Biol Fertil Soils 4: 145–150

Domsch KH (1984) Effects of pesticides and heavy metals on biological processes in soil. Plant Soil 76: 367–378

Entry JA, Rose CL, Cromack K (1991) Litter decomposition and nutrient release in ectomycorrhizal mat soils of a Douglas fir ecosystem. Soil Biol Biochem 23: 285–290

Finlay RD, Frostergård Å, Sonnerfeldt A-M (1992) Utilization of organic and inorganic nitrogen sources by ectomycorrhizal fungi in pure culture and in symbiosis with *Pinus contorta* Dougl. ex Loud. New Phytol 120: 105–115

Firestone MK, Davidson EA (1989) Microbiological basis of NO and N_2O production and consumption in soil. In: Andreae MO, Schimel DS (eds) Exchange of trace gases between terrestrial ecosystems and the atmosphere. Wiley, New York, pp 7–21

Fitter AH, Saunders IR (1992) Interactions with the soil fauna. In: Allen MF (ed) Mycorrhizal functioning. Chapman and Hall, New York, pp 333–354

Gadgil RL, Gadgil PD (1975) Suppression of litter decomposition by mycorrhizal roots of *Pinus radiata*. NZJ For Sci 5: 33–41

Gersper PL, Alexander V, Barkley SA, Barsdate RJ, Flint PS (1980) The soils and their nutrients. In: Brown J, Miller PC, Tieszen LL, Bunnell FL (eds) An Arctic ecosystem. Dowden, Hutchinson & Ross, Stroudsburg, pp 219–254

Gold MH, Alic M (1993) Molecular biology of the lignin-degrading basidiomycete *Phanerochaete chrysosporum*. Microbiol Rev 57: 605–622

Grant RF, Juma NG, McGill WB (1993) Simulation of carbon and nitrogen transformations in soil: mineralization. Soil Biol Biochem 25: 1317–1329

Griffiths RP, Caldwell BA, Cromack K, Morita RY (1990) Douglas-fir forest soils colonized by ectomycorrhizal mats. I. Seasonal variation in nitrogen chemistry and nitrogen cycle transformation rates. Can J For Res 20: 211–218

Halvorson JJ, Smith JL, Franz EH (1991) Lupine influence on soil carbon, nitrogen and microbial activity in developing ecosystems at Mount St. Helens. Oecologia 87: 162–170

Harley JL, Smith SE (1983) Mycorrhizal symbiosis. Academic Press, London

Harris RF (1981) Effect of water potential on microbial growth and activity. In: Parr JF, Gardner WR, Elliott LF (eds) Water potential relations in soil microbiology. American Society of Agronomy, Madison, pp 23–95

Haselwandter K, Bobleter O, Read DJ (1990) Degradation of ^{14}C-labeled lignin and dehydropolymer of coniferyl alcohol by ericoid and ectomycorrhizal fungi. Arch Microbiol 153: 352–354

Ho I (1987) Comparison of eight *Pisolithus tinctorius* isolates for growth rate, enzyme activity, and phytohormone production. Can J For Res 17: 31–35

Ho I, Zak B (1979) Acid phosphatase activity of six ectomycorrhizal fungi. Can J Bot 57: 1203–1205

Hunt HW, Ingham ER, Coleman DC, Elliott ET, Reid CPP (1988) Nitrogen limitation of production and decomposition in prairie, mountain meadow, and pine forest. Ecology 69: 1009–1016

Ingham ER, Coleman DC, Moore JC (1989) An analysis of food-web structure and function in a shortgrass prairie, a mountain meadow, and a lodgepole pine forest. Biol Fertil Soils 8: 29–37

Jones WJ (1991) Diversity and physiology of methanogens. In: Rogers JE, Whitman WB (eds) Microbial production and consumption of greenhouse gases: methane, nitrogen oxides, and halomethanes. American Society for Microbiology, Washington, DC, pp 39–55

Kaal EEJ, de Jong E, Field JA (1993) Stimulation of lignolytic peroxidase activity by nitrogen nutrients in the white rot fungus *Bjerkandera* sp. strain BOS55. Appl Environ Microbiol 59: 4031–4036

Kieft TL, Soroker E, Firestone MK (1987) Microbial biomass response to a rapid increase in water potential when dry soil is wetted. Soil Biol Biochem 19: 119–126

Lawrey JD (1989) Lichen secondary compounds: evidence for a correspondence between antiherbivore and antimicrobial function. Bryologist 92: 326–328

Leahy JG, Colwell RR (1990) Microbial degradation of hydrocarbons in the environment. Microbiol Rev 54: 305–315

Lidstrom ME (1992) The aerobic methylotrophic bacteria. In: Balows A, Trüper HG, Dworkin M, Harder W, Schleifer K-H (eds) The prokaryotes, 2nd edn Springer, Berlin Heidelberg, New York

Ljungdahl LG, Eriksson K-E (1985) Ecology of microbial cellulose degradation. Adv Microb Ecol 8: 237–299

McClaugherty CA, Pastor J, Aber JD, Melillo JM (1985) Forest litter decomposition in relation to soil nitrogen dynamics and litter quality. Ecology 66: 266–275

Meyer O (1993) Functional groups of microorganisms. In: Schultze E-D, Mooney HA (eds) Biodiversity and ecosystem function. Ecological Studies 99. Springer, Berlin Heidelberg, New York, pp 67–96

Miller SL, Allen EB (1992) Mycorrhizae, nutrient translocation, and interactions between plants. In: Allen MF (ed) Mycorrhizal functioning. Chapman and Hall, New York, pp 301–332

Molina R, Massicotte H, Trappe JM (1992) Specificity phenomena in mycorrhizal symbioses: community ecological consequences and practical implications. In: Allen MF (ed) Mycorrhizal functioning. Chapman and Hall, New York, pp 357–423

Nihlgärd B (1985) The ammonium hypothesis—an additional explanation to the forest dieback in Europe. Ambio 14: 2–8

Périé FI, Gold MH (1991) Manganese regulation of manganese peroxidase expression and lignin degradation by the white rot fungus *Dichomitus squalens*. Appl Environ Microbiol 57: 2240–2245

Perry DA, Margolis H, Choquene C, Molina R, Trappe JM (1989) Ectomycorrhizal mediation of competition between coniferous tree species. New Phytol 112: 501–511

Rastetter EB, McKane RB, Shaver GR, Melillo JM (1992) Changes in C storage by terrestrial ecosystems: how C-N interactions restrict responses to CO_2 and temperature. Water Air Soil Pollot 64: 327–344

Read DJ (1992) The mycorrhizal mycelium. In: Allen MF (ed) Mycorrhiza functioning. Chapman and Hall, New York, pp 102–133

Read DJ (1993) Plant-microbe mutualisms and community structure. In: Schultze E-D, Mooney HA (eds) Biodiversity and ecosystem/Function, Springer, Berlin Heidelberg, New York, pp. 181–209

Rice EL (1984) Allelopathy, 2nd edn. Academic Press, New York

Salonius PO (1981) Metabolic capabilities of forest soil microbial populations with reduced species diversity. Soil Biol Biochem 13: 1–10

Schimel JP, Firestone MK (1989) Inorganic nitrogen incorporation by coniferous forest floor material. Soil Biol Biochem 21: 41–46

Schimel JP, Scott W, Killham K (1989) Changes in cytoplasmic carbon and nitrogen pools in a soil bacterium and a fungus in response to salt stress. Appl Environ Microbiol 55: 1635–1637

Schimel JP, Kielland K, Chapin FS III (1994) Nutrient availability and uptake by tundra plants. In: Reynolds JF, Tenhunen JD (eds) Landscape function implication for ecosystem response to disturbance; a case study in arctic tundra. Springer, Berlin Heidelberg, New York, (in press)

Schulze E-D, Mooney HA (1993) Biodiversity and ecosystem function, Ecological Studies 99. Springer Berlin Heidelberg, New York

Schuster M, Conrad R (1992) Metabolism of nitric oxide and nitrous oxide during nitrification and denitrification in soil at different incubation conditions. FEMS Microbiol Ecol 101: 133–143

Skogland T, Lomeland S, Goksoyr J (1988) Respiratory burst after freezing and thawing of soil: experiments with soil bacteria. Soil Biol Biochem 20: 851–866

Skopp J, Jawson MD, Doran JW (1990) Steady-state aerobic microbial activity as a function of soil water content. Soil Sci Soc Am J 54: 1619–1625

Song H-G, Bartha R (1990) Effects of jet fuel spills on the microbial community of soil. Appl Environ Microbiol 56: 646–651

Stephenson SL (1986) Changes in a former chestnut-dominated forest after a half century of succession. Am Midl Nat 116: 173–179

Sugai SF, Schimel JP (1993) Decomposition and biomass incorporation of ^{14}C-labeled glucose and phenolics in taiga forest floor: effect of substrate quality, successional state, and season. Soil Biol Biochem 25: 1379–1389

Tiedje JM (1988) Ecology of denitrification and dissimilatory nitrate reduction to ammonium. In: Zehnder JB (ed) Biology of anaerobic microorganisms. John Wiley, New York, pp 179–244

Trappe JM, Maser C (1977) Ectomycorrhizal fungi: interactions of mushrooms and truffles with beasts and trees. In: Walters T (ed) Mushrooms and man, an interdisciplinary approach to mycology. Linn-Benton Community College, Albany, OR, pp 165–179

Valentine DW, Holland EA, Schimel DS (1994), Ecological controls over methane and carbon dioxide fluxes along a successional gradient. J Geophys Res 99: 1563–1571

Van Gestel M, Ladd JN, Amato M (1991) Carbon and nitrogen mineralization from two soils of contrasting texture and microaggregate stability: influence of sequential fumi- gation, drying and storage. Soil Biol Biochem 23: 313–322

Van Veen JA, Ladd JN, Frissel MJ (1984) Modeling C and N turnover through the micro- bial biomass in soil. Plant Soil 76: 257–274

Visser S, Parkinson D (1989) Microbial respiration and biomass in soil of a lodgepole pine stand acidified with elemental sulfur. Can J For Res 19: 955–961

White CS (1986) Volatile and water-soluble inhibitors of nitrogen mineralization and nitrification in a ponderosa pine ecosystem. Biol Fertil Soils 2: 97–104

Zumft WG (1992) The denitrifying prokaryotes. In: Balows A, Trüper HG, Dworkin M, Harder W, Schleifer K-H (eds) The prokaryotes. Springer, Berlin Heidelberg, New York, pp 554–582

18 Diversity of Biomass and Nitrogen Distribution Among Plant Species in Arctic and Alpine Tundra Ecosystems

J. PASTOR

18.1 Introduction

The decreased number of species encountered as one travels poleward from the equator or upward towards a mountain summit is well known and has led to the generalization that tundra communities are not diverse. Yet, the number of growth forms that can be packed into a square meter (Korner, this Vol.; Shaver and Chapin 1991; Chapin 1993), the differing palatibilities of these to herbivores (Wiegolaski 1975a; Garre and Skogland 1975; Jefferies et al. 1992), the wide ranging responses of their species to resources (Chapin and Shaver 1985, 1989; Shaver and Chapin 1991) as well as the wide ranging effect these species have on resource availability (Hobbie, this Vol.) suggest that functional diversity related to ecosystem properties may be quite high in tundra ecosystems.

This observation that tundra ecosystems have low species diversity at a taxonomic level but high functional diversity at an ecosystem level may at first appear paradoxical. While the total amounts of biomass or nutrients in a piece of landscape are ecosystem properties, the distribution of that biomass or nutrients among species is one form of diversity. Functional diversity increases to the extent that species do things differently with those nutrients. Furthermore, as nutrients and carbon are distributed among species, the future availabilities of those nutrients change because of species differences in allocation of nutrients and carbon in various tissues (Shaver, this Vol.; Chapin, this Vol.), the decay and nutrient release rates from those tissues when returned to the soil as litter (Hobbie, this Vol.), and the palatability of those tissues to herbivores (Bryant and Jefferies, this Vol.). Therefore, the relationship between diversity and ecosystem properties is highly interactive and dynamic.

The purpose of this paper is to synthesize some patterns of the distribution of carbon and nitrogen among species in arctic and alpine tundra communities, to review some experimental evidence for the origins and dynamics of these patterns, and to suggest elements of a process model to account for such patterns.

Natural Resources Research Institute, University of Minnesota, Duluth, MN 55811, USA

Ecological Studies, Vol. 113
Chapin/Körner (eds) Arctic and Alpine Biodiversity

18.2 Concepts of Diversity

Measurements of diversity traditionally depend on counting the number of species or the number of individuals within species in a given area or ecosystem. Species richness is the simplest measure of diversity, but richness per se would have little to do with ecosystem functioning if all species are functionally equivalent or if resources are distributed randomly among species. Quantifying the distribution of resources among species and functional groups is equally important. This requires quantifying relative dominance as well as richness. Measurements of diversity that incorporate dominance and richness include the Shannon-Weaver index and dominance-diversity curves, among others (Whittaker 1975; Magurran 1988). These measures of diversity have traditionally been employed using population data, such as number of individuals within a species. Fortunately, these same measures of diversity can be employed using the distribution of carbon and nutrients among species within communities. This provides a link between traditional measures of diversity and the ecosystem processes responsible for nutrient capital and its distribution.

Diversity is mathematically and conceptually equivalent to uncertainty (MacArthur 1965). If one is walking through the forest and hears a bird which one then identifies, one should ask oneself "Based on what I just heard, what will be the identity of the next bird I shall hear?" The fewer the species and the more dominant is one species over the others, the more likely you will hear the same species again and the lower the diversity of the bird community. Similarly, given a gram of tissue taken at random from a plant community, one can ask to which species should I assign the carbon and nitrogen in this sample, and to which species should I assign the carbon and nitrogen from the next gram of plant tissue. The species to which both this and the next gram of carbon and nitrogen are assigned has great consequences for ecosystem functioning to the extent that each species does something different with that gram of carbon and nitrogen.

Thus, to determine the relationship between diversity and ecosystem properties, we need to examine not only the number of species within a community, but also the relative distribution of carbon and nutrients among those species and their functional forms. The most common and parsimonious way to estimate the uncertainty in the distribution of nitrogen and biomass among plant species is the familiar Shannon-Weaver index:

$$H' = -\sum_{i=1}^{S} p_i(\log_e p_i), \tag{1}$$

where p_i is the proportion of total biomass or nitrogen in the ith of S species. Normally, this index is calculated using the proportion of the total number of individuals within each species; however, there are no mathematical or conceptual difficulties involved in using the proportion of total community carbon or nutrients found within each species. Therefore, we do not need to invent different measures to examine diversity in an ecosystem sense as opposed to a population or community sense.

In addition, one could plot the proportion of total community biomass or nitrogen contained by each species against its rank order compared with all other species (Whittaker 1972). These plots, known as dominance-diversity curves, can be described by several equations encapsulating hypotheses regarding processes responsible for the patterns.

One hypothesis, suggested by MacArthur (1960), states that diversity will be greatest in communities where resources are randomly divided among species. This is otherwise known as the "broken-stick" hypothesis because if one randomly breaks a line or stick (representing total biomass or nutrient) into small pieces (representing the allocation to each species), then the proportion of total nutrient or biomass, p_r, assigned to each species, r, is described as:

$$p_r = \frac{100}{S} \sum_{i=1}^{r} \frac{1}{S-i+1}, \tag{2}$$

where S is the total number of species. Alternative formulations for the broken-stick distribution have also been suggested (Cohen 1968; DeVita 1979), but MacArthur's hypothesis has not been disproven, although it is not a unique explanation for this distribution (Cohen 1968; Pielou 1981) . In any case, the broken-stick distribution is a limiting case of high diversity of resource distribution among constituent species within a community.

An alternative model representing the opposite limiting case is known as the geometric model and is derived from the resource preemption hypothesis (Whittaker 1972). This hypothesis states that the most dominant species obtains or preempts the proportion of total community biomass or nutrient pool it requires, and each succeeding species obtains the same proportion of the remainder. Thus, the resources available to any given species are determined by the amount preempted by the more dominant species; how much each species in turn preempts determines availability to less dominant species. When the proportion of total community biomass or nitrogen content is plotted logarithmically against species rank, the resulting line is straight with slope equal to the resource preemption rate, c. Mathematically, p_r is described as:

$$p_r = c\left(1 - \sum_{i=1}^{r} p_i\right) \tag{3}$$

What value should be assigned to the resource-preemption rate? This can be solved in one of two ways. First, the data can be simply fit to the model and a statistically best estimate of c can be calculated using regression. However, this estimate is a weighted average across all species that may not pertain to any one species. Alternatively, the preemption coefficient for each species could be initially hypothesized as being equal to that of the most dominant species. That is, if the most dominant species preempted 50% of total community nitrogen, then each successive species also preempts 50% of the remainder in sequence. While there may be no a priori reason for this hypothesis, it provides a useful benchmark because divergences of actual species importance from predictions indicate whether a given species is more successful at obtaining its share of remaining resources than the most dominant species (data falls above the predicted model line) or less successful (data falls below the model line). It should be noted that predictions of these two models will converge when the number of species in a community is low and when the proportion of resources obtained by the dominant species is low.

18.3 Patterns of Diversity of Carbon and Nitrogen Distribution

I shall now use these equations to synthesize some literature data on the distribution of biomass and nitrogen among different communities in arctic and alpine tundra. An advantage of the above approach is that it can synthesize a wealth of data already produced by ecosystem studies in arctic and alpine tundra, beginning with the International Biological Program (IBP). As examples, I shall compare four alpine communities at Hardangervidda, Norway, with four arctic communities at Toolik Lake, Alaska (Figs. 1, 2).

There is a surprising range of diversity of biomass and nitrogen distribution among species in both arctic and alpine tundra communities (Figs. 1, 2). The Shannon-Weaver index of diversity of biomass distribution is low in alpine lichen heath ($H' = 0.317$), mainly because lichens comprised greater than 80% of total biomass. At the other extreme, diversity was highest in alpine dry meadow ($H' = 2.151$). The range of diversity of biomass distribution was somewhat less in the arctic communities, ranging from 1.029 in the wet meadow to 1.923 in the tussock tundra. The wider range of diversity in alpine tundra could reflect the greater heterogeneity of the mountainous landscape compared to the relatively flat arctic tundra. These values are comparable to more traditional calculations of diversity using numbers of individuals reported in temperate forests (Auclair and Goff 1971) and even some tropical communities (Whittaker 1975). Therefore, not only can diversity be surprisingly high in some tundra ecosystems, the range of diversity across the landscape is also significant.

The diversity of the distribution of nitrogen among species is nearly equal to that of biomass distribution, except where lichens are the dominant species. In both arctic and alpine lichen heaths, the diversity of nitrogen distribution is much greater than that of biomass distribution. This is because while lichens comprise the bulk of total community biomass, they comprise a much lower proportion of total community nitrogen owing to their low nitrogen concentrations (approximately 0.5%). Nitrogen in this community is thus shared more equally than the distribution of biomass would indicate.

Dominance-diversity curves provide a basis for hypothesizing causes of the different patterns of diversity among communities (Figs. 1, 2). The broken-stick model predicts a more even distribution of biomass or nitrogen among species than does the geometric model, but where the number of species was low (i.e., wet meadow) or where dominance by one species was low (i.e., dry meadow), the theoretical predictions of the models did not differ greatly. However, in most cases, predictions of the geometric models correspond most closely to data (Figs. 1, 2), suggesting that preemption of resources by dominant species is a

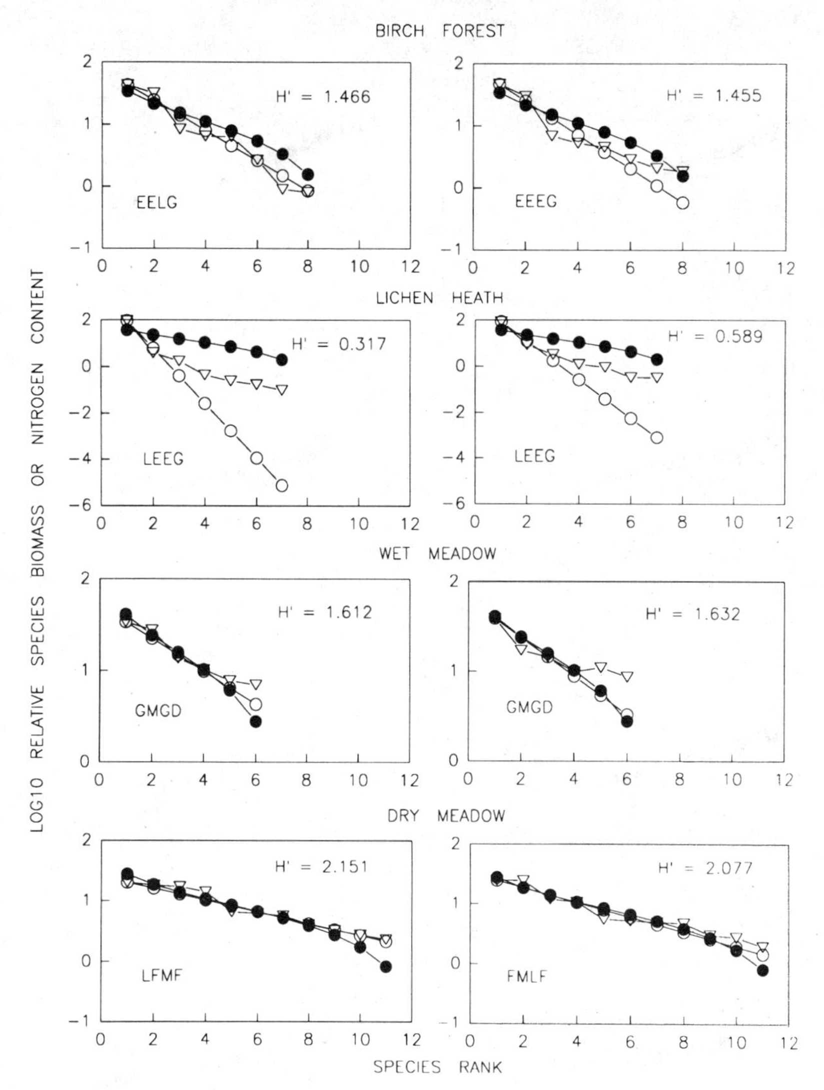

Fig. 1. Diversity of biomass and nitrogen distribution among species in alpine tree-line birch forest, lichen heath, wet meadow, and dry meadow at Hardangervidda, Norway (Wiegolaski 1975a). The distribution of biomass among constituent species was obtained from Kjelvik and Karenlampi (1975) for the birch forest and lichen heath and from Wiegolaski (1975b) for the wet and dry meadows. Nitrogen concentrations of the current growth (leaves and twigs) of the dominant plant species were obtained from Wiegolaski and Kjelvik (1973); for less dominant species, the concentrations of a congener were substituted. Three dominance-diversity curves are plotted for both biomass and nitrogen in each community: *filled circles* predictions of the broken-stick model of resource partitioning [Eq. (2)]; *open circles* predictions of a geometric model of resource preemption given the preemption rate of the dominant species in each community [Eq. (3)]; *triangles* data. The Shannon-Weaver index of biomass or nitrogen distribution [Eq. (1)] for each community is shown in the *upper right corner*. The growth forms of the four most dominant species are shown in the *lower left*: *E* evergreen; *D* deciduous; *L* lichen; *M* moss; *F* forbs; *G* graminoid (includes sedges and grasses)

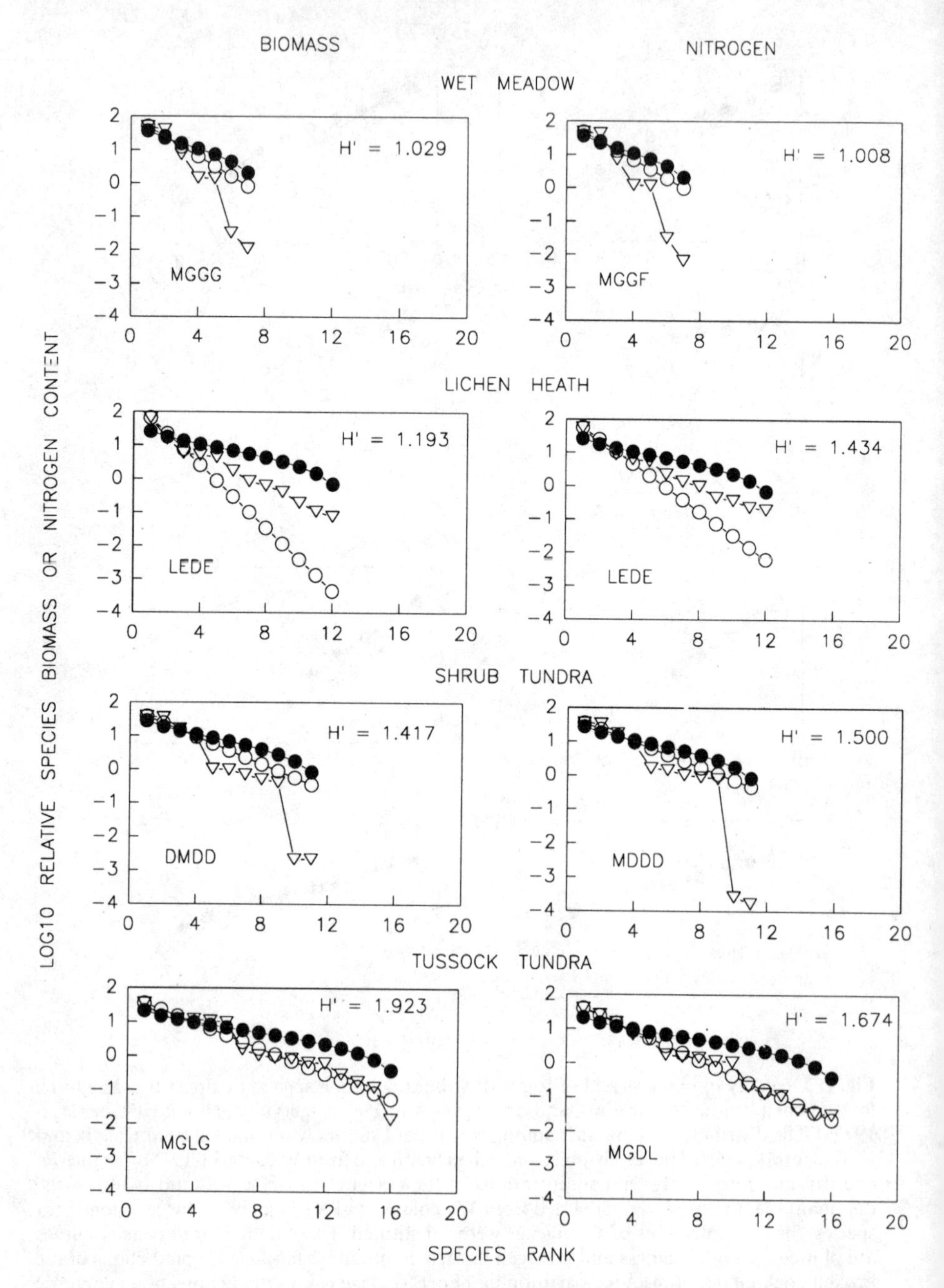

Fig. 2. Diversity of carbon and nitrogen distribution among species in wet meadow, lichen heath, shrub tundra, and tussock tundra at the Toolik Lake LTER site in the foothills north of the Brooks Range in Arctic Alaska (Shaver and Chapin 1991). Explanation of *symbols* as in Fig. 1

major process underlying diversity in tundra ecosystems. In these cases, the conformity to the geometric model is most apparent in the distribution of biomass or nitrogen among the rarer species, rather than among the dominant ones.

The divergence of the data from model predictions also generates interesting hypotheses. There were three communities in which data diverged significantly from model predictions. One such community is the lichen heath at both arctic and alpine sites. In these communities, the dominance-diversity curves for both biomass and nitrogen distribution fall between predictions of the broken-stick and geometric models. All successive species after lichens have a greater share of remaining total community biomass or nitrogen than obtained by lichens despite the high dominance of lichens (> 85% of total community biomass or nitrogen). For the nitrogen dominance-diversity curves, this reflects the low nitrogen concentrations of lichens: proportionally greater amounts of nitrogen remain "unused" by lichens per unit biomass. For the biomass curves, this pattern reflects the prostrate growth form of lichens compared with the upright growth form of grasses, forbs, ericads, and deciduous shrubs: lichens compete for space on the rock or soil surface while vascular plants compete for vertical space above (canopy structure) and below (rooting structure) the soil surface.

In contrast, the arctic wet meadow and shrub tundra show the opposite pattern because data fell below that predicted by the geometric model. Mosses dominate both biomass and nitrogen pools in these communities. The proportion of nitrogen and biomass obtained by all other species was less than that predicted by the geometric model. All successive species after mosses therefore obtain a lower share of remaining resources than obtained by the dominant mosses. In contrast to the lichens, this may reflect the relatively high nitrogen concentrations in mosses (1.7%) as well as their high biomass, thus greatly preempting this limiting resource and greatly depressing the ability of successive species to obtain it.

A pattern of particular significance for issues of biodiversity, and one that may be unique to tundra ecosystems, is that no one growth form dominates these ecosystems (Shaver and Chapin 1991). The four most dominant species, which together comprise > 90% of total community biomass or nitrogen, belong to at least two and in most cases three or even four growth forms (Figs. 1, 2). Shaver (this Vol.) and Chapin (this Vol.) present compelling evidence that arctic tundra ecosystems have high functional diversity despite low numbers of species because of this lack of dominance by one growth form and the fact that each growth form affects nutrient and carbon cycles in different ways. For example, ericads slow rates of nitrogen cycling because of their evergreen habit and low litter quality, whereas deciduous plants enhance it (Hobbie, this Vol.) Examining the diversity of nutrient distribution among growth forms using dominance-diversity curves provides another means to quantify and test such hypotheses.

To some extent, this conclusion must be predicated with the warning that mosses and lichens were not always delineated by species, and so were lumped in this analysis. This would tend to inflate the importance of these growth forms relative to others. Nonetheless, where species were delineated, only one species tended to dominate these growth forms – *Cladonia* comprised virtually all the lichen biomass at the alpine sites (Kjelvok and Karenlampi 1975) and approximately 50% of lichen biomass at the tussock tundra site (Shaver and Chapin 1991). The remaining lichen species each contributed less than 12% of remaining lichen biomass (Shaver and Chapin 1991), which is less than each of the other four dominant species in these communities. With regard to mosses, half their biomass in tussock tundra is attributable to *Dicranum* spp., with the remaining species each comprising less than 20% of the remaining moss biomass. In the shrub tundra, *Hylocomium splendens* comprised 90% of total moss biomass with the remaining species contributing very small proportions of community biomass (Shaver and Chapin 1991). Therefore, although incomplete resolution to species does not distort the pattern of diversity of growth form dominance seen across these communities, identification to species level should be a goal of all future investigations of diversity in tundra ecosystems to avoid possible artifacts.

There was no linear increase or decrease in diversity along biomass or nitrogen gradients, nor any covariation of diversity along soil moisture gradients implied by the site descriptions. In contrast, there was a suggestion that diversity was greatest in communities with intermediate biomass or nitrogen content, with both more and less productive communities having lower diversity. This parabolic relationship between total community resources and diversity, suggested theoretically by Tilman (1982), was more apparent in the arctic tundra than the alpine. In addition, the maximum diversity occurred at greater biomass and nitrogen contents in arctic tundra compared with alpine tundra.

18.4 Experimental Tests of Controls over Diversity

These differences among tundra ecosystems suggest that diversity may be controlled in part by landscape characteristics, such as topography or disturbance regime. Further observations of patterns of diversity along topographic or productivity gradients are needed to more precisely define the broad patterns seen in nature that must be explained by experiments and models.

However, diversity can also change rapidly upon altering rates of nutrient cycling and herbivory. Experimental manipulation of ecosystem processes can be used to evaluate controls over diversity (Tilman 1993). For example, in the alpine tundra at Hardangervidda, both fertilization and grazing increased dominance by one or a few species, resulting in divergence of the two model predictions and tighter conformity to the geometric model (Figs. 3, 4). The changes in the dominance-diversity curves under these experimental manipulations occurred relatively rapidly, within a few years of experimental treatment. Significantly, without grazing, the dominance-diversity curve conformed remarkably well to the broken-stick model rather than the geometric model (Fig. 3). Therefore, the absence of grazing resulted in a different curve than normally seen across the landscape (Fig. 1). In contrast, fertilization did not alter the shape of the domi-

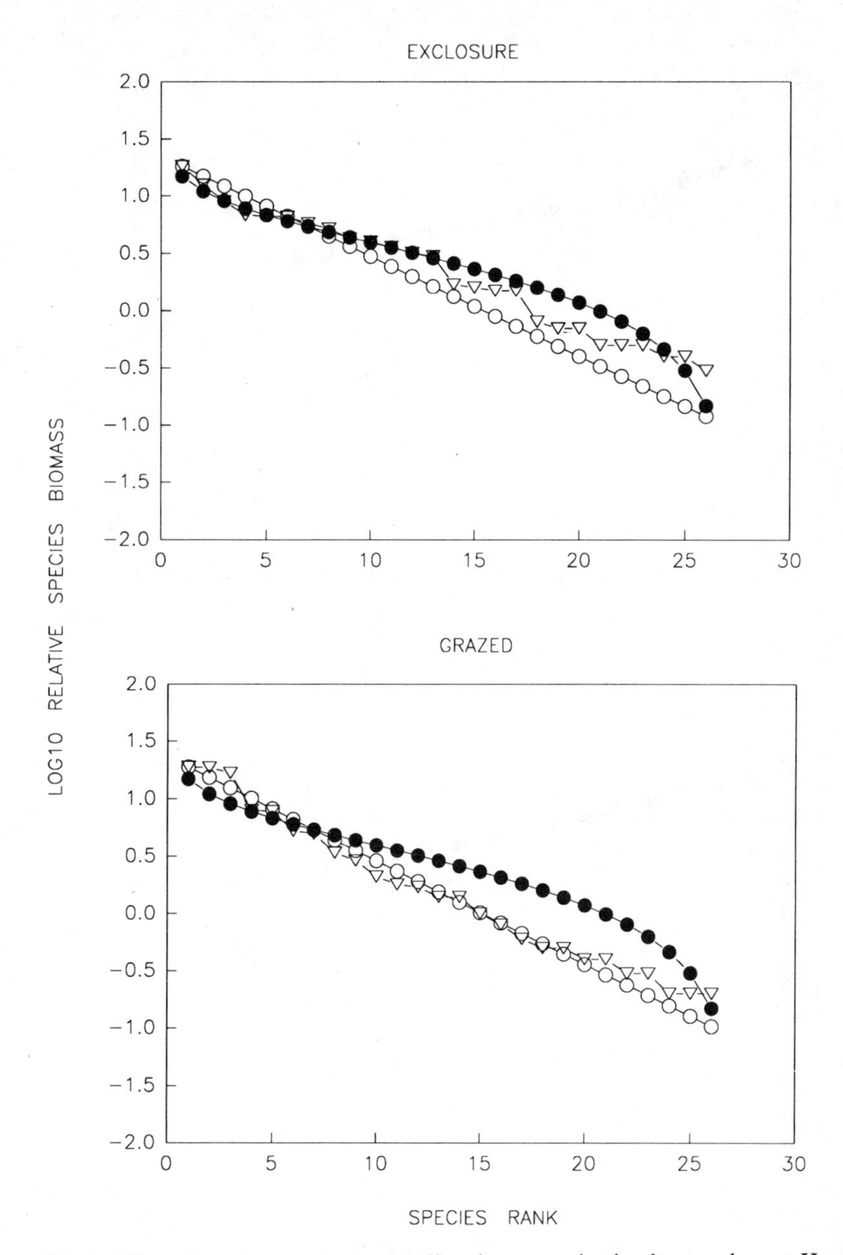

Fig. 3. Effect of grazing on dominance-diversity curves in the dry meadow at Hardangervidda, Norway. *Symbols* as in Fig. 1. Sheep were introduced into an area of the dry meadow that was devoid of grazers and allowed to graze for 3 years while a control plot was fenced and remained ungrazed. (Wiegolaski 1975c)

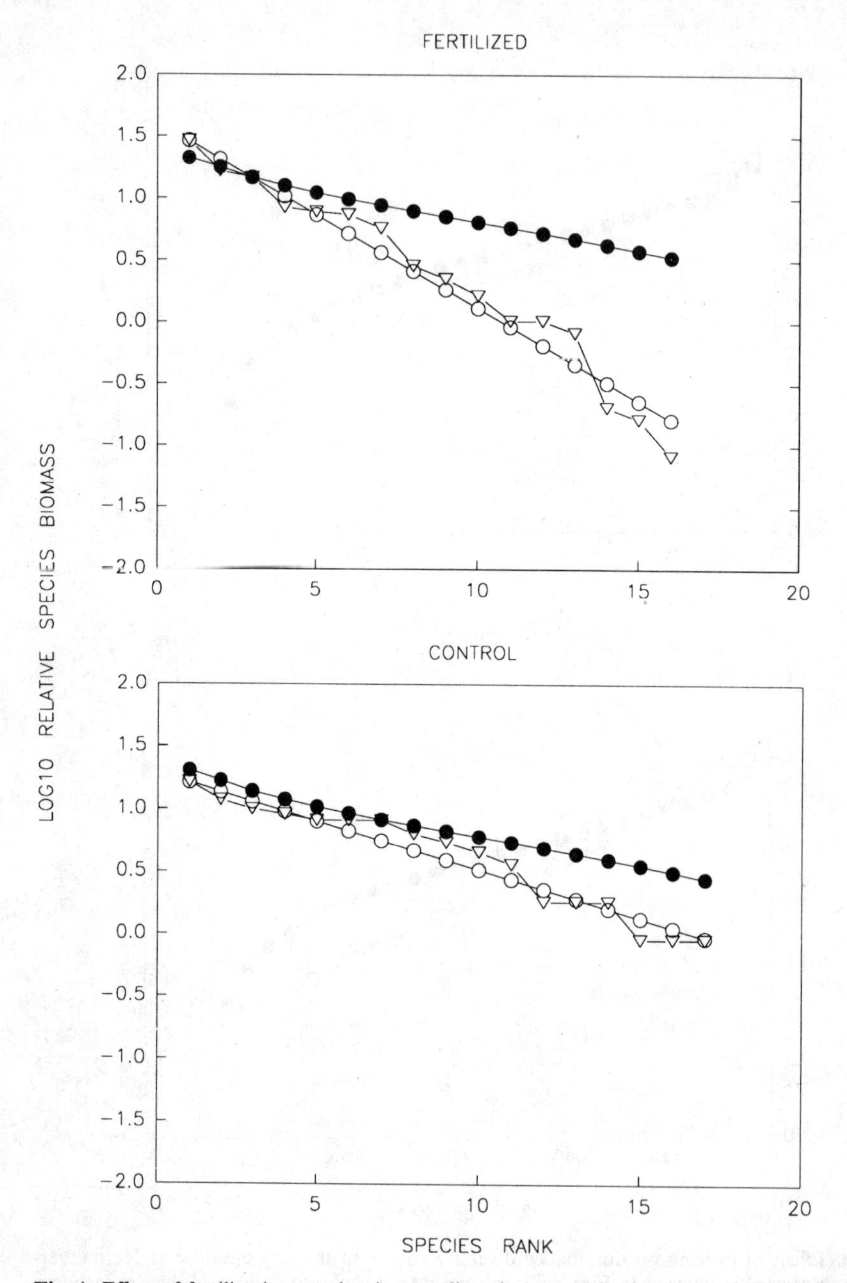

Fig. 4. Effect of fertilization on dominance-diversity curves in the dry meadow at Hardangervidda, Norway, in which 100 g/m^2 of a mixed fertilizer containing 11.5% N, 5% P, 14.5% K, 1.2% Mg, 7.5% S, and 0.02% B was applied annually for 3 years (Wielgolaski 1975b). *Symbols* as in Fig. 1

nance-diversity curve, merely strengthened conformity to the existing geometric model. It appears therefore that grazing is a more importance determinant of the diversity of biomass and nutrient distribution among species, although the rate of nutrient availability is also significant. This suggests that the species that can preempt nitrogen, phosphorus, or light while the system is being grazed will dominate this community. This also suggests an interaction between the availability of soil nutrients and herbivory in determining patterns of diversity.

Experimental manipulation of factors hypothesized to control diversity should be a goal of future investigations. Dominance-diversity curves and the distribution of species and growth forms along them could be used to measure the effects of these processes on both total nutrient pools and their partitioning among species.

18.5 Towards a Dynamic Model of Diversity in Tundra Ecosystems

Such patterns and dynamics as noted above cannot be fully explained by comparative studies or manipulative experiments, valuable and necessary though these are. These dynamics continue over time scales longer than most experiments. Furthermore, the analysis of variance traditionally used to evaluate experimental data cannot elucidate the nonlinearities underlying such dynamics.

One alternative approach would be to simply develop regressions of various diversity indices against environmental variables. However, such an approach is static and does not capture the interactions between plants, resources, and herbivores that control the changes in diversity noted in the experiments cited above. Such regression models are descriptive, not explanatory.

To obtain a full view of the functioning of tundra ecosystems and the relationships between species life characteristics, diversity, and rates of element flows, these data need to be assembled into a simulation model. Such a model should encapsulate the sequence of events and important flows of information and nutrients that comprise feedbacks between organisms and their environment. Chapin and coworkers (Chapin 1986; Bryant and Chapin 1986; Chapin 1993; Chapin et al. 1993) have suggested graphically these feedbacks, flows, and important controls in tundra ecosystems.

An explanatory model should simulate growth, nutrient uptake, herbivory, and return in litterfall at the level of the individual plant because these important processes determine the dynamics of diversity and occur at this level (Chapin and Shaver 1985; Chapin, this Vol.; Shaver, this Vol.; Jefferies and Bryant; this Vol.; Hobbie, this Vol.). This exact approach has been used in individual-based models of forest growth (Shugart 1984; Pastor and Mladenoff 1993), but has not yet been applied to tundra. Furthermore, if simulations are done at the level of individual plants, then diversity indices could be calculated using Eqs. (1)–(3) summing the simulated nitrogen and carbon contents of all individuals within a species as well as by the more traditional methods based on simulated number of individuals within a species or population. Such a model could therefore provide a common mathematical framework to integrate traditional measures of diversity with ecosystem measures.

The basic plant growth equations of the model are two, one for net changes in biomass (dB/dt), the other for net changes in nitrogen content (dN/dt):

$$dB/dt = g - m; \tag{4}$$

$$dN/dt = u - l; \tag{5}$$

where g is relative growth rate, m is mortality (including respiration, litter, and loss to herbivores), u is uptake, and l is loss of nitrogen (again to litterfall or herbivores). Chapin (this Vol.), Shaver (this Vol.), and Hobbie (this Vol.) describe experiments that can be used to parameterize these equations.

We also need equations describing the fate of carbon and nitrogen in litter once it is shed: dead cohorts of litter are still part of the population of plants, merely detached and subject to other influences. We need this because if the dominance of a given species is altered by global climate change or other factors, then so is the array of litter types returned to the soil, the availability of nutrients in the soil, and hence the future growth of individuals. Changes in the diversity of a plant community are thereby propagated through the limiting soil resource by decay of litter; this, in turn, will affect the diversity of future communities.

Two candidate equations are the exponential decay equation from Olson (1963) for mass or carbon loss and an equation for the change in nitrogen with respect to changes in carbon noted by Aber and Melillo (1982) and Bosatta and Staaf (1982):

$$dC/dt = -kC; \tag{6}$$

$$dN/dt = \frac{N}{C}(-kC) + ake^{-kt}C; \tag{7}$$

where k is the instantaneous decay rate of carbon from litter (C respired per unit carbon in tissue per unit time) and a is the instantaneous change in N/C ratio in the litter-microbe complex per unit carbon respired. The first term on the right-hand side of Eq. (7) depicts gross N mineralization relative to carbon respired, while the second term depicts nitrogen immobilization rate as a function of carbon remaining. Both k and a differ among genera and species (Aber and Melillo 1982; Pastor and Post 1986; Geng et al. 1993; Hobbie, this Vol.) and at a whole ecosystem level they would therefore be affected by changes in diversity. Decomposition experiments that can be used to parameterize these equations are described by Hobbie (this Vol.), Shaver (this Vol.), and Callaghan and Jonasson (this Vol.).

These four equations couple the flows of carbon and nitrogen through live individuals and their shed parts; when integrated and summed over all individuals and species, they describe the distribution of carbon and nitrogen among species within a community subject to the feedback of carbon and nitrogen between individual plants and soils.

We can now decompose any of the elements of the right hand side of the equations into quantitative hypotheses regarding controlling factors. For example, if allocation to different organs (roots, shoots, leaves, and reproductive parts) is hypothesized to control establishment, spread, or survival (Chapin, this Vol.), then the growth and mortality terms in Eq. (4) need to be partitioned into a sum of the allocation to different tissues. These partition coefficients are, in turn, functions of environmental factors. When these parts are shed as litter, their decay must be described by different parameterizations of Eq. (6) and (7).

Similarly, if decay rate depends on soil temperature, then k in Eq. (6) can be represented as a function of temperature. This, in turn, is a function not only of climate, landscape position, and physical features of the soil, but also of plant cover, particularly that of *Sphagnum* (Chapin 1993). Thus, there is a feedback between decay rate, k, in Eq. (6) and the biomass of *Sphagnum*, which is the integral of dB/dt for *Sphagnum* in Eq. (4). The extreme dominance of mosses and low diversity noted in arctic wet meadows (Fig. 2) could arise from this positive feedback.

Such a model could therefore prove useful in predicting the patterns discussed in this chapter as functions of the interactions of individuals with their environment. This model could simultaneously predict the number of species in a given environment, the distribution of carbon and nitrogen among them, and total ecosystem carbon storage and flux. It thus provides a framework to investigate the complex interactions between material flows and biodiversity.

18.6 Conclusions

1. With respect to ecosystem properties, the diversity of carbon and nitrogen distribution among species and functional groups is more important than species richness.
2. Traditional measures of diversity used in population ecology, such as the Shannon-Weaver index and dominance-diversity curves, can also be used to quantify diversity of carbon and nitrogen distribution among species.
3. Topographic complexity, resource preemption, herbivory, and soil nutrient availability are major factors responsible for the general patterns of diversity of carbon and nitrogen distribution among species in tundra ecosystems.
4. Because of strong feedbacks between species, resource availability, and herbivory, simulation models are required to examine the relationships between diversity and ecosystem processes.

Acknowledgments. I have learned much about the tundra from discussions over the years with Terry Chapin, Gus Shaver, Anne Hershey, Sarah Hobbie, Knute Nadelhoffer, and many others; these discussions were the basis of this paper. Terry Chapin kindly provided additional data for this paper as well. The presentation of this paper at Kongsvold, Norway, was made possible by a grant from the National Science Foundation to Terry Chapin and Christian

Korner and by a grant from the National Science Foundation to Scott Bridgham, Jan Janssens, Tom Malterer, and myself. The assistance of this organization is greatly appreciated.

References

Aber JD, Melillo JM (1982) Nitrogen immobilization in decaying hardwood leaf litter as a function of initial nitrogen and lignin content. Can J Bot 60: 2263–2269

Auclair AN, Goff FG (1971) Diversity relations of upland forests in the Western Great Lakes area. Am Nat 105: 499–528

Bosatta E, Staaf H (1982) The control of nitrogen turn-over in forest litter. Oikos 39: 143–151

Bryant JP, Chapin FS (1986) Browsing-woody plant interactions during boreal forest plant succession. In: Van Cleve K, Chapin FS, Flanagan PW, Viereck LA, Dyrness CT (eds) Forest ecosystems in the Alaskan taiga. Springer, Berlin Heidelberg New York, pp 213–225

Chapin FS (1986) Controls over growth and nutrient use by taiga forest trees. Pages. In: Van Cleve K, Chapin FS, Flanagan PW, Viereck LA, Dyrness CT (eds) Forest Ecosystems in the Alaskan taiga. Springer, Berlin Heidelberg New York, pp 96–111

Chapin FS (1993) Functional role of growth forms in ecosystem and global processes. In: Schimel DS (ed) Scaling physiological processes: leaf to globe. Academic Press, New York, pp 287–312

Chapin FS, Shaver GR (1985) Individualistic growth response of tundra plant species to manipulation of light, temperature, and nutrients in a field experiment. Ecology 66: 564–576

Chapin FS, Shaver GR (1989) Differences in growth and nutrient use among arctic plant growth forms. Funct Ecol 3: 73–80

Chapin FS, Autumn K, Pugnaire F (1993) Evolution of suites of traits in response to environmental stress. Am Nat 142 (Suppl): 78–92

Cohen JE (1968) Alternative derivations of a species-abundance relation. Am Nat 102: 165–172

De Vita J (1979) Niche separation and the broken-stick model. Am Nat 114: 171–178

Gaare E, Skogland T (1975) Wild reindeer food habits and range use at Hardangervidda. In: Wielgolaski FE (ed) Fennoscandian tundra ecosystems, part 2. Springer, Berlin Heidelberg New York, pp 206–216

Geng X, Pastor J, Dewey B (1993) Decay and nitrogen dynamics of litter from disjunct, congeneric tree species in old-growth stands in northeastern China and Wisconsin. Can J Bot 71: 693–699

Jefferies RL, Svoboda J, Henry G, Raillard M, Reuss R (1992) Tundra grazing systems and climatic change. In: Chapin FS, Jefferies RL, Shaver GR, Svoboda J (eds) Arctic physiological processes in a changing climate. Academic Press, New York, pp 391–412

Kjelvik S, Karenlampi L (1975) Plant biomass and primary production of fennoscandian subarctic and subalpine forests and of alpine willow and heath ecosystems. In: Wielgolaski FE (ed) Fennoscandian tundra ecosystems, part 1. Springer, Berlin Heidelberg New York pp 111–120

MacArthur RH (1960) On the relative abundance of species. Am Nat 94: 25–36

MacArthur RH (1965) Patterns of species diversity. Biol Rev 40: 510–533

Magurran AE (1988) Ecological diversity and its measurement. Princeton University Press, Princeton

Olson JS (1963) Energy storage and the balance of producers and decomposers in ecological systems. Ecology 44: 322–331

Pastor J, Post WM (1986) The influence of climate, soil moisture, and succession on forest carbon and nitrogen cycles. Biogeochemistry 2: 3–27

Pastor J, Mladenoff DJ (1993) Modeling the effects of timber management on population dynamics, diversity, and ecosystem processes. In: LeMaster DC (ed) Modeling sustainable forest ecosystems. American Forests, Washington DC, pp 16–29

Pielou EC (1981) The broken-stick model: a common misunderstanding. Am Nat 117: 609–610

Shaver GR, Chapin FS (1991) Production: biomass relationships and element cycling in contrasting arctic vegetation types. Ecol Monogr 61: 1–31

Shugart HH (1984) A theory of forest dynamics. Springer, Berlin Heidelberg New York

Tilman D (1982) Resource competition and community structure. Princeton Monographs in Population Biology 17. Princeton University Press, Princeton

Tilman D (1993) How productivity limits colonization and species richness in grasslands. Ecology 74: 2179–2191

Whittaker RH (1972) Evolution and measurement of species diversity. Taxon 21: 213–251

Whittaker RH (1975) Communities and ecosystems. Macmillan, New York

Wiegolaski FE (ed) (1975a) Fennoscandian tundra ecosystems, parts 1 and 2. Springer, Berlin Heidelberg New York

Wiegolaski FE (1975b) Primary production of alpine meadow communities. In: Wiegolaski FE (ed) Fennoscandian tundra ecosystems, part 1. Springer, Berlin Heidelberg New York, pp 121–128

Wiegolaski FE (1975c) Comparison of plant structure on grazed and ungrazed tundra meadows. In: Wiegolaski FE (ed) Fennoscandian tundra ecosystems, part 1. Springer, Berlin Heidelberg New York, pp 86–93

Wiegolaski FE, Kjelvik S (1973) Mineral elements and energy of plants at Hardangervidda, Norway. In: Bliss LC, Wiegolaski FE (eds) Primary, production and production processes, tundra biome. Proc Conf in Dublin, Ireland, Tundra Biome Steering Committee, Stockholm, Sweden, pp 231–238 20

19 The Plant-Vertebrate Herbivore Interface in Arctic Ecosystems

R. L. JEFFERIES[1] and J.P. BRYANT[2]

19.1 Introduction

Both natural and anthropogenic disturbances affect ecosystem integrity and biodiversity. Nearly all natural disturbances in arctic regions and elsewhere are directly or indirectly driven by climate (Walker and Walker 1991). Human disturbances also operate over large spatial and temporal scales and produce immediate direct effects and numerous indirect effects (Harte et al. 1992). The effects of these two types of disturbances on ecosystem function and biodiversity may be synergistic. Each disturbance reinforces the other, so that there is mutual amplification of their respective effects (Myers 1992). Organisms often respond nonlinearly to environmental variability (DeAngelis 1992). The responses are linked, to positive and negative feedback processes which involve both abiotic and biotic components of the environment. The environmental changes associated with these feedbacks, which are not easily predictable, may lead to a loss of ecosystem resilience and biodiversity.

The theme of the mutual amplification of the effects of different disturbances on biodiversity is developed in the remainder of the chapter. We begin by describing patterns of vegetation and vertebrate herbivory in arctic tundra, and then consider how expected responses of northern vegetation to warming might affect interactions between plants and vertebrate herbivores in ways that change biodiversity in tundra ecosystems. We conclude by focusing on a well-documented study of how anthropogenic influences are affecting both top-down and bottom-up trophic processes in arctic coastal habitats, leading to loss of biodiversity and ecosystem integrity. These influences are linked to recent weather patterns and provide a potential climatic scenario for some of the predicted ecological effects associated with global warming.

19.2 Patterns of Vegetation and Vertebrate Herbivory in Arctic Tundra

Without good knowledge of floristic and vegetation patterns within the vast northern regions of Eurasia and North America (approximately 7 500 000 km^2),

[1] Department of Botany, University of Toronto, 25 Wilcocks Street, Toronto, Ontario, M5S 3B2, Canada
[2] Institute for Arctic Biology, University of Alaska, Fairbanks, AK 99775-0180, USA

Ecological Studies, Vol. 113
Chapin/Körner (eds) Arctic and Alpine Biodiversity

studies of effects of ecosystem processes on biodiversity are difficult (Bliss and Matveyeva 1992). Tundra vegetation varies at several spatial scales (Walker and Walker 1991; Walker et al. 1993). From the forest-tundra ecotone to the High Arctic, woody vegetation decreases in importance, graminoids and forbs increase in importance, and both the standing crop of plant biomass and aboveground net primary production decline (Bliss 1988; Bliss and Matveyeva 1992). Within each latitudinal zone tundra vegetation varies as a result of geology and history. For example, "tundra steppe" plant communities occur on Wrangel Island and locally on the Chukotka mainland. These communities are dominated by grasses and a rich herbaceous flora and are considered a relic from the cold, dry Beringian Steppe of the Pleistocene (Yurtsev 1974). In low arctic regions, chemical defenses of shrubs such as willow and birch that are important foods of vertebrate herbivores (Bryant and Kuropat 1980) vary longitudinally (Bryant et al. 1989). Although unresolved, the reason for this biogeographical variation may be a response by vegetation to selective browsing in the Holocene (Bryant et al. 1992). At the landscape level topography, geomorphology, and patterns of snow accumulation and melt, and wildfires interact to generate vegetation mosaics exploited by vertebrate herbivores in their annual cycle of foraging.

Vertebrate herbivores of the arctic tundra all have catholic diets. However, the mixture of graminoids and woody species normally eaten varies among species. Some species, for example, geese (Jefferies 1988a, b) and brown lemming (*Lemmus sibericus*) (Batzli and Jung 1980), are primarily grazers, whereas others such as caribou (*Rangifer tarandus*) (Kuropat 1984), musk-oxen (*Ovibos moschatus*) (Robus 1981), and tundra vole (*Microtus oeconomus*) (Batzli and Jung 1980) eat substantial quantities of both graminoids and woody browse, and other species such as ptarmigan (*Lagopus*) (Williams et al. 1980), collard lemming (*Dicrostonyx torquatus*) (Batzli and Jung 1980), and moose (*Alces alces*) (Mould 1977) are primarily browsers. Few species feed heavily on nonvascular plants, but there are notable exceptions such as the dependence of caribou and reindeer on lichens in winter. Thus, climate change would affect the plant-herbivore interface by altering absolute and relative abundances of plant growth forms.

19.3 Vertebrate Herbivory in a Warming Climate: Implications for Tundra Biodiversity

In high-latitude ecosystems, warming is expected to result in northward migration of vegetation (Davis 1988; Pastor and Post 1988; Bryant and Reichardt 1992). This migration would (1) shift a modified forest-tundra ecotone northward, (2) increase the area dominated by shrub tundra, and (3) restrict graminoid-dominated tundra either to a latitudinal zone bounded on the north by the Arctic Ocean, the High Arctic islands or to the higher elevations of mountains. Paleoecological evidence (cf. Hopkins et al. 1982) and recent experiments (Chapin and Shaver 1985; Barta et al. 1989) indicate such a reduction in the diversity of tundra vegetation could reduce the abundance and diversity of tundra vertebrate herbivores.

During the last glaciation, a vast expanse of Arctic Asia and Alaska (Beringia) was unglaciated (Hopkins et al. 1982). Although there is debate about the absolute abundance of vegetation and herbivores in Pleistocene Beringia (Guthrie 1982, 1985, 1990; Ritchie and Cwyner 1982), there is general agreement that the Beringian landscape was a complex mosaic of steppe and polar desert (Ager 1982; Ritchie and Cwyner 1982; Schweger 1982; Brubaker et al. 1983; Lozhikin et al. 1993; Brubaker et al. this Vol.) and megaherbivore grazers dominated the vertebrate herbivore fauna (Guthrie 1982, 1984, 1985, 1990). Early in the Holocene warming event there was a very rapid shift in vegetation to dominance by shrub birch (e.g., *Betula exilis*) over much of Beringia (Ager 1982; Ritchie and Cwyner 1982; Schweger 1982; Brubaker et al. 1983; Lozhikin et al. 1993; Brubaker et al. this Vol.). This catastrophic vegetation change appears to have destabilized the Beringian ecosystem (Young 1982) and contributed significantly to the extinction of Beringian megaherbivore grazers (Guthrie 1984), because it is unlikely that the grazers could subsist on a low diversity diet of chemically defended woody species, such as shrub birch. Is there evidence that present-day tundra ecosystems could respond similarly to the expected future global warming?

In arctic ecosystems,the expected rise in surface temperature in response to global warming is about 3–6 K (Mitchell et al. 1989; Maxwell 1992). Chapin and Shaver (1985) found that a temperature increase of this magnitude can affect tundra vegetation. In tussock tundra dominated by the graminoid, *Eriophorum vaginatum*, 2 years of warming increased the growth of dominant shrubs such as birch, but not the growth of *E. vaginatum* and other graminoids. These observations indicate warming in the Arctic could again result in rapid replacement of graminoids by more chemically defended shrubs to the detriment of tundra grazers (Bryant and Reichardt 1992). For example, decreased production of inflorescences of *E. vaginatum* and other graminoid and herbaceous plants could adversely affect caribou populations. At calving time *E. vaginatum* inflorescences are an extremely important forage of cow caribou, because these flowers are often the only abundant source of chemically undefended, highly digestible protein (Kuropat 1984). It is very likely that loss of this protein supply at calving time and early in lactation would reduce the productivity of caribou herds (White 1983; Kuropat 1984). Such a loss of caribou would be a severe blow to the food economies of some tundra predators and some indigenous peoples of the Arctic.

A migration of taiga vegetation to tundra in response to warming could have a further negative effect on populations of tundra herbivores. Populations of taiga woody species may be more chemically defended against browsing by vertebrates than populations of tundra woody species (Bryant and Reichardt 1992). Thus, forced consumption of relatively heavily defended taiga woody vegetation by vertebrate herbivores of the tundra could result in starvation, thereby reducing the abundance and diversity of tundra herbivores. For example, in a recent study of competition between a tundra browser, the arctic hare (*Lepus arcticus*), and a taiga browser, the snowshoe hare (*L. americanus*), in Newfoundland, Barta et al. (1989) have shown that confinement of arctic hare to forested islands resulted in death by starvation in 1–3 months. Such a response occurred during winter and early spring, irrespective of the presence of potential competitors such as the snowshoe hare. This is the period of the year when taiga woody plants are most heavily defended chemically against browsing by snowshoe hares (Reichardt et al. 1990; Bryant et al. 1992). In contrast, arctic hares confined to their typical tundra barrens habitat maintained weight and survived well. Thus, starvation in forests may have been a result of low tolerance by arctic hares for chemical defenses of taiga woody plants. In contrast, snowshoe hares were able to maintain weight in either their typical forest habitat or in tundra barrens. This would be expected if tolerance of snowshoe hares for phytotoxins of taiga woody plants is higher than that of arctic hares. Bryant et al. (1989) have provided circumstantial evidence that snowshoe hares may be able to tolerate higher intakes of taiga woody plant toxins than other high-latitude hares.

Selective herbivory by vertebrate herbivores could amplify effects of warming on the biodiversity and functioning of tundra ecosystems. Currently, intense grazing in tundra ecosystems is not widespread (Jefferies et al. 1995). However, this situation could change in a warming event, because compression of tundra graminoids into a smaller area is likely to be accompanied by a commensurate increase in grazing intensity. This increase in grazing could accelerate shrub encroachment. The majority of evidence indicates that moderate to severe grazing frequently results in conversion of grassland to shrubland and loss of plant species diversity (Crawley 1983; Milchunas et al. 1988; Milchunas and Lauenroth 1993). In tundra ecosystems, such a grazing-facilitated shift in the abundance of plant functional groups is likely to affect rates of element cycling (Pastor this Vol.).

19.4 Effects of Anthropogenic Disturbances on Trophic Interactions – A Potential Climatic Change Scenario

Terrestrial ecosystems dominated by one species at each trophic level often occur in extreme or disturbed environments. Such systems are relatively susceptible to catastrophic loss of the dominant species brought about by biotic agents (Vitousek and Hooper 1993). As a first approximation, the soil-vegetation-goose-human interactions described below can be represented as a trophic ladder, in which different trophic levels may influence and be influenced by climatic events. Within the coastal ecosystems of Hudson Bay lowlands, the lesser snow goose (*Anser caerulescens caerulescens*) is the only major herbivore and populations are increasing rapidly as a result of human intervention. Increased use of agricultural crops (rice, corn, and spring wheat) on the wintering grounds and along the flyways, together with reduced hunting pressure (Francis et al. 1992) and the establishment of refugia along flyways probably account for the buildup of numbers of geese. These processes affecting the goose populations, which are brought about by human agencies, operate from both the top and the bottom of the trophic ladder simultaneously. Conservative estimates indicate that in the eastern Canadian Arctic more than 2 million birds were present in 1980 (Boyd et al. 1982). Since that time, birds have probably increased further but accurate census data are unavailable for the region as a whole. At La Pérouse Bay, Manitoba, the population has risen from 1800 pairs in 1968 to 23 000 pairs in 1991 (Cooke et al. 1982; R.H. Kerbes, unpubl. data). Such an increase in goose numbers places considerable demands on the environment for food resources.

Direct and indirect deleterious effects on plant growth brought about by adverse spring and summer weather conditions during the last decade have also affected goose populations. The decade has been marked by a series of late springs, especially in northern areas of Hudson Bay, the result of a net negative temperature anomaly during the period 1980–1989 compared with the prevailing temperatures from 1950 to 1979 (Findlay and Deptuch-Stapf 1991). As the authors indicate, these low temperatures, which have occurred especially in winter and spring, although paradoxical, could be a stage in global warming as climate systems adjust to a change in forcing. In years when the low temperature anomaly was marked (1978, 1983, 1986, 1987, 1992) snowmelt occurred later in the season in northern areas of Hudson Bay (H. Boyd, unpubl. results) and the northward migration of birds in spring was delayed, so that birds remained on the southern coasts of Hudson Bay.

The effects of large numbers of birds and low spring temperatures on the vegetation of the southern coasts have been considerable. In early spring, immediately after snow melt, but before aboveground plant growth begins, lesser snow geese grub for roots and rhizomes of graminoid plants, creating patches where little vegetation remains (Jefferies 1988a, b). The grubbing is carried out not only by the breeding population of geese, but also by birds that are forced to stage in the area, but which breed further north (Foxe Basin, Baffin Island, and northern area of Hudson Bay). Where grubbing is extensive, the patches of bare sediment increase in size and the vegetation fails to recover. Approximately 50% of salt-marsh graminoid swards have been destroyed by geese at La Pérouse Bay since 1985 (R.L. Jefferies, unpubl. data), largely as a result of grubbing, but also where roosting birds deposit piles of feces in spring. Hence, large numbers of birds and poor spring weather have combined to limit the area of available summer forage for the breeding population. Both of these changes are linked directly or indirectly to anthropogenic causes.

The change in arctic salt-marsh ecosystems initiated directly by the geese and indirectly by weather conditions acts as a trigger for a further series of changes that amplifies the scale of destruction. The reduced area of vegetation and the high numbers of adults and goslings of these colonial populations result in intense foraging during the summer months. At a number of sites on the southern Hudson Bay coast, the aboveground standing crop of grazed swards may be only 15 g dry wt. m^{-2} or less, in contrast to 40 g dry wt. m^{-2} for undamaged grazed swards (Cargill and Jefferies 1984; Cooch et al. 1992; Williams et al. 1993). The effect of these processes is to reduce the thickness of the vegetation mat (live and dead material) that insulates the underlying sediments from the air. Rates of evaporation from surface sediments increase and inorganic salts from the underlying marine clays produce hypersaline conditions (32–120 g of dissolved solids per liter) at the surface (Iacobelli and Jefferies 1991; Srivastava 1993). These salinities reduce the growth of the preferred forage plants (*Puccinellia phryganodes* and *Carex subspathacea*) in summer and this together with the intense foraging lead to sward destruction and the development of mudflats via a destructive, positive feedback process (Srivastava 1993).

The potential for the development of hypersaline conditions has been enhanced during the last decade because summer temperatures along the southwestern coasts of Hudson Bay under the influence of continental praire weather systems have been above the long-term average, leading to hot, dry conditions. In 1991, for example, 13 daily records of temperature were broken in July and August at Churchill, Manitoba. One of the predictions of climate change based on a 2 × CO_2 climate scenario is that temperature will increase substantially in the Prairie Provinces from June to August, and in particular there will be substantial decreases in available soil moisture for plant growth (Manabe and Wetherald 1987).

The weather patterns during the last decade provide a potential climatic scenario for some of the effects that may be predicted in the event of global climatic change brought about by anthropogenic influences. The amplification of the effects of weather and high numbers of geese lead to a deleterious synergistic effect on the growth of forage plants.

What is the outcome of these changes on the integrity of these coastal ecosystems around the Hudson Bay and elsewhere in the Arctic? In addition, to the destruction of salt-marsh swards, the geese pull up shoots of fresh-water sedges and eat the swollen bases. Extensive moss carpets have now developed where once sedge communities dominated. The geese are changing the character of these coastal plant communities and the abundance of the dominant species. These changes have had an adverse effect on the breeding population of lesser snow geese. Cooch et al. (1991b) have shown that gosling size has decreased significantly between 1976 and 1988. Likewise, there has been a 13% decrease in adult size for cohorts between 1969 and 1986; small goslings have become small adults. In addition, goslings that hatch early in the season are larger than later-hatched goslings when size is corrected for age (Cooch et al. 1991a), presumably because families arriving earlier at the grazing sites deplete the food source and later families have less, or lower quality food available. Goslings that left the traditional feeding area and walked at least 15 km to alternate areas where supplies of preferred forage were available were significantly larger than goslings that remained in the damaged area (Cooch et al. 1992). Francis et al. (1992) have shown that in years when gosling size is reduced, gosling survivorship is also reduced and that increased mortality occurs between the time the birds are banded and the time the birds are hunted. The use of alternative forage sources by goslings and adult birds may contribute to the decline in weight of birds. The results of short-term feeding trials using captive goslings indicate that due to a suite of nutritional and architectural characteristics of plants, birds acquire maximal amounts of essential nutrients and show the greatest growth when grazing on salt-marsh graminoids in comparison to alternative forages (Gadallah 1993). Hence, the necessity for adequate supplies of preferred salt-marsh forage to sustain the populations of geese.

These changes in edaphic conditions and in vegetation at a number of coastal sites across the North American Arctic, the outcome of this positive degenerative feedback process, adversely affect other populations besides geese and salt-marsh plants. As a result of the decreasing organic base in sediments brought about by destruction of swards, overall rates of soil nitrogen mineralization are appreciably lower in grubbed areas devoid of vegetation, thus affecting the supply of this essential element for plant growth in a nitrogen-limited environment (Wilson 1993). Coincident with this, there has been a decline in populations of shore birds and probably soil invertebrates and some species of duck, such as widgeon (*Anas americana*) which are grazers (F. Cooke, unpubl. data). The adverse effect of this trophic cascade brought about by the mutual amplification of the influence of high goose numbers and recent weather patterns ramifies through the entire coastal ecosystem.

19.5 Conclusions

At sites where large populations of lesser snow geese breed and where there has been destruction or partial destruction of plant communities and loss of ecosystem integrity, there is a marked synchrony between the reestablishment of the vegetative sward and the life expectancy of geese and other animals. It is unlikely that these damaged and fragmented plant communities will reestablish in their original form (the so-called Humpty-Dumpty effect, Pimm 1993), because of increased salinity and erosion of bare sediments in the short term, and geomorphological changes at the landscape level associated with isostatic uplift over the long term. These relatively simple coastal communities are dependent on a few species, the product of short assembly processes (Pimm 1993). As Strong (1992) has argued, trophic cascades are restricted to habitats with a low species diversity, in which runaway consumption will occur in the absence of a highly reticulate, species-rich food web which buffers a system against change. In the scenario presented above, anthropogenic-triggered events amplify top-down and bottom-up processes associated with this simple trophic ladder of a few species, to produce a precipitous change in these coastal ecosystems. Sudden shifts in populations (and biodiversity) occur as feedbacks result in environmental thresholds being exceeded in a trophic cascade.

These examples of coupled feedback processes discussed in this chapter in which the effects on components of these different ecosystems share many similar characteristics, in spite of the very different spatial and temporal scales involved. One is the product of events that occurred at the end of the last glaciation over a broad geographical area, the other is a consequence of changes that have occurred during the last two decades in a specific ecosystem at a regional level. The types of synergisms discussed are applicable across a wide range of spatial and temporal scales. In both examples, climate acts as a trigger that initiates a sequence of changes as a result of its effects on other processes. In the first example, the detailed mechanistic basis leading to the changes in herbivore numbers is unknown, hence, the formulation of the hypothesis linking climate, vegetational change, herbivory, and the production of secondary chemicals by trees. In the second example, two very unrelated types of disturbance, both of which may well be a consequence of human activities reinforce one another to produce a trophic cascade in which trophic interactions are severely modified. The return to the original pattern of plant-animal interactions as well as the original plant assemblages is likely to be protracted. What is particularly of concern is the rapidity with which these changes have occurred. In less than a decade there have been large shifts in community assemblages. Recognition of further coupled feedback cycles leading to synergistic effects on ecosystems is urgent. One such example is the potential effect on vegetation of late snow clearance in spring associated with increased snow depth. This may limit the availability of inflorescences of plants such as *Eriophorum vaginatum* at a time when lactating female caribou actively seek flowers as a prime food source. The likely outcome is that survivorship of calves will be adversely affected.

19.6 Summary

1. The mutual amplification of the effects of interactions between climate, vegetation, and herbivores on ecosystem function and biodiversity in tundra and boreal forest regions are examined.
2. Vertebrate herbivores, which have catholic diets, exploit the mosaic of vegetation types in which species differ in their chemical defenses and nutritional qualities. Climatic change may be expected to affect plant-herbivore relationships by altering the absolute/relative abundances of plant growth forms.
3. Increases in abundance of shrubs as the forest-tundra ecotone moves north will lead to a loss of existing tundra vegetation which, in turn, may result in the extinction of vertebrate herbivore populations. The shrubs may be rich in secondary compounds that deter herbivores. The destabilization of the Beringia ecosystem and the extinction of megaherbivore population serve as a model for the likely sequence of events following climate change.
4. Recent rapid increases in breeding populations of lesser snow geese, probably as a result of anthropogenic influences, have led to the partial destruction of tundra coastal ecosystems and loss of biodiversity. The destruction has been enhanced by the direct and indirect effects of weather patterns during the last decade which may be linked to global climate change.
5. These anthropogenic-triggered events amplify top-down and bottom-up trophic processes, resulting in sudden shifts in sizes of populations (and loss of

biodiversity) in these relatively simple coastal communities dependent on a few species.

6. In spite of the different spatial and temporal scales involved in the two examples discussed, climate acts as a trigger that initiates a sequence of changes as a result of its effect on other processes. Recognition of further examples of coupled feedback cycles which result in synergistic effects on ecosystems and changes in biodlverslty is urgent.

Acknowledgments. One of us (J.P.B.) acknowledges N.S.F. grant BSR-8702672 for long-term ecological research in the Alaskan taiga. The other (R.L.J.) thanks N.S.E.R.C. for financial support for projects on the coast of Hudson Bay. As always, Mrs. C. Siu typed the manuscript against a series of her and our deadlines.

References

Ager TA (1982) Vegetation history of western Alaska during the Wisconsin glacial interval and the Holocene. In: Hopkins DM, Matthews JV Jr, Schweger CE, Young SB (eds) Paleoecology of Beringia. Academic Press, New York, pp 75–94

Barta RM, Keith LB, Fitzgerald SM (1989) Demography of sympatric arctic and snowshoe hare populations: an experimental assessment of interspecific competition. Can J Zool 67: 2762–2775

Batzli GO, Jung HG (1980) Nutritional ecology of microtine rodents: resource utilization near Atkasook, Alaska. Arct Alp Res 12: 483–499

Bliss LC (1988) Arctic tundra and polar desert biome. In: Barbour MG, Billings WD (eds) North American terrestrial vegetation. Cambridge University Press, Cambridge, pp 1–32

Bliss LC, Matveyeva NV (1992) Circumpolar arctic vegetation. In: Chapin FS III, Jefferies RL, Reynolds JF, Shaver GR, Svoboda J (eds) Arctic ecosystems in a changing climate. Academic Press, New York, pp 59–89

Boyd H, Smith GEJ, Cooch FG (1982) The lesser snow geese of the eastern Canadian Arctic. Canadian Wildlife Service Occasional Paper No 46, Ottawa, Canada

Brubaker LB, Garfinkel HL, Edwards ME (1983) A late Wisconsin and Holocene vegetation history from the Central Brooks Range: implications for Alaskan palaeoecology. Quat Res 20: 194–214

Bryant JP, Kuropat PJ (1980) Selection of winter forage by subarctic browsing vertebrates: the role of plant chemistry. Annu Rev Ecol Syst 11: 261–285

Bryant JP, Reichardt PB (1992) Controls over secondary metabolite production by arctic woody plants. In: Chapin FS III, Jefferies RL, Reynolds JF, Shaver GR, Svoboda J (eds) Arctic ecosystems in a changing climate. Academic Press, New York, pp 377–390

Bryant JP, Tahvanainen J, Sulkinoja M, Julkunen-Tiitto R, Reichardt P, Green T (1989) Biogeographic evidence for the evolution of chemical defense by boreal birch and willow against mammalian browsing. Am Nat 134: 20–34

Bryant JP, Reichardt PB, Clausen TP, Provenza FD, Kuropat PJ (1992) Woody plant-mammal interactions. In: Rosenthall GA, Bernbaum MR (eds) Herbivores: their interactions with secondary plant metabolites. Academic Press, New York, pp 344–371

Cargill SM, Jefferies RL (1984) The effects of grazing by lesser snow geese on the vegetation of a sub-arctic salt marsh. J Appl Ecol 21: 669–686

Chapin FS III, Shaver GR (1985) Individualistic growth response of tundra plant species to environmental manipulations in the field. Ecology 66: 564–576

Cooch EG, Lank DB, Dzubin A, Rockwell RF, Cooke F (1991a) Body size variation in lesser snow geese: environmental plasticity in gosling growth rates. Ecology 72: 503–512

Cooch EG, Lank DB, Rockwell RF, Cooke F (1991b) Long-term decline in body size on a snow goose population: evidence of environmental degradation? J Anim Ecol 60: 483–496
Cooch EG, Jefferies RL, Rockwell RF, Cooke F (1992) Environmental change and the cost of philopatry: an example in the lesser snow goose. Oecologia 93: 1–7
Cooke F, Abraham KF, Davies JC, Findlay CS, Healey RF, Sadura A, Seguin RJ (1982) The La Pérouse Bay snow, goose project – a 13 year report. Department of Biology, Queen's University, Kingston, Canada, 194 pp
Crawley MJ (1983) Herbivory. Studies in Ecology 10. University of California Press, Berkeley
Davis MB (1988) Ecological systems and dynamics. In: Toward an understanding of global change. National Academy Press, Washington DC, pp 69–105
DeAngelis DL (1992) Dynamics of nutrient cycling and food webs. Chapman and Hall, London
Findlay BF, Deptuch-Stapf A (1991) Colder than normal temperatures over north eastern Canada during the 1980's. Climat Perspect 13 (April Issue): 9–12
Francis CM, Richards MH, Cooke F, Rockwell RF (1992) Long-term changes in survival of lesser snow geese. Ecology 73: 1346–1362
Gadallah F (1993) Forage quality and gosling nutrition in the lesser snow goose. MSc Thesis, University of Toronto, Toronto, 122 pp
Guthrie RD (1982) Mammals of the mammoth steppe as paleoenvironmental indicators. In: Hopkins DM, Matthews JV Jr, Schweger CE, Young SB (eds) Paleoecology of Beringia. Academic Press, New York, pp 207–328
Guthrie RD (1984) Mosaics, allelochemics, and nutrients: an ecological theory of late Pleistocene megafaunal extinctions. In: Martin PS, Klein RG (eds) Quaternary extinctions. University of Arizona Press, Tucson, pp 259–298
Guthrie RD (1985) Woolly arguments against the mammoth steppe: a new look at the palynological data. Q Rev Arch 6: 9–16
Guthrie RD (1990) Frozen fauna of the Mammoth Steppe. University of Chicago Press, Chicago
Hopkins DM, Matthews JV Jr, Schweger CE, Young SB (1982) Paleoecology and Beringia. Academic Press, New York
Harte J, Torn M, Jensen D (1992) The nature and consequences of indirect linkages between climate change and biological diversity. In: Peters RL, Lovejoy TE (eds) Global warming and biological diversity. Yale University Press, New Haven, pp 325–343
Iacobelli A, Jefferies RL (1991) Inverse salinity gradients in coastal marshes and the death of stands of *Salix*: the effects of grubbing by geese. J Ecol 79: 61–73
Jefferies RL (1988a) Vegetation mosaics, plant-animal interactions and resources for plant growth. In: Gottlieb LD, Jain SK (eds) Plant evolutionary biology. Chapman and Hall, London, pp 340–361
Jefferies RL (1988b) Pattern and process in arctic coastal vegetation in response to foraging by lesser snow geese. In: Weger MJA, van der Aart PJM, During HJ, Verhoeven JTA (eds) plant form and vegetational structure adaptation, plasticity and relationship to herbivory. SPB Academic Publishing, The Hague, pp 281–300
Jefferies RL, Klein DR, Shaver GR (1994) Herbivores and northern plant communities: reciprocal influences and responses. Oikos 71: 193–206
Kuropat PJ (1984) Foraging behavior of caribou on a calving ground in northwestern Alaska. M Sc Thesis, University of Alaska, Fairbanks, 94 pp
Lozhikin AV, Anderson PM, Eisner WR, Ravako LG, Hopkins DM, Brubaker LB, Colinvaux PA, Miller MC (1993) Late Quaternary lacustrine pollen records from southerwestern Beringia. Quat Res 39: 314–324
Manabe S, Wetherald RT (1987) Large scale changes in soil wetness induced by an increase in atmospheric carbon dioxide. J Atmos Sci 44: 1211–1235
Maxwell B (1992) Arctic climate: potential change under global warming. In: Chapin FS III, Jefferies RL, Reynolds JF, Shaver GR, Svoboda J (eds) Arctic ecosystems in a changing climate. Academic Press, New York, pp 11–34
Milchunas DG, Lauenroth WK (1993) Quantitative effects of grazing on vegetation and soils over a global range of environments. Ecol Monogr 63: 327–366

Milchunas DG, Sala OE, Lauenroth WK (1988) A generalized model of the effects of grazing by large herbivores on grassland community structure. Am Nat 132: 87–106

Mitchell JFB, Senior CA, Ingrahm WJ (1989) CO_2: a missing feedback? Nature 341: 132–134

Mould ED (1977) Movement patterns of moose in the Colville River area, Alaska. M Sc Thesis, University of Alaska, Fairbanks, 82 pp

Myers N (1992) Synergisms: joint effects of climate change and other forms of habitat destruction. In: Peter RL, Lovejoy TE (eds) Global warming and biological diversity. Yale University Press, New Haven, pp 344–354

Pastor J, Post WM (1988) Response of northern forests to CO_2-induced climate change. Nature 334: 55–58

Pimm SL (1993) Biodiversity and the balance of nature. In: Schulze E-D, Mooney HA (eds) Biodiversity and ecosystem function. Ecological studies 99. Springer, Berlin Heidelberg New York, pp 348–359

Reichardt PB, Bryant JP, Mattes BR, Clausen TP, Chapin FS III, Meyer M (1990) Winter chemical defense of Alaskan balsam poplar against snowshoe hares. J Chem Ecol 16: 1941–1960

Ritchie JC, Cwynar LC (1982) The late Quaternary vegetation of the north Yukon. In: Hopkins DM, Matthews JV Jr, Schweger CE, Young SB (eds) Paleoecology of Beringia. Academic Press, New York, pp 113–126

Robus M (1981) Foraging behavior of muskoxen in Arctic Alaska. M Sc Thesis, University of Alaska, Fairbanks, 78 pp

Schweger CE (1982) Late pleistocene vegetation of eastern Beringia: pollen analysis of dated alluvium. In: Hopkins DM, Matthews JV Jr, Schweger CE, Young SB (eds) Paleoecology of Beringia. Academic Press, New York, pp 95–112

Srivastava DS (1993) The role of lesser snow geese in positive, degenerative feedback processes resulting in the destruction of salt-marsh swards. M Sc Thesis, University of Toronto, Toronto, 244 pp

Strong DR (1992) Are trophic cascades all wet? Differentiation and donor-control in speciose ecosystems. Ecology 73: 747–754

Vitousek PM, Hooper DU (1993) Biological diversity and terrestrial ecosystem biogeochemistry. In: Schulze E-D, Mooney HA (eds) Biodiversity and ecosystem function. Ecological studies 99. Springer, Berlin Heidelberg New York, pp 3–14

Walker DA, Walker MD (1991) History and pattern of disturbance in Alaskan arctic terrestrial ecosystems: a hierarchical approach to analyzing landscape change. J Appl Ecol 28: 244–276

Walker DA, Halfpenny JC, Walker MD, Wessman CA (1993) Long-term studies of snow-vegetation interactions. Bioscience 43: 287–301

White RG (1983) Foraging patterns and their multiplier effects on productivity of northern ungulates. Oikos 40: 377–384

Williams JB, Best D, Warford C (1980) Foraging ecology of ptarmigan at Meade River, Alaska. Wilson Bull 92: 341–351

Williams TD, Cooch EG, Jefferies RL, Cooke F (1993) Environmental degradation, food limitation and reproductive output: juvenile survival in lesser snow geese. J Anim Ecol 62: 766–777

Wilson DJ (1993) Nitrogen mineralization in a grazed sub-arctic salt marsh. M Sc Thesis, University of Toronto, Toronto, 226 pp

Young SB (1982) The vegetation of the Bering Land Bridge. In: Hopkins DM, Matthews JV Jr, Schweger CE, Young SB (eds) Paleoecology of Beringia. Academic Press, New York, pp 179–194

Yurtsev BA (1974) Problems of botanical geography in north eastern Asia. Nauka, St Petersburg (in Russian)

20 Insect Diversity, Life History, and Trophic Dynamics in Arctic Streams, with Particular Emphasis on Black Flies (Diptera: Simuliidae)

A.E. HERSHEY[1], R.W. MERRITT[2], and M.C. MILLER[3]

20.1 Introduction

Streams in Arctic Alaska typically have very low levels of inorganic nutrients, low levels of coarse particulate organic matter (CPOM), no influence of woody debris, and high concentrations of dissolved organic matter due to leaching from the peat soils which they drain. Overall stream habitat diversity in tundra regions is fairly low, mainly consisting of first–fourth order streams flowing over boulder, cobble, and peat substrata. These habitats are not shaded, except for some dwarf willows along the riparian corridors.

The arctic insect fauna is depauperate compared to that of temperate regions (Oliver 1968; Danks 1978). Low insect diversity probably results from a combination of life history and habitat characteristics, rather than biogeographic constraints. In comparison, arctic fish diversity is constrained biogeographically, as well as by life history factors (Deegan and Peterson 1992). Because most arctic streams freeze solid for many months of the year, only stream insects which have physiological adaptations to survive while encased in ice (see Olsson) 19813 are found there. Insects with life cycles longer than 1 year are subjected to greater stress because of the need to overwinter more than once, possibly in more than one life stage. Although this is reasonably common for lentic chironomids in the arctic (Butler 1982; Hershey 1985), it is less common among lotic species.

In this chapter, we discuss diversity of the lotic insect fauna of arctic regions, with emphasis on Arctic Alaska in the vicinity of the Toolik Lake Arctic Research Station (68°38′N, 149°36′W). We focus particularly on the family Simuliidae (black flies), which often dominate secondary production of Alaskan arctic streams. We present data illustrating the trophic significance of black fly feeding in arctic streams with respect to the available organic matter pools. We then discuss the relationship between taxonomic diversity and trophic function.

20.2 Arctic Insect Diversity

Although aquatic insects are somewhat thermally buffered, only aquatic Diptera are well represented in the Arctic, due particularly to the success of the

[1] Department of Biology, University of Minnesota – Duluth, Duluth, MN 55812, USA
[2] Department of Entomology, Michigan State University, East Lansing, MI, USA
[3] Department of Biological Sciences, University of Cincinnati, Cincinnati, OH, USA

Ecological Studies, Vol. 113
Chapin/Körner (eds) Arctic and Alpine Biodiversity

Chironomidae (Oliver 1968) and Simuliidae (this Chap.). In temperate streams, mayflies (Ephemeroptera), stoneflies (Plecoptera), caddisflies (Trichoptera), and aquatic beetles (Coleoptera) are typically species-rich and often account for the highest portion of insect secondary production (e.g., Benke et al. 1984; Smock et al. 1985). However, these important groups lose diversity with latitude (Oliver 1968; Danks 1978) and become less important trophically, while Diptera increase in importance.

20.2.1 Black Fly Diversity

Two summer seasons of sampling 20–25 stream, river, and lake-outlet sites covering approximately 19 km^2 yielded approximatley 16 species (Table 1). Habitats included cobbles and boulders in larger streams, or sedges and grasses in smaller streams. Farther south in Alaska, aquatic habitats are more varied and species richness for black flies is greater than in the northern tundra region (Sommerman et al. 1955; Table 2). However, the number of species in the Alaskan tundra is comparable with the total number of black fly species recorded from Nebraska in the USA (Pruess and Peterson 1987). Nebraska streams also are fairly uniform in their channel morphology and pattern, consisting of sand-bottom streams that have often been channelized and contain trailing vegetation for black fly attachment sites (Table 2).

Table 1. Black fly taxa collected from streams in the North Slope Borough, Alaska, in 1992 and 1993

Taxon	1st-order tundra	Lake outlet	Higher order
Prosimulium neomacropyga Peterson		X	X
Prosimulium perspicuum Sommerman			X
Gymnopais holopticus Stone			X
Metacnephia pallipes (Fries)	X	X	X
Metacnephia sommermanae (Stone)	X		
Cnephia eremites Shewell		X	X
Stegopterna emergens (Stone)	X	X	X
Stegopterna sp.			X
Simulium decolletum Adler and Currie		X	X
Simulium vernum group	X	X	X
Simulium decorum cytospecies nr. *churchill*		X	
Simulium truncatum Lundstroem (= S. *venustum* cytospecies EFG/C)	X	X	
Simulium rostratum Lundstroem (= *S. verecundum* cytospecies ACD)		X	
Simulium tuberosum cytospecies FGI	X	X	X
Simulium (Schoenbaueria) sp.	X		X
Sinmulium (Hellichiella) sp.		X	

Table 2. Comparison of black fly species richness in arctic, subarctic, and temperate streams

Location	Latitude	Habitats	Area km^2	Estimated No. of species	Reference
Toolik Lake area Arctic Alaska, USA	67°N	Tundra streams, peat and rocky substrates; trailing vegetation	Approx. 168	16	This Chapter
Southern Alaska	60°–66°N	Diverse mountain and boreal forest streams	310 800	38	Sommerman et al. (1955)
Nebraska, USA	40°–43°N	Sand-bottom channelized streams; trailing vegetation	198 495	16	Pruess and Peterson (1987)
Michigan, USA	42°–47°N	Lower peninsula: low gradient, deciduous /forest streams on glacial moraine outwash terrain Upper peninsula: boreal forest (with some northern hardwoods) streams of bedrock with low to moderate relief and including outwash/ lacustrine plains	250 738	50–55	Merritt et al. (1978); R. W. Merritt (unpubl.)
South Carolina, USA	32–35°N	Sandy, silt mud-bottomed streams at lower elevations; swift, rocky-bottomed streams at high elevations	78 236	50	P. Adler (pers. comm.)
Mono Co., California, USA	38°N	Diverse stream habitats at various altitudes with predominantly rock-bottoms; some silt-bottom meadow streams with high concentrations of organic matter	8037	55–60	P. Adler (pers. comm.)
Pennsylvania, USA	40–42°N	Slower moving sandy-bottom streams at low elevations; swift and clear rocky-bottom streams at high elevations	116 270	45	Alder and Kim (1986)
Ammarnäs, central Swedish Lapland	66°N	Boreal or mixed hardwood forest streams with rocky bottoms: birch forests at tree line	Approx. 40 km along 3 stream reaches	13	Ulfstrand (1968)

Stream habitat diversity in the states of Michigan and South Carolina (Table 2) is considerably greater due to different geomorphic and vegetational patterns. The number of black fly species ranges from 50–55 in these states (Merritt et al. 1978, unpubl. data; P. Adler, pers. comm.). Similarly, in Mono County, California, one finds approximately 55–60 species of black flies occurring in a mosaic of different stream habitat types from desert to mountain elevations (P. Adler, pers. comm.). These regional comparisons suggest that stream habitat heterogenity is positively related to species richness in black flies; however, other factors, such as lower stream temperatures and a reduced larval developmental period, also may limit species richness.

The 16 species of black flies belong to 6 different genera (Table 1). Five of these genera (*Gymnopais*, *Cnephia*, *Stegopterna*, *Metacnephia*, and *Prosimulium*) are mainly northern and, according to Crosskey (1990), belong to the tribe Prosimulinii, which constitutes approximately 45% of the Nearctic black fly fauna. They are primarily Holarctic in origin and absent from the transtropical faunal regions (Crosskey 1990). The larvae of this group generally live in cool streams and often emerge as adults earlier than those from other black fly genera. Among these, the genus *Prosimulium* has traditionally been considered the most primitive (plesiomorphic) in black flies, based on pupal and adult characters (Crosskey 1990). The majority of species found in our study area that belong to the genus *Simulium* also have Holarctic distributions (Crosskey 1990; P. Adler, pers. comm.).

20.2.2 Diversity of Other Insect Taxa

Mayflies tend to be species-poor at high latitudes and high altitudes (Brittain 1982). For example, Jacobi and Benke (1991) reported 28 species of mayflies on snag habitats of the Ogeechee River, a blackwater coastal plain river in Georgia, USA, but in Arctic Alaska (68°N), the fourth-order Kuparuk River had only 5 or 6 (Hershey et al. 1995), and no mayflies were found in nearby first-order Imnavait Creek (Miller et al. 1986). At least two of the Kuparuk River species belong to the genus *Baetis*, which has high tolerance to environmental stress and high life cycle plasticity (Brittain 1982). Caribou Poker Creek, a fourth-order subarctic Alaskan stream, had five mayfly species (Miller et al. 1986).

Stoneflies also are represented by very few species in arctic and subarctic streams. One shredder and one predator stonefly are known from Arctic Alaska (Miller et al. 1986; Hershey et al. 1994). Danks (1978) reported that stoneflies are rare in the high Arctic, and that most species of stoneflies are temperate. In subarctic Monument Creek, which drains Alaskan boreal forest, shredder stoneflies are numerically abundant even though their diversity is low (Cowan and Oswood 1984).

Trichoptera also rank high in diversity and abundance in temperate streams, which has been attributed to their ability to construct cases and retreats using silk, and their high trophic diversity (Mackay and Wiggins 1979). However,

Harper (1981) reported them to be rare or absent from arctic streams. At high latitudes, caddisfly life cycles are typically longer than 1 year (see Hershey and Hiltner 1988), and survivorship encased in ice was not as high as for Chironomidae (Olsson 1981). We sampled four species of caddisflies in the Kuparuk River, none of which were abundant, and one from Imnavait Creek (Miller et al. 1986). Although not species rich, they can be important in the trophic dynamics of arctic streams due to their large size and aggressive behavior. Experimental fertilization of the Kuparuk River resulted in a dramatic increase in the filter-feeding caddisfly, *Brachycentrus americanus* (Banks), which aggressively displaced black flies, decreasing the abundance and trophic importance of that group (Hershey and Hiltner 1988). Also, predation by the predatory caddisfly, *Rhyacophila* sp., accounted for a significant portion of the seasonal mortality of black flies in Oksrukuyik Creek, also near the Toolik Lake Arctic Research Station (Wheeler and Hershey 1992).

Aquatic beetles were not found in larger, rocky streams such as the Kuparuk River (Hershey et al. 1994). Danks (1978) reported them to be virtually absent from the Arctic. However, small beaded streams, such as Imnavait Creek, were found to contain a few species of beetles belonging to the Dytiscidae and Hydrophilidae (Miller et al. 1986). Representatives of these families also are found in arctic ponds (A.E. Hershey, unpubl.).

Chironomidae are very successful in the Arctic, where they are nearly as species rich as in temperate streams (Table 3). In streams, most arctic chironomids

Table 3. Comparison of species richness within the Chironomidae in selected arctic and temperature streams

Stream	Latitude	Habitat types	No. of species	Reference
Kuparuk River, Alaska, USA	68°N	Tundra; Cobble, boulder, and peat substrates	> 27	This Chapter
Imnavait Creek, Alaska USA	68°N	Tundra; peat and cobble substrates	> 40	Miller et al. (1986)
Caribou-Poker Creek, Alaska, USA	65°N	Boreal forest, diverse habitats	> 35	Miller et al. (1986)
Linding A, Western Jutland, Denmark	56°N	Cobble, gravel, sand, and mud substrates with dense vegetation	64	Lindegaard-Petersen (1972)
Hunt Creek, Michigan, USA	45°N	Sand, clay, silt, and woody debris substrates	> 55	Wiley (1978)
North Fork of Bigoray River, Alberta, Canada	54°N	Diverse habitats with high macrophyte density	32	Boerger et al. (1982)
Cedar Creek, South Carolina, USA	32°N	Second-order black-water stream; snags and sand substrates	27	Smock et al. (1985)

require a full year for development, although the same species may complete several generations per year in temperate regions. For example, in the Fort Simpson area of the subarctic, Northwest Territories, *Cricotopus bicinctus* (Meigen) has subpopulations with two and three generations per year (Rosenberg et al. 1977), whereas the same species in the Kuparuk River has only one generation (A.E. Hershey, unpubl.).

20.3 Black Fly Life History Features

It appears that most tundra black fly species have one generation a year (univoltine), with eggs serving as the overwintering stage (Sommerman et al. 1955; Peterson 1970). Eggs hatch shortly after ice breakup, with larval and pupal development lasting 3–6 weeks. However, some species of *Simulium* (e.g., members of the *decorum*, *tuberosum*, and *venustum* species complexes), have been reported to have at least two to three annual generations (multivoltine) in southern Alaska and in Alberta, Canada (Sommerman et al. 1955; Adler 1986; Currie 1986). Three species of *Simulium* (*decorum*, *rostratum*, *truncatum*) have been primarily associated with lake outlets (Burger 1987; Wotton 1987; Crosskey 1990). Others are generalists that often occur at these sites, but can be found farther downstream as well (Crosskey 1990; McCreadie and Colbo 1991). While the feeding mode of most larval black flies utilizes a pair of labral fans to capture small particulate organic material in flowing water (Wallace and Merritt 1980), larvae of the genus *Gymnopais* have lost the labral fans of their presumed *Prosimulium* ancestors and have evolved mouth parts to scrape algae and associated materials from the surfaces of stones in glacier-fed streams (Currie and Craig 1987).

Since most species are filter feeders, they are probably very similar trophically, and a species-level understanding is not needed to understand trophic function. However, insect species within a functional feeding group often have slightly different phenologies. For example, the two dominant black fly species in the Kuparuk River in Arctic Alaska have emergence periods separated by 3–4 weeks (Hershey and Hiltner 1988). Thus, although similar trophically, temporal separation of development extends the period of organic matter processing by black flies in arctic Alaskan streams; diversity contributes to trophic function on a temporal scale.

Biting activity of adult Alaskan black flies varies among different species and genera. Downes (1971) noted that autogeny (i.e., the ability to produce a batch of eggs without requiring a blood meal) is a common adaptation in black flies associated with successful colonization of unpredictable terrestrial environments, where density-independent factors associated with harsh weather can be responsible for catastrophic mortalities. Most of these species undergo only one gonotrophic cycle (Anderson 1987). Adults of the *Prosimulium* spp. and *Gymnopais* sp. recorded from our study area are reported to be nonblood-feeding species; eggs mature from nutrients carried over from the larval stage (Anderson

1987). Other species may be obligately anautogenous (i.e., requiring a blood meal to produce their first batch of eggs), feeding on birds [e.g., *S. vernum* species complex, *S.* (*hellichiella*) sp.], small mammals (e.g., *S. tuberosum* species complex), and large mammals (e.g., *M. pallipes*, *S. truncatum*, *S. rostratum*, *S. venustum* species complex) (Sommerman et al. 1955; Abdelnur 1968; Peterson 1970; Cupp and Gordon 1983; Crosskey 1990; Hunter 1990; Table 1).

20.4 Trophic Dynamics

Low diversity in several taxonomic groups coupled to high diversity among the dipteran families Chironomidae and Simuliidae is reflected in the trophic diversity of arctic streams. Most arctic chironomids are unspecialized collectors, although *Orthocladius rivulorum* Kieffer grazes on algae on its tube (Hershey et al. 1988). These chironomids share the collector-gatherer niche with the few species of mayflies found in the arctic, particularly *Baetis* spp. (Merritt and Cummins 1984). Black flies, on the other hand, are the only common group of filter-feeders in arctic streams.

Fine particulate organic matter (FPOM), operationally defined as particles < 1 mm but > 0.45 μm, results from the breakdown of coarse particulate organic matter (CPOM), epilithic sloughing, flocculation of dissolved and colloidal (< 0.45 μm) organic matter (DOM), and as feces produced by all functional feeding groups (Cummins and Klug 1979). This organic matter source dominates macroconsumer food webs in a wide range of running water types (e.g., Cummins 1974; Naiman and Sedell 1979; Vannote et al. 1980; Wallace et al. 1982). FPOM is the major food source for filter-feeding collectors, as well as the collector-gatherer consumers which use FPOM on the substratum (Wallace and Merritt 1980). In arctic streams, the major terrestrial source of FPOM is peat, thus, to a large extent, aquatic invertebrate collectors represent secondary production that is delayed from terrestrial primary production by thousands of years (Schell 1983).

Filter-feeding black flies strongly dominate insect secondary production in the arctic rivers we have studied, ranging from one-third to two-thirds of total insect production in the Kupaurk River (see Peterson et al. 1993; Hershey et al. 1994), and nearly 100% in the E-1 outlet stream. As indicated above, grazer and collector-gatherers are represented by a fairly high diversity of Chironomidae (Diptera), and a few Ephemeroptera (Hershey et al. 1994). During years of high chironomid density, these collector-gatherers control algal biomass in the Kuparuk River (Gibeau 1992; Peterson et al. 1993). Shredder and predator functional feeding groups are very underrepresented (Hershey et al. 1995). The low amount of coarse plant detritus present in unforested streams probably accounts for the low abundance of shredders, although life history constraints are undoubtedly also a factor in low shredder diversity (Cowan and Oswood 1984). Predators, which tend to have much longer life cycles than their prey, are most likely limited by overwintering conditions. Thus, although Diptera are well represented by filter-feeding black flies and grazer/collector-gatherer Chironomidae, other insect orders have few representatives.

In the Kupaurk River, the chironomid, *Orthocladius rivulorum*, represents 6–20% of the total insect secondary production, but the remainder of the chironomid community contributes $<3\%$ to secondary production (estimated for 1984 only), which is very low compared to chironomids in temperate streams (Hauer and Benke 1991), and reflects their small size and slow turnover. Since black flies contribute 40 to nearly 100% to secondary production, the contributions of other groups to stream trophic dynamics is generally very small. Thus, trophic function is maintained by a smaller array of taxonomic groups than in temperate streams, and trophic diversity is skewed toward collectors, especially black flies, which are filter-feeding collectors.

20.5 Organic Matter Processing by Black Flies in an Arctic Lake Outlet

20.5.1 Gut Analyses of Larval Black Flies

In the Lake E-1 outlet stream, near the Toolik Lake Arctic Research Station, black fly gut contents contained four major categories of FPOM particles, ranging in size from 1–280 µm: algae, solitary bacteria, bacterial clusters, and detritus (Table 4). The most common algae found in the E-1 outlet larval guts were the colonial yellow-green species, *Dinobryon divergens*, and clusters of blue greens (Table 4). Bacterial clusters were abundant. Detritus comprised 10.9%, while total bacteria in guts of E-1 outlet black flies comprised over 80% of the mean area/gut (Table 4). These results show that black flies ingest a wide array of FPOM particle types and sizes, thus the filter-feeding trophic function is quite broad.

20.5.2 Black Fly Feeding on Dissolved Organic Matter

Dense aggregations of filter-feeders, which often include Simuliidae, Hydropsychidae, and some Chironomidae, occur at lake outlets and some other stream

Table 4. Gut contents of black flies from the E-1 outlet stream, North Slope Borough, Alaska

Particle category	Percent by number	Mean size (μm^2)	Percent by area
Algae[a]	0.16	47.31	7.51
Bacteria[b]	98.17	0.74	70.03
Bacterial clusters[c]	1.37	8.71	11.56
Detritus	0.30	37.90	10.90

[a] Includes, but is not necessarily limited to, *Cyclotella, Navicula, Dinobryon,* and *Coelosphaerium.*
[b] Consists of rods and cocci.
[c] Consists of clusters of cocci.

habitats (e.g., Wotton 1992; Lake and Burger 1983), and have very high secondary production (Richardson and Mackay 1991). In arctic streams, where other filter-feeders are uncommon, such aggregations are limited to black flies. In addition to FPOM, large amounts of the organic matter in streams is in dissolved form (see Fisher and Likens 1973; Meyer 1990). Labile fractions of DOM fuel stream microbes (e.g., Lock et al. 1984; Edwards and Meyer 1987; Meyer 1990), but much DOM is refractory, thus is generally unavailable to consumers (Meyer 1990). However, DOM becomes incorporated into the FPOM pool by physical and chemical processes, and by microbial uptake of labile DOM components (Wotton 1991). This material is subsequently used by filtering collectors, such as black flies, and collector-gatherers, such as chironomids.

Biogenic flocculation of DOM occurs on mucus-coated structures of some marine filter-feeders, and is trophically important to those animals (Stephens 1982). Similar processes also have been hypothesized to also occur in streams (Wotton 1988; Cibrowski et al. 1993). In laboratory experiments using black flies from the E-1 outlet stream, we showed that the larvae ingest significant quantities of DOM (Fig. 1), in addition to their FPOM diet. Ingestion of ^{14}C-labeled DOM

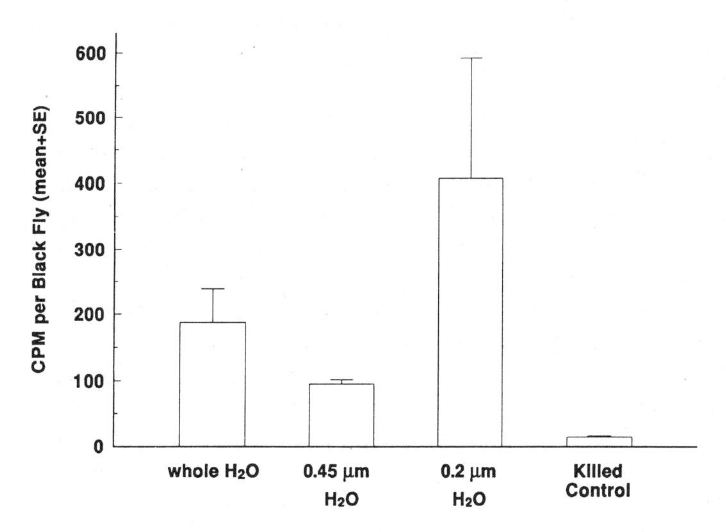

Fig. 1. Results from a laboratory experiment demonstrating that larval black flies feed on dissolved organic matter (DOM) in an arctic stream. To test the hypothesis that black flies feed on DOM, we offered ^{14}C-labeled DOC to black flies in recirculating stream microcosms (see Hershey and Dodson 1987; Hershey 1987). To provide a source of ^{14}C-labeled DOC, cultured *Chalmydomonas* were inocculated with ^{14}C bicarbonate. After incorporation of the label, algae were homogenized to disrupt cells, then filtered through either a 0.45-µm or a 0.2-µm filter to produce two size fractions of labeled DOM. Microcosms containing black flies and either unfiltered E-1 Outlet water (whole water), 0.45-µm filtered water, or 0.2-µm water, were all inocculated with labeled DOC (either 0.45-µm DOC for whole water and 0.45-µm filtered water, or 0.2-µm DOC for 0.2-µm filtered water). In all treatments, live black fly larvae incorporated significantly more of the ^{14}C label than the dead larvae used as controls for surface adsorption ($p < 0.01$).These results strongly supported the hypothesis that black flies feed on DOM as well as FPOM

occurred regardless of whether particles were also present in the stream water (Fig. 1). Although the mechanism for DOM feeding is unknown, it seems unlikely that black flies are able to ingest it directly, but rather that they are feeding on material that either precipitates on their fans or is formed in the ambient water column. In both cases, they would be feeding on DOM material through indirect rather than direct means.

20.6 Summary and Conclusions

1. Diversity of insects in Alaskan arctic streams is low for most groups, but relatively high for Diptera, especially in the families Simuliidae (black flies) and Chironomidae. Black fly and chironomid species richness is comparable to temperate streams and is very high relative to nondipteran taxonomic groups. This most likely reflects the fact that members of these families typically have short life cycles, usually overwinter as eggs or cold-hardy larval forms, and are able to expolit the food resources and habitats available.
2. The very low diversity of predators is probably due to their relatively large size and long life cycles, which makes overwintering difficult. Low diversity of shredders probably reflects a combination of life-cycle constraints and low availability of coarse detritus.
3. Most Simuliidae and Chironomidae are collectors, which use fine particulate organic matter (FPOM), an abundant resource in most tundra streams. Simuliidae also use DOM, which is also abundant in Alaskan arctic streams draining peat tundra. Other functional feeding groups are poorly represented.
4. Black flies are the only abundant group of filter-feeding collectors in most Alaskan arctic streams, thus occupying an important position in terms of trophic diversity. The dynamics of FPOM and DOM are linked to black fly feeding and overall trophic diversity. In arctic streams, the role of black flies in FPOM and DOM dynamics may be especially important to stream ecosystem function because other filter-feeder groups are absent.
5. Although black fly species overlap trophically, they have temporally distinct life cycles. Thus, high black fly diversity contributes to trophic function on a temporal scale.

Acknowledgments. We would like to thank R.S. Wotton, D.L. Lawson, and M.E. McDonald for comments on various aspects of this research; J. McCrea, W.T. Morgan, T.L. Galarowicz, T. Elton, and J.R. Wheeler for technical assistance; and Pat Mulholland for the loan of the stream microcosms used in laboratory feeding experiments. We would like to thank J. McCreadie and P. Adler for reviewing a portion of the manuscript, and the latter for assistance with simuliid species identifications. The research was supported in part by NSF awards BSR 87-02328 and DPP 90-24188 to John Hobbie with subcontracts to AEH and MCM, BSR 91-19417 to AEH, and by Northeast Regional Black Fly Project NE-118 to RWM.

References

Abdelnur OM (1968) The biology of some black flies (Diptera: Simuliidae) of Alberta. Quaest Entomol 4: 113–174.

Alder PH (1986) Ecology and cytology of some Alberta black flies (Diptera: Simuliidae). Quaest Entomol 22: 1–18

Adler, PH, Kim KC (1986) The black flies of Pennsylvania (Simuliidae, Diptera): bionomics, taxonomy, and distribution. Penn State Agric Exp Stn Bull No 856, 87 pp

Anderson JR (1987) Reproductive strategies and gonotrophic cycles of black flies. In: Kim KC, Merritt, RW (eds) Black flies: ecology, population management, and annotated world list. Penn State Univ Press, University Park, pp 276–293

Benke AC, Van Arsdall TC Jr, Gillespie DM, Parrish FK (1984) Invertebrate productivity in a subtropical blackwater river: the importance of habitat and life history. Ecol Monogr 54: 25–63

Boerger HJ, Clifford HF, Davies RW (1982) Density and microdistribution of chironomid larvae in an Alberta brown-water stream. Can J Zool 60: 913–920

Brittain JE (1982) Biology of mayflies. Annu Rev Entomol 27: 119–147

Burger JF (1987) Specialized habitat selection by black flies. In: Kim KC, Merritt RW (eds) Black flies: ecology, population management, and annotated world list. Penn State Univ Press, University Park, pp 129–145

Butler MG (1982) A 7-year life cycle for two *Chironomus* species in arctic Alaskan ponds (Diptera: Chironomidae). Can J Zool 60: 58–70

Cibrowski JH, Craig DA, Fry K (1993) Can black fly larvae (Diptera: Simuliidae) ingest dissolved organic matter? – Laboratory evidence. Bull N Am Benthol Soc 10(1): 83 (Abstr)

Cowan CA, Oswood MW (1984) Spatial and seasonal associations of benthic macroinvertebrates and detritus in an Alaskan subarctic stream. Polar Biol 3: 211–215

Crosskey RW (1990) The natural history of blackflies. Wiley, Chichester, 711 pp

Cummins KW (1974) Structure and function of stream ecosystems. BioScience 24: 631–641

Cummins KW, Klug MJ (1979) Feeding ecology of stream invertebrates. Annu Rev Ecol Syst 10: 147–172

Cupp EW, Gordon AE (eds) (1983) Notes on the systematics, distribution, and bionomics of black flies (Diptera: Simuliidae) in the northeastern United States. Search Agric 25: 1–75

Currie DC (1986) An annotated list of keys to the immature black flies of Alberta (Diptera: Simuliidae). Mem Entomol Soc Can 134: 1–90

Currie DC, Craig DA (1987) Feeding strategies of larval black flies. In: Kim KC, Merritt RW (eds) Black flies: ecology, population management and annotated world list. Penn State Univ Press, University Park, pp 155–170

Danks HV (1978) Modes of seasonal adaptation in the insects. I. Winter survival. Can Entomol 110: 1167–1205

Deegan LA, Peterson BJ (1992) Whole river fertilization stimulates fish production in an arctic tundra river. Can J Fish Aquat Sci 49: 1890–1901

Downes JA (1971) The ecology of blood-sucking Diptera: an evolutionary perspective. In: Fallis AM (ed) Ecology and physiology of parasites. Adam Hilger, London pp 232–258

Edwards RT, Meyer JL (1987) Bacteria as a food source for black fly larvae in a blackwater river. J N Am Benthol Soc 6: 241–250

Fisher SG, Likens GW (1973) Energy flow in Bear Brook, New Hampshire: an integrative approach to stream ecosystem metabolism. Ecol Mongr 43: 421–429

Gibeau G (1992) Grazer and nutrient limitation of algal biomass in an arctic tundra river. MS Thesis, University of Cincinnati, Cincinnati

Harper PP (1981) Ecology of streams at high latitudes. In: Lock MA, Williams A (eds) Perspectives in running water ecology. Plenum Press, New York, pp 313–337

Hauer FR, Benke AC (1991) Rapid growth of snag-dwelling chironomids in a blackwater river: the influence of temperature and discharge. J N Am Benthol Soc 10: 154–164

Hershey AE (1985) Littoral chironomid communities in an arctic Alaskan lake. Holarct Ecol 8: 39–48

Hershey AE (1987) Tubes and foraging behavior in larval Chironomidae: implications for predator avoidance. Oecologia 73: 236–241

Hershey AE, Dodson SL (1987) Predator avoidance by *Cricotopus*: cyclomorphosis and the importance of being big and hairy. Ecology 68(4): 913–920
Hershey AE, Hiltner AL (1988) Effects of caddisfly activity on black fly density: interspecific interactions outweigh food limitation. J N Am Benthol Soc 7: 188–196
Hershey AE, Hiltner AL, Hullar MAJ, Miller MC, Vestal JR, Lock MA, Rundle S, Peterson BJ (1988) Nutrient influence on a stream grazer: *Orthocladius* microcommunities track nutrient input. Ecology 69: 1383–1392
Hershey AE, Bowden WB, Deegan LA, Hobbie JE, Peterson BJ, Kipphut GW, Kling GW, Lock MA, Merritt RW, Miller MC, Vestal JR, Schuldt JA (1994) The Kuparuk River: a long term study of biological and chemical processes in an arctic river. In: Milner A, Oswood M (eds) Alaskan freshwaters. Springer, Berlin Heidelberg New York (in press)
Hunter FF (1990) Ecological, morphological and behavioral correlates to cytospecies in the *Simulium venustum/verecundum* complex. PhD Thesis, Queen's University, Kingston, Ontario
Jacobi DI, Benke AC (1991) Life histories and abundance patterns of snag-dwelling mayflies in a blackwater coastal plain river. J N Am Benthol Soc 10: 372–387
Kling GW, Kipphut GW, Miller MC (1991) Arctic lakes and streams as gas conduits to the atmosphere: implications for tundra carbon budgets. Science 251: 298–301
Lake DJ, Burger JF (1983) Larval distribution and succession of outlet-breeding blackflies (Diptera: Simuliidae) in New Hampshire. Can J Zool 61: 2519–2533
Lindegaard-Petersen C (1972) The chironomid fauna in a lowland stream in Denmark compared with other European streams. Verh Int Verein Limnol 18: 726–729
Lock MA, Wallace RR, Costerton JW, Ventullo RM, Charlton SE (1984) River epilithon: toward a structural-functional model. Oikos 42: 10–22
Mackay RJ, Wiggins GB (1979) Ecological diversity in Trichoptera. Annu Rev Entomol 24: 185–208
McCreadie JW, Colbo MH (1991) Spatial distribution patterns of larval cytotypes of the *Simulium venustum/verecundum* complex (Diptera: Simuliidae) on the Avalon Peninsula, Newfoundland: factors associated with occurrence. Can J Zool 69: 2651–2659
Merritt RW, Cummins KW (eds) (1984) An introduction to the aquatic insects of North America. Kendall/Hunt, Dubuque, IA
Merritt RW, Ross DH, Peterson BV (1978) Larval ecology of some lower Michigan black flies (Diptera: Simuliidae) with keys to the immature stages. Great Lakes Entomol 4: 34–39
Meyer JL (1990) A blackwater perspective on riverine ecosystems. BioScience 40: 643–51
Miller MC, Stout JR, Alexander V (1986) Effects of a controlled under-ice oil spill on invertebrates of an arctic and subarctic stream. Environ Pollut 42: 99–132
Naiman RJ, Sedell JR (1979) Benthic organic matter as a function of stream order in Oregon. Arch Hydrobiol 87: 404–422
Oliver DR (1968) Adaptations of arctic Chironomidae. Ann Zool Fenn 5: 111–118
Olsson TI (1981) Overwintering of benthic macroinvertebrates in ice and frozen sediment in a North Swedish river. Holarct Ecol 4: 161–166
Oswood MW, Everett KR, Schell DM (1989) Some physical and chemical characteristics of an arctic beaded stream. Holarct Ecol 12: 290–295
Peterson BJ, Hobbie JE, Corliss TL (1986) Carbon flow in a tundra stream ecosystem. Can J Fish Aquat Sci 43: 1259–1270
Peterson BJ, Deegan L, Helfrich J, Hobbie J, Hullar M, Moller B, Ford T, Hershey A, Hiltner A, Kipphut G, Lock MA, Fiebig DM, McKinley V, Miller MC, Vestal JR, Ventullo R, Volk G (1993) Biological responses of a tundra river to fertilization. Ecology 74: 653–672
Peterson BV (1970) The *Prosimulium* of Canada and Alaska. Mem Entomol Soc Can 69: 1–216
Pruess KP, Peterson BV (1987) Black flies (Diptera: Simuliidae) of Nebraska: an annotated list. J Kansas Entomol Soc 60: 528–534
Richardson JS, Mackay RJ (1991) Lake outlets and the distribution of filter feeders: an assessment of hypotheses. Oikos 62: 370–380
Rosenberg DM, Wiens AP, Saether OA (1977) Life histories of *Cricotopus* (*Cricotopus*) *bincinctus* and *C. (C.) mackenziensis* (Diptera: Chironomidae) in the Fort Simpson area, Northwest Territories. J Fish Res Board Can 34: 247–253

Ross DH, Merritt RW (1987) Factors affecting larval black fly distributions and population dynamics. In: Kim KC, Merritt RW (eds) Black flies: ecology, population management and annotated world list. Penn State Univ Press, University Park, pp 90–108
Schell DM (1983) Carbon-13 and carbon-14 abundances in Alaskan aquatic organisms: delayed production from peat in arctic food webs. Science 219: 1068–1071
Smock LA, Gilinsky E, Stoneburner DL (1985) Macroinvertebrate production in a southeastern United States blackwater stream. Ecology 66: 1491–1503
Sommerman KM, Sailer RI, Esselbaugh CO (1955) Biology of Alaskan black flies (Simuliidae, Diptera). Ecol Monogr 25: 345–385
Stephens GC (1982) The role of uptake of organic solutes in nutrition of marine organisms. Am Zool 22: 611–620
Ulfstrand S (1968) Life cycles of benthic insects in Lapland streams (Ephemeroptera, Plecoptera, Trichoptera, Diptera Simuliidae). Oikos 19: 167–190
Vannote RL, Minshall GW, Cummins KW, Sedell JR, Cushing CE (1980) The river continuum concept. Can J Fish Aquat Sci 37: 130–137
Wallace JB, Merritt RW (1980) Filter-feeding ecology of aquatic insects. Annu Rev Entomol 25: 103–132
Wallace JB, Ross DH, Meyer JL (1982) Seston and dissolved organic carbon dynamics in a southern Appalachian stream. Ecology 63: 824–838
Wheeler J, Hershey AE (1992) The role of *Rhyacophila* predation in an arctic stream: results from experimental stream microcosms and field sampling studies. Bull N Am Benthol Soc 9: 103
Wiley M (1978) The biology of some Michigan trout stream chironomids (Diptera: Chironomidae). Mich Acad 11: 193–209
Wotton RS (1987) The ecology of lake-outlet black flies. In Kim KC, Merritt RW (eds) Black flies: ecology, population management, and annotated world list. Pennsylvania State University Press, University Park, pp 146–154
Wotton RS (1988) Dissolved organic material and trophic dynamics. BioScience 38: 172–177
Wotton RS (1991) Particulate and dissolved organic material as food. In: Wotton RS (ed) The biology of particles in aquatic systems. CRC Press, Boca Raton, pp 213–216
Wotton RS (1992) Animals that exploit lake outlets. Freshwater Forum 2: 62–67

21 Land-Water Interactions: The Influence of Terrestrial Diversity on Aquatic Ecosystems

G. W. KLING

21.1 Introduction

As ecologists, we have learned much about the effects of species diversity on the biological aspects of communities. In an effort to expand our understanding to broader levels of organization, there has been much recent attention given to the role of species diversity in the workings of ecosystems, including abiotic as well as biological effects (see Schulze and Mooney 1993). Despite this recent attention, and despite an increasing awareness of the importance of landscape heterogeneity (Hansen and di Castri 1992), there has been little explicit study of the interactions between land and water as they relate to ecological diversity. An analysis of the role of biological and landscape diversity in the interaction between terrestrial and aquatic ecosystems requires information about the following questions: (1) which inputs from land regulate aquatic systems?, (2) which of these inputs is related to terrestrial species diversity or landscape diversity?, and (3) what controls the transfer of important inputs across the land-water boundary? This chapter reviews the information available and needed from tundra systems to answer these questions within the general framework of diversity and land-water interactions.

21.2 Inputs from Land and the Regulation of Aquatic Systems

Inputs of materials from land strongly regulate the functioning of lake and river ecosystems. These inputs include water, nutrients, and organic matter, which form the building blocks of aquatic food webs. The amounts, chemical forms, and timing of delivery of materials from land to water are critical, and thus the chemical transformations of material in terrestrial environments can have an important impact on aquatic systems. Although there are many indirect effects of terrestrial inputs, such as the support of grazing fish by algal production that is in turn driven by nutrient availability, it is the direct effects on aquatic systems that are most related to inputs from land. The export of water itself is a special case, because not only is it related to the total amount of materials delivered to aquatic systems, it is also related to the productivity of stream systems in that small

Department of Biology, University of Michigan, Ann Arbor, MI 48109–1048, USA

Ecological Studies, Vol. 113
Chapin/Körner (eds) Arctic and Alpine Biodiversity

streams may dry completely during mid-summer and lowered water level in large streams may affect production and fish habitat. The direct effects of terrestrial exports on aquatic diversity and on the functions of aquatic productivity and decomposition are discussed below.

21.2.1 Productivity

Aquatic productivity is driven and often limited by the amounts and forms of available nutrients. Inputs of nutrients in streams or groundwater are the ultimate source of nutrients for lakes; direct input from the atmosphere is the only other source and is unimportant in all but the largest lakes. Algal productivity in arctic surface waters is light-limited in winter but nutrient-limited during the ice-free season (Hobbie 1980; Peterson et al. 1985; Miller et al. 1986). Co-limitation by nitrogen (N) and phosphorus (P) is more common in alpine and arctic lakes (Miller et al. 1986; Morris and Lewis 1988) than in other systems, where one or the other nutrient is limiting; P usually limits temperate-zone lakes and N sometimes limits tropical lakes (Hecky and Kilham 1988). P and especially N are strongly retained by terrestrial tundra ecosystems (Chapin et al. 1980; Dowding et al. 1981), and this may explain the tendency for aquatic tundra ecosystems to be co-limited by N and P.

The timing of nutrient inputs is also important, although in different ways for lakes and streams. In Toolik Lake, for example, over 80% of the nutrient load that is available for primary production for the whole year is delivered to the lake during spring snowmelt, and the highest rates of production are just following snowmelt even though the lake is still ice covered (Whalen and Cornwell 1985; Miller et al. 1986). By contrast, during the spring snowmelt, the bottom of streams may be encased in ice and the water is turbid, creating poor conditions for stream algal production. Consequently, primary production in streams is more dependent on the continuous inputs of nutrients in the seepage of groundwater from land during the growing season.

Finally, the chemical forms of nutrients are important for algal production. For example, the uptake of NH_4^+ requires less energy than the uptake of NO_3^-, and thus NH_4^+ is preferred by phytoplankton. Because of this, the conversion of NH_4^+ to NO_3^- by nitrification in terrestrial environments can have an important impact on aquatic systems. In addition, there is increasing evidence that nutrients in the form of dissolved organic nitrogen and phosphorus (DON, DOP) may be important for phytoplankton nutrition and growth in temperate (Palenik and Morel 1990) and arctic (Kling 1994) systems, especially in nutrient-limited situations. The mass balance of nutrient budgets in many arctic lakes and streams is dominated by dissolved organic nutrients (Whalen and Cornwell 1985; Peterson et al. 1986;), and in Toolik Lake up to 30% of the annual nutrient demand by phytoplankton must be supplied by DON and DOP (Whalen and Cornwell 1985).

The direct effect of terrestrial inputs on secondary production in aquatic environments is related more to organic matter than to inorganic nutrients for both bacteria (discussed below) and animals. Stream ecologists have long recognized the powerful influence of dissolved and particulate organic matter (DOM, POM) inputs from land on the function and structure of running waters (Ross 1963; Vannote et al. 1980). Similar influences occur in lakes, where POM is most important for benthic or benthic-feeding animals (Schell 1983) and POM plus DOM is most important for pelagic animals (Salonen et al. 1992). For example, arctic zooplankton may rely on terrestrial DOM and detritus for nutrition. This reliance can be very strong in northern Scandinavian lakes (Hessen et al. 1990), but less important in foothill lakes of the North Slope of Alaska (Kling 1994). The strength and importance of this linkage in the Arctic are related to the ratio of organic to inorganic carbon in the lake, and to the ratio of the amount of terrestrial organic matter present compared to the amount of organic matter derived from in-lake production by algae (Kling 1994).

21.2.2 Bacteria and Decomposition

Despite some debate, it appears that decomposition through the activity of bacteria in aquatic systems is limited by the supply of organic substrate rather than by grazing of bacteria by protozoans (Cole et al. 1988; Hobbie and Helfrich 1988; Stockner and Porter 1988) in all but the most eutrophic systems (Riemann 1985; Pace 1993). Some of this organic substrate originates in terrestrial environments and some is produced within aquatic environments as exudates from algae and zooplankton. The relative importance of these two sources to bacterial activity varies, in part as a result of the amounts of DOM available from each source (Cole et al. 1988; Sundh and Bell 1992; Wetzel 1992), and in part as a result of the quality of the DOM. The general perception that DOM leached from terrestrial systems is of low quality and unavailable for further bacterial use in surface waters is being challenged (Meyer and Edwards 1990; Jones 1992; Tranvik 1992; Kaplan and Newbold 1993). There is evidence from Toolik Lake in arctic Alaska that terrestrial DOM leached from surface plant litter contains labile compounds and strongly stimulates bacterial activity in the lake (Fig. 1). Because most arctic lakes are oligotrophic and in-lake sources of DOM are limited, it is probable that bacterial activity and decomposition is regulated by the inputs of terrestrial DOM. There is also evidence that in arctic lakes this terrestrial material initially incorporated by bacterial secondary production is made available to higher trophic levels through trophic interactions in the microbial food web (Rublee 1992; Kling 1994), as has been found in other aquatic systems (Azam et al. 1983; Stockner and Porter 1988).

21.2.3 Aquatic Diversity

Aquatic species diversity is controlled by abiotic, biotic, and biogeographical factors (Kling et al. 1992a; Hershey et al. this Vol.). In relation to terrestrial diversity, the chemical nature of terrestrial inputs is most important in directly shaping aquatic diversity. For example, the ratios and absolute amounts of nutrient elements such as N, P, and silica can alter the diversity of algal species in

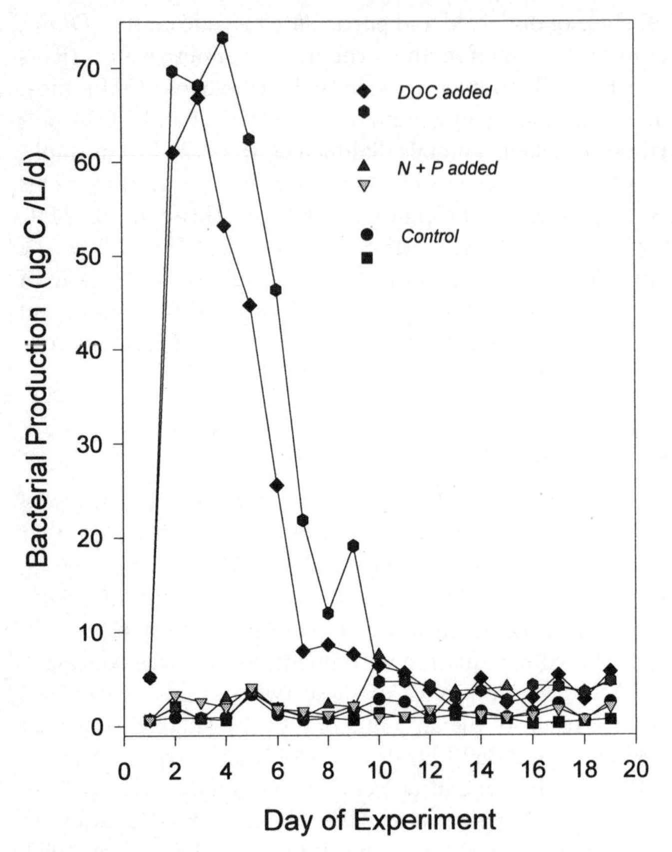

Fig. 1. Bacterial productivity in 10 m^3 limnocorrals in Toolik Lake, arctic Alaska. Additions of 1 mg/l DOC (to a background of 6 mg/l) leached from terrestrial plant detritus at the soil surface enhanced productivity in light and dark corrals compared to control corrals and to corrals with added inorganic N and P. (J.E. Hobbie, G.W. Kling unpubl.)

arctic lakes and streams (Welch et al. 1989; Miller et al. 1992). In addition, the amounts of major ions such as Ca^{2+} and HCO_3^- delivered from land can directly control the diversity and species distribution of mollusks in alpine lakes due to constraints on shell construction (Saunders and Kling 1990). Species diversity in zooplankton and aquatic insects appears to be little related to the direct effects of inorganic water chemistry arctic Alaska (Kling et al. 1992a), and the effects of nontoxic organic materials on diversity of these groups of organisms are unstudied.

21.3 Terrestrial Diversity

Having identified water, nutrients, and organic matter as the major inputs from terrestrial systems that affect aquatic systems, it is necessary to relate the biogeochemistry of these materials to particular vegetation groups and to particular habitats on land. Although the amounts, forms, and timing of terrestrial exports have been studied extensively in several ecosystems, we understand little about the specific plant and soil origins of these materials or the controls on production and delivery, especially in the Arctic.

21.3.1 The Effects of Species Diversity

As water moves across a landscape its chemistry is modified by vegetation and soil processes. We understand a good deal about the importance of vegetation type on the uptake of inorganic nutrients from soil waters (Chapin 1980; Goldberg 1990), including the role of riparian vegetation in some temperate systems (Lowrance et al. 1984; Peterjohn and Correll 1984). The most complete research on the role of vegetation in transforming and transferring inorganic nutrients across terrestrial landscapes was done in the Arctic at Imnaviat Creek (Marion and Everett 1989) and at the Sagavanirktok River near Toolik (Giblin et al. 1991; Shaver et al. 1991). This work showed that different vegetation types and locations on the landscape differ strongly in their chemical interactions with soil water and groundwater, and thus have highly variable effects on the chemistry of water exported from land. Unfortunately, in these arctic studies the fluxes between land and water across the riparian zone were not examined. There are also substantial gaps in our knowledge of DOM in groundwaters, especially of dissolved organic carbon and nitrogen (DOC, DON). What little information exists for temperate terrestrial systems about DOC (Sollins and McCorison 1981; Cronan and Aiken 1985; McDowell and Likens 1988; Kaplan and Newbold 1993) or about DON (Yavitt and Fahey 1986; Qualls et al. 1991; Binkley et al. 1992) suggests that vegetation type and soil composition play a major role in determining the concentrations and element ratios. Even though DOM tends to dominate the surface water budgets of C, N, and P in the Arctic, there are apparently no published studies on the production, dynamics, or transfers of groundwater DOC, and one reference to DON (Kielland and Chapin 1992) in environments of the far north. There is evidence from a small catchment near Toolik Lake in arctic Alaska that the concentrations of groundwater DOC are related in part to the type of overlying vegetation (Fig. 2).

21.3.2 The Effects of Landscape Diversity

The heterogeneity of landscapes may also play a role in the production and export of nutrients and organic matter, although the specific mechanisms involved and even the general patterns need clarification. This heterogeneity includes the local effects of water track or stream gradient and size, and the regional effects of landscape age, geology and geomorphology, vegetation, and soil composition. At the local level, the effect of chemical transformation of NH_4^+ to NO_3^- in the last few meters of the riparian zone is shown for an arctic setting in Fig 3. This chemical

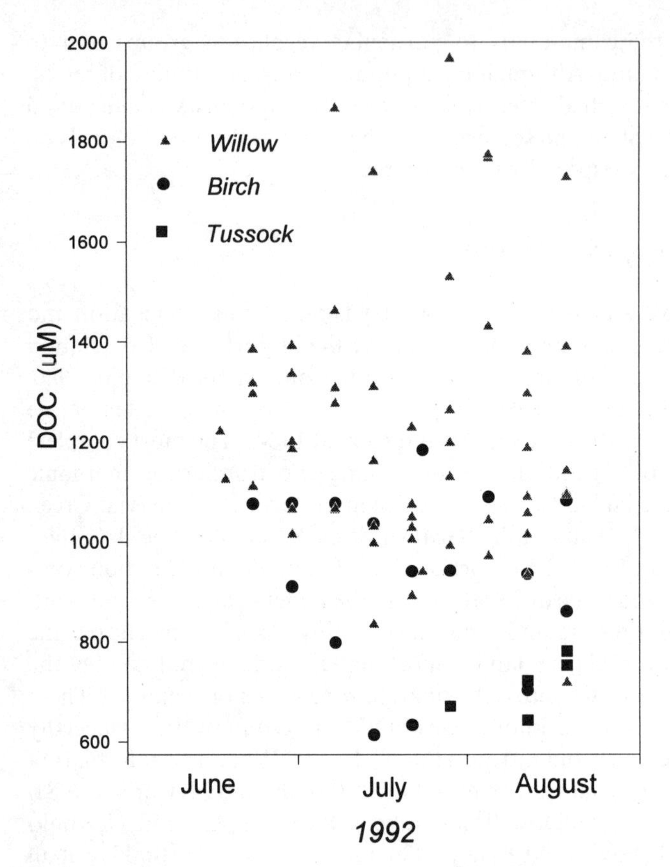

Fig. 2. Groundwater concentrations of DOC in a small catchment (8 ha) located near Toolik Lake, under predominantly birch, willow, or tussock vegetation

transformation occurs in response to increased exposure of groundwater to oxygen through the cracked and eroding peat banks near the river edge. Again, at the local level, there appears to be a consistent relationship between position on the drainage slope and groundwater DOC concentrations in a small arctic catchment (Fig. 4). The relationship is independent of known effects of vegetation type on DOC (Fig. 2), although the processes responsible are as yet unknown.

At the regional level, the age of the land surface influences the extent of weathering, and there are large differences in chemistry of water draining various landscapes in the foothills of arctic Alaska that were deglaciated within the last 10 000–20 000 years, versus those deglaciated more than 100 000 years ago (Kling et al. 1992a). In addition, the effects of slope, aspect, and wind-blown materials such as alkaline loess create very different acid-base conditions in arctic soils (Walker and Webber 1979; Walker et al. 1993). It remains to be determined exactly how these local and regional components of landscape heterogeneity are controlling land-water linkages and aquatic systems.

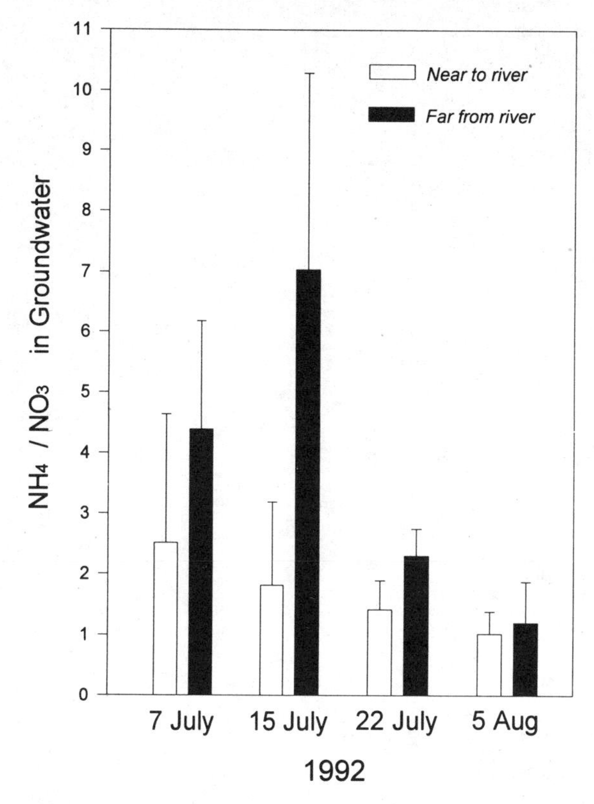

Fig. 3. Molar ratios of NH_4^+ to NO_3^- (mean $\pm$ 1 SE) in groundwaters near to (within 10 m) and far from (> 10 m) the bank of the Kuparuk River, arctic Alaska

21.4 Movement of Material from Land to Water

Research at the scale of catchments has shown that material export from land to water is primarily a function of water flow within specific basins, and of landscape heterogeneity, such as geologic setting and vegetation type, among different basins. Secondary controls are a function of disturbance (e.g. clear-cutting) and biological transformations of organic matter and nutrients (Likens and Bormann 1974; Likens et al. 1977). Similar studies have established these generalities for other temperate and tropical systems, especially with regard to water flow (Swank and Crossley 1988; Lewis and Saunders 1989). However, there are only a few references describing the association between water flow and the export of materials or discharge-solute relationships in the Arctic (Peterson et al. 1986, 1992; Everett et al. 1989; Marion and Everett 1989; Cornwell 1992). Water flow and the hydrologic cycle in the Arctic have unique characteristics (Woo et al. 1981; Kane et al. 1989). Most importantly, infiltration and groundwater movement is confined to shallow (0.5 m) zones by permafrost, and the seasonal

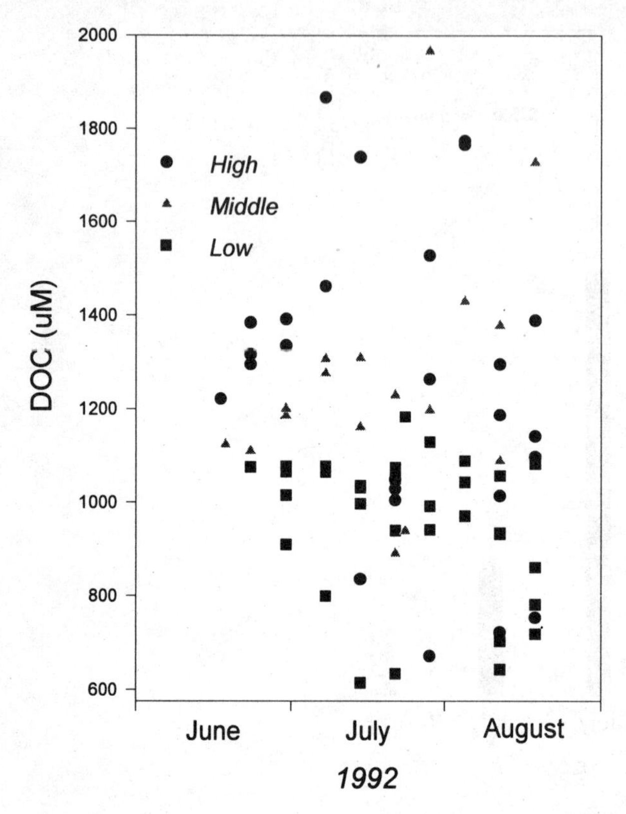

Fig. 4. Groundwater concentrations of DOC in a small catchment near Toolik Lake at various locations in the catchment. *High* indicates samples at the top of the catchment, and *low* indicates samples taken about 400 m lower in the catchment

hydrologic regime is dominated by freezing and thawing processes. Because extensive areas of the Arctic have uniform topography, moderate changes in hydrological conditions can lead to large changes in the area of surface water, the amount of soil moisture, and the amount of ground ice near the land surface (Lachenbruch and Marshall 1986). These changes, in turn, will have important feedbacks on the transfer of materials from land to surface waters.

The transport of nutrients across tundra may be strongly controlled by the spatial and temporal relation between water pathways and vegetation uptake. For example, data from the North Slope of Alaska show that because of rapid biological uptake, nutrients added to tundra have limited movement downslope in groundwater; this movement is at most only tens of meters per year, in some cases even if water flow is also increased (Kummerow et al. 1987; Chapin et al. 1988; Marion and Everett 1989). The significance of these findings is that the expected downslope movement in a natural setting is small compared to the normal interfluvial distances. Because of this, much of the upland tundra may be chemically isolated from nearby water tracks during dry periods, and the role of intermittent rain events will be magnified in the transport of materials from land to water. While such rain events are known to have a flushing or diluting effect on the export of major elements, nutrients, and DOC in the Arctic (Peterson et al. 1986, 1992; Everett et al. 1989; Marion and Everett 1989), there has been little specific research on the linkage between different landscape patches and on the processes that determine the interaction between soils, vegetation, and water flow during transport of dissolved materials across the tundra.

The final segment in the transfer of materials from land is the land-water interface. These interfaces are among the most distinctive and persistent of landscape boundaries, with the exception of that between the earth and atmosphere. Recent advances in the theory of boundaries (Naiman et al. 1988; Forman and Moore 1992; Weins 1992) describe useful conceptual frameworks to define the determinants of flows across interfaces. Although the concepts include those of permeability and the role of "filtering", it is clear from the above discussion that in a biogeochemical perspective the concepts of chemical exchange and transformation must also be included (e.g., Fig. 3). These processes are important both at the boundaries between landscape patches and within the patches themselves.

21.4.1 Feedbacks

For the most part, the movement of nutrients and organic matter is unidirectional from land to water over geologically short time scales. Some notable exceptions include the harvest of fish, and the emergence of insects from streams (Jackson and Fisher 1986; Hershey et al. 1993). However, perhaps the most important feedback from water to land, especially in the Arctic, involves the cycling of carbon gases. The cycle begins by fixation of atmospheric CO_2 by tundra vegetation, and the subsequent respiration of plant organic matter in the soil to produce CO_2 and CH_4. These gases then dissolve in groundwaters and are transported to lakes and streams where they are subsequently released to the atmosphere to complete the cycle (Fig. 5). The flux to the atmosphere resulting from excess CO_2 and CH_4 in surface waters is a consistent feature of arctic Alaska (Kling et al. 1992b), and probably of other tundra areas (Table 1). This feedback of terrestrially produced carbon to the atmosphere from aquatic systems represents an important flux in the global carbon cycling of tundra environments (Kling et al. 1991), and is related in part to the diversity of terrestrial vegetation and landscapes.

21.5 Conclusions

1. Given our current limited knowledge of land-water interactions, it is apparent that three main factors are likely to regulate the transformations of terrestrial-derived materials and their transfers to surface waters: (a) water flow, (b) vegetation and soil uptake and release, and (c) landscape heterogeneity.

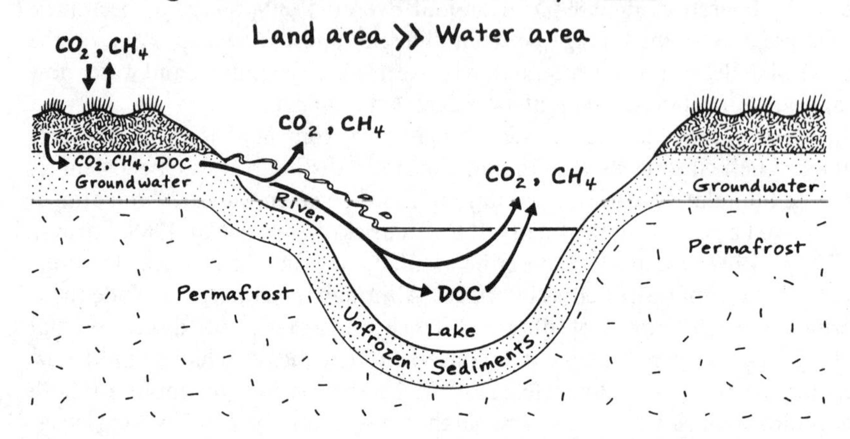

Fig. 5. Conceptual diagram of arctic carbon cycle showing the uptake of CO_2 by plants, the transport of carbon in groundwaters to rivers and lakes (either as CO_2, CH_4, DOC, or POC), and the release of carbon gases from surface waters to the atmosphere directly or after in-lake metabolism of DOC or POC

Table 1. Dissolved gas concentrations in small lakes and streams in West Siberia (between 66° and 67°N and 79° and 80°E; elevations < 40 m a.s.l.) showing that most waters contain excess CO_2 (relative to atmospheric pressure, 365 μatm) that is produced in terrestrial environments. Conductivity, alkalinity, pH, and P_{CO2} measured or calculated following Kling et al. (1992b)

Site	Date (1991)	Temp. (°C)	Cond. (μS/cm)	pH	Alk. (μeq/l)	P_{CO2} (μatm)
1	1 Aug	14.0	98.9	7.53	800	1274
2	2 Aug	17.5	12.0	6.48	37	720
3	2 Aug	18.5	18.1	6.99	116	702
4	4 Aug	14.0	33.8	7.10	300	1316
5	4 Aug	15.5	19.5	7.45	130	262
6	4 Aug	15.5	12.5	6.05	59	3007
7	6 Aug	17.0	9.4	5.48	21	4073
8	6 Aug	17.0	10.4	5.85	29	2395
9	9 Aug	11.5	23.7	7.25	163	493
10	9 Aug	11.5	11.4	6.63	53	676
11	10 Aug	12.0	24.7	7.52	174	284
12	10 Aug	12.5	16.1	6.69	62	694
13	10 Aug	12.5	16.4	6.77	89	829
14	10 Aug	12.0	11.8	5.54	5	790
15	10 Aug	12.0	20.0	7.12	129	531

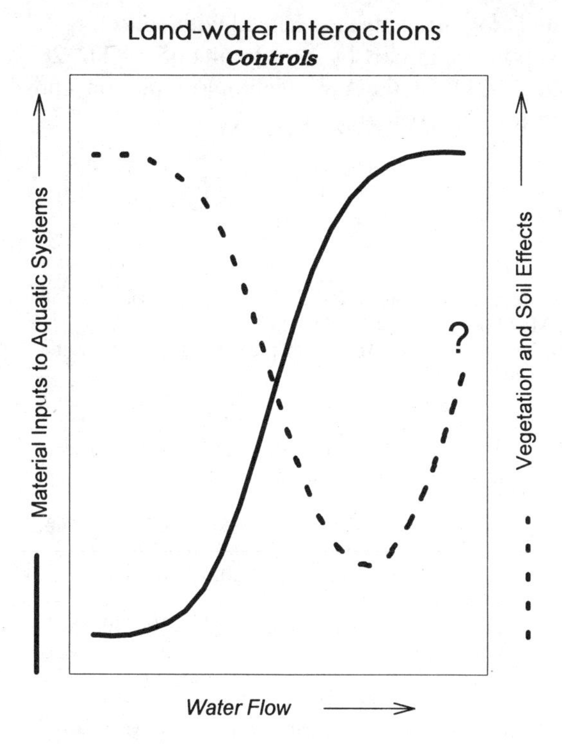

Fig. 6. Conceptual diagram showing major controls of terrestrial material production and export. Material flux increases with water flow, while vegetation and soil effects decrease due to reduced interaction time in soils. Highest water flows may increase vegetation and soil effects (e.g., for DOC, see Schindler et al. 1992). The shape of these curves may be different for given landscapes with different characteristic slope, age or vegetation

2. A first step will be to determine the relationships between water flow and vegetation and soil effects as represented by the curves shown in Fig. 6. Secondly, we must determine the relative importance of these factors in controlling the material fluxes arising from different vegetation types and landscape units.
3. Finally, research on determining the processes and controls of material production, transformation, and transfers from land to water is needed to develop models that predict how changing environmental conditions will alter land-water interactions and, consequently, the structure and function of tundra terrestrial and aquatic ecosystems.

Acknowledgments. I thank John Hobbie, George Kipphut, Mike Miller, Bruce Peterson, Ed Rastetter and Gregory Vilchek for critcial discussions and for generously providing unpublished data. I also appreciate the comments of Gus Shaver and Terry Chapin as reviewers. Peder Yurista, Tammy Gage, Bonnie

Kwaitkowski, Mark Brahce, and Tanya Buchatskaya helped with collecting or processing samples. Work was supported in part by NSF grants BSR-8702328, DPP-8722015, DPP-8320544, DEB-9211775, the A.W. Mellon Foundation, and the Institute of Geography of the Russian Academy of Sciences.

References

Azam F, Fenchel T, Field JG, Meyer-Reil LA, Thingstad F (1983) The ecological role of water-column microbes in the sea. Mar Ecol Prog Ser 10: 257–263

Binkley D, Sollins P, Bell R, Sachs D, Myrold D (1992) Biogeochemistry of adjacenet conifer and alder-conifer stands. Ecology 73: 2022–2033

Chapin FS III (1980) The mineral nutrition of wild plants. Annu Rev Ecol Syst 11: 233–260

Chapin FS III, Miller PC, Billings WD, Coyne PI (1980) Carbon and nutrient budgets and their control in coastal tundra. In: Brown J, Miller PC, Tieszen LL Bunnell FL (eds) An arctic ecosystem: the coastal tundra at Barrow. Dowden, Hutchinson & Ross, Stroudsburg, pp 458–482

Chapin FS III, Fetcher N, Kielland K, Everett KR, Linkens AE (1988) Productivity and nutrient cycling of Alaskan Tundra: enhancement by flowing soil water. Ecology 69: 693–702

Cole JJ, Findlay S, Pace ML (1988) Bacterial production in fresh and saltwater ecosystems: a cross-system overview. Mar Ecol Prog Ser 43: 1–10

Cornwell JC (1992) Cation export from Alaskan arctic watersheds. hydrobiologia 240: 15–22

Cronan CS, Aiken GR (1985) Chemistry and transport of soluble humic substances in forested watersheds of the Adirondack Park, New York. Geochim Cosmochim Acta 49: 1697–1705

Dowding P, Chapin FS III, Weilgolaski FE, Kilfeather P (1981) Nutrients in tundra ecosystems In: Bliss LC, Heal OW, Moore JJ (eds) Tundra ecosystems, a comparative analysis. Cambridge University Press, Cambridge, pp 647–683

Everett KR, Marion GM, Kane DL (1989) Seasonal geochemistry of an arctic tundra drainage basin. Holarct Ecol 12:279–289

Forman RT T, Moore PN (1992) Theoretical foundations for understanding boundaries in landscape mosaics. In: Hansen AJ, di Castri F (eds) Landscape boundaries. Springer, Berlin, Heidelberg, New York, pp 236–258

Giblin AE, Nadelhoffer KJ, Shaver GR, Laundre JA, McKerrow AJ (1991) Biogeochemical diversity along a riverside toposequence in arctic Alaska. Ecol Monogr 61: 415–435

Goldberg DE (1990) Components of resource competition in plant communities. In: Grace JB, Tilman D (eds) Perspectives on plant competition. Academic Press, New York, pp 27–49

Hansen AJ, di Castri F (1992) Landscape boundaries. Springer, Berling Heidelberg, New York, 452 pp

Hecky RE, Kilham P (1988) Nutrient limitation of phytoplankton in freshwater and marine environments: a review of recent evidence on the effects of enrichment. Limnol Oceanogr 33: 796–822

Hershey AE, Pastor J, Peterson BJ, Kling GW (1993) Stable isotopes resolve the drift paradox for *Baetis* mayflies in an arctic river. Ecology 74: 2315–2325

Hessen DO, Andersen T, Lyche A (1990) Carbon metabolism in a humic lake: pool sizes and cycling through zooplankton. Limnol Oceanogr 35: 84–99

Hobbie JE (1980) Limnology of Tundra Ponds. Dowden, Hutchinson & Ross, Stroudsburg

Hobbie JE, Helfrich JVK III (1988) The effect of grazing by microprotozoans on production of bacteria. Archiv Hydrobiol Beih 31: 281–288

Jackson JK, Fisher SG (1986) Secondary production, emergence, and export of aquatic insects of a Sonoran desert stream. Ecology 67: 629–638

Jones RI (1992) The influence of humic substances on lacustrine planktonic food chains. Hydrobiologia 229: 73–91

Kane DL, Hinzman LD, Benson CS, Everett KR (1989) Hydrology of Imnavait Creek, and arctic watershed. Holarct Ecol 12: 262–269

Kaplan LA, Newbold JD (1993) Biogeochemistry of dissolved organic carbon entering streams. In:Ford TE (ed) Aquatic microbiology, an ecological approach. Blackwell, Cambridge, MA, pp 139–165

Kielland K, Chapin FS III (1992) Nutrient absorption and accumulation in arctic plants. In: Chapin FS, Jeffries RL, Reynolds JF, Shaver GR, Suoboda J (eds) Arctic ecosystems in a changing climate. Academic Press, Sah Diego, pp 321–336

Kling GW (1994) Ecosystem-scale experiments in freshwaters: the use of stable isotopes. In; Baker LA (ed) Environmental chemistry of lakes and reservoirs. Advances in Chemistry Series 237. American Chemical Society, Washington DC, pp 91–120

Kling GW, Kipphut GW, Miller MC (1991) Arctic lakes and rivers as gas conduits to the atmosphere: implications for tundra carbon budgets. Science 251: 298–301

Kling GW, O'Brien WJ, Miller MC Hershey AE (1992a) The biogeochemistry and zoogeography of lakes and rivers in arctic Alaska. Hydrobiologia 240: 1–14

Kling GW, Kipphut GW, Miller MC (1992b) The flux of carbon dioxide and methane from lakes and rivers in arctic Alaska. Hydrobiologia 240: 23–36

Kummerow J, Mills JN, Ellis BA, Hastings SJ, Kummerow A (1987) Downslope fertilizer movement in arctic tussock tundra. Holarct Ecol 10: 312–319

Lachenbruch AH, Marshall BV (1986) Changing climate: geothermal evidence from permafrost in the Alaskan arctic. Science 234: 689–696

Lewis WM Jr, Saunders JF III (1989) Concentration and transport of dissolved and suspended substances in the Orinoco River. Biogeochemistry 7: 203–240

Likens GE, Bormann FH (1974) Linkages betwen terrestrial and aquatic ecosystems. Bioscience 24: 447–456

Likens GE, Bormann FH, Pierce SR, Eaton SJ, Johnson MN (1977) Biogeochemistry of a forested ecosystem. Springer, Berlin Heidelberg, New York

Lowrance R, Todd R, Fail J Jr, hendrickson O Jr, Leonard R, Asmussen L (1984) Riparian forests as nutrient filters in agricultural watersheds. Bioscience 34: 374–377

Marion GM, Everett KR (1989) The effect of nutrient and water additions on element mobility through small tundra watersheds. Holarc Ecol 12: 317–323

McDowell WH, Likens GE (1988) origin, composition, and flux of dissolved organic carbon in the Hubbard Brook Valley. Ecol Monogr 58: 177–195

Meyer J, Edwards RT (1990) Ecosystem metabolism and turnover of organic carbon along a blackwater river continuum. Ecology 71: 668–677

Miller MC, Hater GR, Spatt P, Westlake P, Yeakel P (1986) Primary production and its control in Toolik Lake, Alaska. Arch Hydrobiol (Suppl) 74: 97–134

Miller MC, DeOliveira P, Gibeau GG (1992) Epilithic diatom community response to years of PO_4 fertilization: Kuparuk River, Alaska (68 °N Lat.). Hydrobiologia 240: 103–119

Morris DP, Lewis WM (1988) Phytoplankton nutrient limitation in Colorado mountain lakes. Freshwater Biol 20:315–327

Naiman RJ, Decamps H, Pastor J, Johnston CA (1988) The potential importance of boundaries to fluvial ecosystems. JN Benthol Soc 7: 289–306

Pace ML (1993) Heterotrophic microbial processes. In; Carpenter SR, Kitchell JF (eds) The trophic cascade in lakes. Cambridge University Press, New York, pp 252–277

Palenik B, Morel FMM (1990) Aminio acid utilization by marine phytoplankton: novel mechanism. Limnol Oceanogr 35: 260–269

Peterjohn WT, Correll DL (1984) Nutrient dynamics in an agricultural watershed: observations on the role of a riparian forest. Ecology 65: 1466–1475

Peterson BJ, Hobbie JE, Hershey AE, Lock ME, Ford TE, Vestal JR, McKinley VL, Hullar MAJ, Ventullo RM, Volk GS (1985) Transformation of a tundra river from heterotrophy to autotrophy by addition of phosphorus. Science 229: 1383–1386

Peterson BJ, Hobbie JE, Corliss TL (1986) Carbon flow in a tundra stream ecosystem. Can J Fish Aquat Sci 43: 1259–1270

Peterson BJ, Corliss TL, Kriet K, Hobbie JE (1992) Nitrogen and phosphorus concentrations and export for the upper Kuparuk on the North Slope of Alaska in 1980. Hydrobiologia 240: 61–69

Qualls RG, Haines BL, Swank WT (1991) Fluxes of dissolved organic nutrients and humic substances in a deciduous forest. Ecology 72: 254–266
Riemann B (1985) Potential importance of fish predation and zooplankton grazing on natural populations of freshwater bacteria. Appl Environ Microbiol 50: 187–193
Ross HH (1963) Stream communities and trerrestrial biomes. Arch Hydrobiol 59: 235–242
Rublee P (1992) Community structure and bottom-up regulation of heterotrophic microplankton in arctic lakes. hydrobiologia 240: 133–141
Salonen K, Kairesalo T, Jones RI (eds) (1992) Dissolved organic matters in lacustrine ecosystems: energy source and system regulator. Hydrobiologia 229: 1–291
Saunders JF, Kling GW (1990) Species distributions and shell characteristics of *Pisidium* (Mollusca: Bivalvia) in the Colorado Front Range: the role of abiotic factors. Freshwater Biol 24: 275–285
Schell DM (1983) ^{13}C and ^{14}C abundances in Alaskan aquatic organisms: delayed consumer production from peat in arctic food webs. Science 219:1068–1071
Schindler DW, Bayley SE, Curtis PJ, Parker BR, Stainton MP, Kelly CA (1992) Natural and man-caused factors affecting the abundance and cycling of dissolved organic substances in Precambrian shield lakes. Hydrobiologia 229: 1–21
Schulze E-D, Mooney HA (eds) (1993) Biodiversity and ecosystem function. Ecological Studies 99, Springer, Berlin, Heidelberg, New York.
Shaver, GR, Nadelhoffer KJ, Giblin AE (1991) Biogeochemical diversity and element transport in a heterogeneous landscape, the North Slope of Alaska. In: Turner MG, Gardner RH (eds), Quantitative methods in landscape ecology. Springer, Berlin, Heidelberg, New York, pp 105–125
Sollins P, Mc Corison FM (1981) Nitrogen and carbon solution chemistry of an old growth coniferous forest watershed before and after cutting. Wat Resour Res 17: 1409–1418
Stockner JG, Porter KG (1988) Microbial food webs in freshwater planktonic ecosystems. In: Carpenter SR (ed), Complex interactions in lake communites. Springer, Berlin, Heidelberg, New York, pp 69–84
Sundh I, Bell RT (1992) Extracellular dissolved organic carbon released from phytoplankton as a source of carbon for heterotrophic bacteria in lakes of different humic content. Hydrobiologia 229: 93–106
Swank WT, Crossley DA (1988) Forest hydrology and ecology at Cowetta. Ecological Studies 66. Springer, Berlin, Heidelberg, New York
Tranvik LJ (1992) Allochthonous dissolved organic matter as an energy source for pelagic bacteria and the concept of the microbial loop. hydrobiologia 229: 107–114
Vannote RL, Minshall GW, Cummins KW, Sedell JR, Cushing CE (1980) The river continuum concept. Can J Fish Aquat Sci 37: 130–137
Walker DA, Webber PJ (1979) Relationships of soil acidity and air temperature to the wind and vegetation of Prudhoe Bay, Alaska. Arctic 32: 224–236
Walker DA, Auerbeach NA, Everett KR, Shippert MM, Walker MD (1993) Large-scale disturbance features in northern Alaska affect regional carbon budgets. Bull Ecol Soc Am 74: 475
Weins JA (1992) Ecological flows across landscape boundaries: a conceptual overview. In: Hansen AJ, di Castri F (eds) Landscape boundaries. Springer, Berlin, Heidelberg, New York, pp 217–235
Welch HE, Legault JA, Kling HJ (1989) Phytoplankton, nutrients, and primary production in fertilized and natural lakes at Saqvaqjuac, N.W.T. Can J Fish Aquat Sci 46: 90–107
Wetzel RG (1992) Gradient-dominated ecosystems: sources and regulatory functions of dissolved organic matter in freshwater ecosystems. Hydrobiologia 229: 181–198
Whalen SC, Cornwell JC (1985) Nitrogen, phosphorus, and organic carbon cycling in an arctic lake. Can J Fish Aquat Sci 42: 797–808
Woo MK, Heron R, Steer P (1981) Catchment hygrology of a high arctic lake. Cold Regions Sci Tech 5: 29–41
Yavitt JB, Fahey TJ (1986) Litter decay and leaching from the forest floor in *Pinus contorta* (Lodgepole pine) ecosystems. J Ecol 74: 525–545

Part IV
A Synthesis

22 Patterns, Causes, Changes, and Consequences of Biodiversity in Arctic and Alpine Ecosystems

F.S. CHAPIN III[1] and CH. KÖRNER[2]

22.1 The Arctic and Alpine Biota

The land area covered by arctic and alpine vegetation is roughly 11 million km^2 or 8% (5% arctic, 3% alpine) of the terrestrial surface of the globe, stretching from 80°N to 67°S and reaching elevations of more than 6000 m in the subtropics. This area is similar to that covered by boreal forests or crops and supports about 4% of the global flora (10 000 alpine and 1500 arctic lowland species; Walker, Chap. 1; Körner, Chap. 4). Most of the fauna of these cold environments occurs below ground and comprises a similar proportion of the world's animal species (Chernov, Chap. 6; Meyer and Thaler, Chap. 7). The local floras of individual mountains (except for isolated volcanic peaks) throughout the world support between 200 and 300 species – a surprisingly constant number. The floras of whole mountain ranges may have over 1000 species in diversity hotspots such as the Caucasus or the mountains of Central Asia (Agakhanjanz and Breckle, Chap. 5) or parts of the subtropical Andes. In most areas of the arctic and alpine zone, less than ten species of higher plants make up more than 90% of the vascular-plant biomass. Perhaps as few as 20 genera account for most of the vascular-plant biomass of the circumpolar arctic (e.g. shrubs of the genera *Betula*, *Rubus*, *Dryas*, *Vaccinium*, *Empetrum*, and *Ledum*, and the sedges *Eriophorum* and *Carex*).

The magnitude of genetic diversity within species does not change with latitude or altitude within either the arctic or the alpine floras (Murray, Chap. 2; McGraw, Chap. 3). Genetic differences among populations result in ecotypic differentiation along both large- and small-scale environmental gradients. Within populations, genetic diversity is created in some plant taxa by frequent hybridization and polyploidy (particularly in deglaciated regions where formerly isolated populations or species come into secondary contact) and in other taxa by recruitment of genetically distinct individuals from the buried seed pool (McGraw, Chap. 3).

Species richness generally declines with increasing latitude and altitude because low temperature and the short growing season are severe environmental filters that exclude species from progressively more severe climates (Walker, Chap. 1; Körner, Chap. 4; Chernov. Chap. 6; Meyer and Thaler, Chap. 7; Grabherr et al., Chap. 12). Under the most extreme conditions, major functional groups of organisms (e.g. tall shrubs; plants or animals with annual life cycles) are

[1] Department of Integrative Biology, University of California, Berkeley, CA 94720, USA
[2] Botanisches Institut, Universität Basel, Schonbeinstrasse 6, 4056 Basel, Switzerland

Ecological Studies, Vol. 113
Chapin/Körner (eds) Arctic and Alpine Biodiversity

absent. There are also predictable changes with latitude and altitude in specific groups of animals (Chernov, Chap. 6; Meyer and Thaler, Chap. 7). For example, the Coleopetra decline more strongly with decreasing temperature than do other groups, resulting in a *relative* increase in the abundance of Colembola (alpine) and Diptera (arctic). Similarly, amphibians and reptiles are missing at high latitude and altitude, whereas the number of species of some bird groups may even increase (Chernov, Chap. 6). In all animal groups, the proportion of species that are carnivorous increases with latitude.

In cold-dominated ecosystems, the balance between the formation of a soil organic mat and disturbance results in an inverse relationship between soil carbon and species diversity. Thus, the Arctic has three fold more soil carbon (55 Pg) than the alpine (20 Pg), but only 13% of the number of plant species. This pattern reflects active accumulation of soil organic matter and a low degree of disturbance in low-arctic compared to high-altitude ecosystems. In the alpine, gravity (1) prevents water accumulation that would reduce decomposition and cause organic accumulation and (2) disrupts the soil organic mat as freeze-thaw action displaces the soil surface down-slope, opening space for many colonizing species. Such slope effects are found in both arctic and alpine regions, so that within each region the greatest species diversity is found on slopes steep enough to minimize soil organic accumulation (Walker, Chap. 1; Körner, Chap. 4). Most of the arctic landscape has a thick organic mat and very low species diversity, whereas areas of topographic relief such as pingos and mountain slopes are hot spots of diversity. Conversely, in alpine regions, where vertical relief is more pronounced, many areas have a high diversity within each square meter, and flat, often peat-covered areas of low diversity are less common. In both the Arctic and the humid alpine region, a substantial part of the regional flora and fauna can be found within 1 km^2 (often within 10 m^2), and very few additional species are added at the mountain-range or regional scale (Körner, Chap. 4; Chernov, Chap. 6).

At a regional and continental scale arctic and alpine organismic diversity is determined by the ancestral (mostly tertiary) stock of species, long-distance migration during the Holocene, and the evolution of new taxa (Murray, Chap. 2; Agakhanjanz and Breckle, Chap. 5; Chernov, Chap. 6; Ammann, Chap. 10). In the Central Asian mountains, the rate of tectonic uplift of mountain systems is similar to the rate of speciation, so that climatic changes caused by uplift are an important selective influence (Agakhanjanz and Breckle, Chap. 5). Following glacial disturbances and extinctions, migration becomes crucial for the rearrangement and diversity of arctic and alpine flora and fauna. Whereas the floristic composition of the Arctic tends to intergrade continuously from four to five centers of floristic richness, the alpine biota are often more discrete, contributing to local speciation and endemism (Murray, Chap. 2; Körner, Chap. 4; Agakhanjanz and Breckle, Chap. 5; Grabherr et al., Chap. 12). Thus, the dominant species of the most widespread ecosystems in any region have a circumpolar distribution and are common throughout the Arctic, whereas each mountain range has a different group of alpine dominants. These geographically discrete patterns of endemism contribute greatly to the overall biotic richness of alpine ecosystems. For example, the alpine flora of New Zealand, with 650 species, shares hardly any species with other mountain areas of the globe.

22.2 Past, Present, and Future Changes in Biodiversity

Climatic changes since the Pleistocene altered the geographic distribution of arctic ecosystems and caused vertical migration of alpine vegetation belts (Ammann, Chap. 10; Agakhanjanz and Breckle, Chap. 5; Brubaker et al., Chap. 8). However, each species typically showed a unique pattern of migration in response to climatic change because of individualistic responses to the environment (Brubaker et al., Chap. 8; Chapin et al., Chap. 16). Consequently, past communities often had quite different species composition than those of today. Extinction of large grazers by human hunting may have contributed strongly to these community changes because of the large effect of herbivores on ecosystem processes (Zimov et al., Chap. 9).

The paleorecord suggests that it will be difficult to predict future patterns of migration. A given species often migrated into quite different ecological communities, indicating that there was no predictable pattern of succession nor any "preparation" (e.g., presence of nitrogen fixers) necessary to allow invasion of new taxa (Brubaker et al., Chap. 8). Migrating populations often produced peak pollen abundances recorded in sediments rapidly after they first appeared, suggesting either rapid migration or rapid onset of reproduction in established clones that were previously not represented in the pollen profile. Very different dominant species (e.g., birch and spruce) were more similar in their ecosystem impacts (e.g., effects on watershed chemistry) than were communities that differed in dominant life form (e.g., herbaceous vs forest communities).

Climatic warming during the past century (0.7 °C) has already caused upward migration of alpine species (Grabherr et al., Chap. 12). If climatic warming continues, taxa restricted to narrow alpine zones at the summits of mountains may disappear (Grabherr et al., Chap. 12). However, the migration is half the rate that would be expected if species had maintained an equilibrium relationship to temperature. Thus, both the rate of individual migration and the movement of ecosystems are slower than would be predicted from change in temperature. This is consistent with recent findings that altitudinal ecotones of forest species move slowly in response to climatic shifts, since their position is strongly determined by species interactions, particularly in the understory (Körner, Chap. 4).

Experimental studies provide a strong basis for predicting how arctic and alpine communities may respond to climatic change. At high latitudes experimental increases in air temperature cause large changes in growth, reproductive output, and clonal expansion, whereas in the mid- and low-Arctic, changes in other factors, such as nutrient supply, are more important (Callaghan and Jonasson, Chap. 11). In the high Arctic, temperature seems to operate directly on the vegetation rather than through soils processes, at least over the first years of experimentation. CO_2 enrichment has little effect on plant growth in arctic and alpine tundra in the short term, perhaps because other factors more strongly restrict growth (Körner, Chap. 4; Callaghan and Jonasson, Chap. 11).

In both arctic and alpine regions, human impact will be the greatest source of environmental change in the coming decades (Young and Chapin, Chap. 13). Although there have been substantial direct impacts associated with resource extraction in the Arctic, changes associated with arctic haze, nitrogen deposition, and altered fire and grazing regimes may have greater impact on arctic biodiversity and ecosystem processes. For example, air pollution from industrial Europe has dramatic effects on the species composition and ecosystem effects of arctic mosses and on lake acidification. In alpine regions, tourist, agricultural, forestry, and hydroelectric developments have caused the most severe impacts. Human impacts depend strongly on economic and social forces outside arctic and alpine regions, and, therefore, feedback loops involving people are relatively insensitive to changes within these ecosystems (Young and Chapin, Chap. 13). People directly influence biodiversity by harvesting targeted species of plants and animals. In some areas, this harvest threatens species because of changes in local social institutions and exogenous forces such as demand for animal products.

22.3 The Significance of Biodiversity for Ecosystem Function

Ecosystem ecologists often equate ecosystem processes with the flow of energy and nutrients. To persists, individuals that comprise populations and species must (1) reproduce and, to achieve this, must (2) acquire resources to maintain themselves and produce biomass. In the process, they create conditions that may be essential or detrimental to the existence of other species. Regardless of the impact of these interactions, the ultimate result is to select for traits that promote persistence of certain genotypes in space and time, not maximazation of production or rates of biogeochemical cycling per se. In some situations, high productivity may promote persistence. For example, following disturbance, rapidly growing species quickly monopolize the available light and nutrient resources. Other species may occupy niches where slow growth and space occupancy lead to greater long-term persistence and reproductive output. In other words, high diversity is not necessarily coupled to a particular rate of production or biogeochemical cycling but may depend on the maintenance of an environmental matrix in which different strategies are favored at different times or places (Körner, Chap. 4). For this reason, the biological feedbacks that maintain the integrity of ecosystems are of greater functional importance in the long term than instantaneous fluxes of matter. The retention of soil resources, the maintanence of structures and pools, and the persistence of interactions among organisms are particularly significant. The long-term persistence of ecosystem functions such as the uptake and cycling of carbon, nutrients, and water will depend on the maintenance of this integrity, which thereby becomes the key component to consider when discussing feedbacks of biodiversity on ecosystem function.

22.3.1 Material and Energy Flow

Current evidence suggests that species diversity within major functional groups has little effect on biogeochemical pools and fluxes of carbon, water, and nutrients in arctic and alpine ecosystems (Shaver, Chap. 14). All plant species are qualitatively similar with respect to the basic processes of resource acquisition and loss and differ mainly in their effect on rates of these processes (Hobbie, Chap. 15; Chapin et al., Chap. 16). In closed communities, any change in abundance of one species which shifts competitive balances will cause opposing changes in resource acquisition by other species, with little overall effect on nutrient cycling or energy flow at the ecosystem level (Chapin et al., Chap. 16). Only when productivity and energy flow are limited by factors other than resources (e.g., rate of seedling establishment in the high Arctic or times since disturbance in sites of frost action) will the species diversity within functional groups be important in governing fluxes or pools of carbon and nutrients. Dominance-diversity curves from closed arctic and alpine plant communities fit a geometric model, suggesting that competitive interactions determining resource preemption best explain the patterns of diversity in these ecosystems (Pastor, Chap. 18). When grazing or nutrient addition alters the abundance of the dominant species, other species change in biomass to acquire the remaining resources. Resource preemption, grazing, and nutrient availability, in that order, appear to be the major processes determining the diversity of carbon and nitrogen distribution among species (Pastor, Chap. 18). This contrasts strikingly with earlier assumptions that physical environment exerted stronger control than biotic interactions over ecosystem processes in cold environments.

In contrast to the situation *within* functional groups, addition or losses of entire plant functional groups can have great effects on ecosystem processes (Hobbie, Chap. 15). For example, mosses, which are sensitive to vascular-plant shading or pollution, reduce decomposition and ecosystem carbon turnover through low litter quality, whereas graminoids, which respond rapidly following disturbance, have high litter quality and substantial rates of tissue turnover, enhancing nutrient cycling. Changes in abundance of plant functional groups that are extreme with respect to litter quality could alter rates of decomposition and the extent to which nutrients are bound up in soil organic matter. Plant functional groups also *indirectly* influence rates of nutrient supply through modification of the environment. For example, mosses, with their low rates of evapotranspiration (leading to waterlogging), and effective insulation (preventing soil warming), inhibit decomposition. By directly or indirectly influencing the rate of supply of limiting resources, changes in abundance of plant functional groups alter the productivity and rates of nutrient cycling of arctic and alpine ecosystems.

In contrast to plant diversity, microbial diversity is difficult to define by species. Many microbial functions (e.g., certain enzymatic potentials related to common substrates, e.g., cellulose, lignin, protein) remain surprisingly constant across a wide range of ecosystem types. Consequently, these enzymatic potentials are ubiquitous, and microbial diversity in these functions has no clear ecosystem consequence (Schimel, Chap. 17). Other microbial processes (e.g., methanogen activity and nitrification) differ strikingly among ecosystem types and are carried out by a relatively small number of species, so that diversity with respect to these processes has profound ecosystem implications (Schimel, Chap. 17). Here, microbial diversity has a clear relationship to trace gas production and biogeochemical cycles. The challenge is to determine how microbial diversity relates to these processes.

22.3.2 Biotic Interactions

In contrast to biogeochemical pools and fluxes, ecosystem processes involving species interactions, such as mycorrhizal interactions, trophic dynamics, and pollination are extremely sensitive to the diversity and identity of species in a landscape. These interactions generally control the ecosystem properties that are of greatest interest to humans and provide a powerful pragmatic basis for understanding and maintaining biodiversity in arctic and alpine ecosystems. Animals play a key role in determining the structure and diversity of arctic and alpine ecosystems. For example, moderate grazing of arctic salt marshes by geese maintains a high productivity (Jefferies and Bryant, Chap. 19), just as grazing by Pleistocene megafauna may have contributed to a productive steppe-grassland in Beringia 20 000 years ago (Zimov et al., Chap. 9). However, recent increase in grain availability in temperate wintering areas has augmented snowgeese populations beyond their summer carrying capacity so that they are destroying widespread areas of summer salt marsh. This illustrates how the impact of human activities outside the Arctic can alter the activity of key arctic organisms.

Browsing mammals are both a product and a cause of plant diversity in cold-dominated ecosystems (Jefferies and Bryant, Chap. 19). Fire and other disturbances create patches of early successional habitat that are essential to the maintenance of populations of browsing mammals. These mammals selectively browse early successional vegetation, speeding its transition to dominance by late-successional species, which are more flammable, increasing the probability of fire and return to early succession. In addition, browsing mammals maintain a mixed diet to minimize intake of any single plant secondary metabolite. One consequence of this mixed feeding strategy is that animals tend to eliminate rare species and avoid those species that are most common, thus reducing plant species diversity. Browsing mammals thus contribute to landscape diversity by speeding succession, but reduce species diversity within individual patches of vegetation. In contrast to browsers, grazers, which are particularly common in alpine regions, tend to increase plant species diversity by reducing competition from the dominant species and creating disturbed microsites for seedling establishment.

Landscape heterogeneity provides an important component of biodiversity in arctic and alpine ecosystems. Those habitats with most plant species and which are used most intensively by mammals are associated with topographic relief. Communities along topographic gradients differ strikingly in groundwater chemistry and, therefore, their effect on aquatic ecosystems (Kling, Chap. 21). For example, riparian shrub communities have strong nitrification potentials, so that groundwater passing through these communities is enriched in nitrate, whereas ammonium dominates the soil solution of many other ecosystem types. Upland heath ecosystems are net sources of nitrogen to groundwater, whereas lowland sedge-meadow communities are net sinks for nitrogen. Because of differences in resource supply and historical factors governing species composition of fish and zooplankton communities, streams and lakes in different landscape positions differ in the relative importance of resource supply and predators in controlling community diversity (Hershey et al., Chap. 20).

22.3.3 Integrity of the Environmental Matrix

Plant functional groups also influence ecosystem processes through effects on energy budgets (Chapin et al., Chap. 16). Tall plants such as trees and tall shrubs, which are absent or uncommon in tundra, reduce the albedo (reflectance) by masking the snow, thereby increasing annual energy inputs to cold-dominated ecosystems. Because decomposition, nutrient supply, and plant growth are potentially sensitive to temperature and length of growing season, changes in the ecosystem energy balance meditated by invasion of new plant functional groups could alter both nutrient and carbon flow within ecosystems. By determining the amount of energy absorbed by an ecosystem, albedo – as controlled by plant functional groups – determines the energy inputs to the atmosphere and, therefore, local and regional climate.

Functional groups which influence the disturbance regime can strongly affect both ecosystem processes and species diversity. For example, certain scree species must be present to consolidate the substrate and open it to other species, which, in turn, establish islands of humus where the quantity and quality of biomass is adequate to support herbivores. The elimination of the deeply anchored pioneers would immediately destabilize the entire system (Körner, Chap. 4). Thus, a few pioneer species of low abundance are critical to slope stability, water retention, and seasonality of water flow to rivers and thus hydroelectric power plants. The degree of disturbance following an avalanche depends on plant traits. Elimination of the forest will not cause erosion if an understory exists with plants of various rooting depths. In the Arctic, invasion of conifers which are highly flammable could greatly alter the fire regime, causing replacement of fire-sensitive by fire-resistant species.

22.3.4 Insurance

Genetic and species diversity provide insurance against loss of ecosystem function. The greater the biotic diversity within a functional group, the less likely that any extinction event resulting from extreme climatic or chemical perturbation will have serious ecosystem consequences (Chapin et al., Chap. 16). For example, catastrophic events such as a snow-free winter situation could eliminate a large fraction of snow-bed species that depend on protection by snow. If the community contains a few non-snow-bed species or genotypes with higher frost resistance, the integrity of the ecosystem can be maintained. For these reasons, genetic and species diversity per se are important to maintain ecosystem structure and function.

22.4 Conclusions

1. Patterns of diversity differ between arctic and alpine ecosystems for both historical and current ecological reasons. Low temperature is an effective filter that limits the number of species that can colonize arctic and alpine environments. Greater isolation and niche differentiation promoted speciation and restricted species migration in alpine regions, resulting in a higher species richness in alpine than in arctic ecosystems.

   Moreover, the greater verticle relief in alpine ecosystems is unfavourable to peat formation and causes more disturbance, leading to high ecological diversity in the alpine ecosystem at the landscape level and a concentration of arctic biodiversity in localized sites of high vertical relief. Because the most widespread communities in the Arctic (and in alpine areas of low relief) have very few species, the loss of even a few species would dramatically alter species diversity.
2. Biodiversity in arctic and alpine ecosystems is currently threatened most strongly by diffuse impacts of human activity. CO_2-induced climatic warming is causing upward migration of alpine species with the possible loss of some alpine ecosystems from low-altitude summits. Input of pollutants from low latitudes and altitudes has a low-level chronic impact on key functional groups. This combined with climatic warming, could alter conditions for establishment and cause an advance of the tree line, leading to changes in the role of arctic and alpine ecosystems in global carbon and energy balance. Because these human impacts are caused by forces outside arctic and alpine areas, there is no clear mechanism by which ecological impacts will alter human activities responsible for the problems.
3. Biogeochemical pools and fluxes are the ecosystem traits and processes that are least sensitive to changes in biodiversity. Only if there are major changes in abundance of functional groups of plants, animals, or microorganisms will biogeochemical processes be strongly altered. By contrast, traits of individual species strongly affect organism interactions such as herbivory, mycorrhizal association, and pollination, so that even subtle changes in biodiversity will affect these interactions and their impact on ecosystem processes.

Subject Index

Ecological Studies

Volumes published since 1989

Volume 74
Inorganic Contaminants in the Vadose Zone (1989)
B. Bar-Yosef, N.J. Barrow, and J. Goldshmid (Eds.)

Volume 75
The Grazing Land Ecosystems of the African Sahel (1989)
H.N. Le Houérou

Volume 76
Vascular Plants as Epiphytes: Evolution and Ecophysiology (1989)
U. Lüttge (Ed.)

Volume 77
Air Pollution and Forest Decline: A Study of Spruce (*Picea abies*) on Acid Soils (1989)
E.-D. Schulze, O.L. Lange, and R. Oren (Eds.)

Volume 78
Agroecology: Researching the Ecological Basis for Sustainable Agriculture (1990)
S.R. Gliessman (Ed.)

Volume 79
Remote Sensing of Biosphere Functioning (1990)
R.J. Hobbs and H.A. Mooney (Eds.)

Volume 80
Plant Biology of the Basin and Range (1990)
B. Osmond, G.M. Hidy, and L. Pitelka (Eds.)

Volume 81
Nitrogen in Terrestrial Ecosystems: Questions of Productivity, Vegetational Changes, and Ecosystem Stability (1991)
C.O. Tamm

Volume 82
Quantitative Methods in Landscape Ecology: The Analysis and Interpretation of Landscape Heterogeneity (1990)
M.G. Turner and R.H. Gardner (Eds.)

Volume 83
The Rivers of Florida (1990)
R.J. Livingston (Ed.)

Volume 84
Fire in the Tropical Biota: Ecosystem Processes and Global Challenges (1990)
J.G. Goldammer (Ed.)

Volume 85
The Mosaic-Cycle Concept of Ecosystems (1991)
H. Remmert (Ed.)

Volume 86
Ecological Heterogeneity (1991)
J. Kolasa and S.T.A. Pickett (Eds.)

Volume 87
Horses and Grasses: The Nutritional Ecology of Equids and Their Impact on the Camargue (1992)
P. Duncan

Volume 88
Pinnipeds and El Niño: Responses to Environmental Stress (1992)
F. Trillmich and K.A. Ono (Eds.)

Volume 89
Plantago: A Multidisciplinary Study (1992)
P.J.C. Kuiper and M. Bos (Eds.)

Volume 90
Biogeochemistry of a Subalpine Ecosystem: Loch Vale Watershed (1992)
J. Baron (Ed.)

Volume 91
Atmospheric Deposition and Forest Nutrient Cycling (1992)
D.W. Johnson and S.E. Lindberg (Eds.)

Volume 92
Landscape Boundaries: Consequences for Biotic Diversity and Ecological Flows (1992)
A.J. Hansen and F. di Castri (Eds.)

Volume 93
Fire in South African Mountain Fynbos: Ecosystem, Community, and Species Response at Swartboskloof (1992)
B.W. van Wilgen et al. (Eds.)

Volume 94
The Ecology of Aquatic Hyphomycetes (1992)
F. Bärlocher (Ed.)

Volume 95
Palms in Forest Ecosystems of Amazonia (1992)
F. Kahn and J.-J. DeGranville

Volume 96
Ecology and Decline of Red Spruce in the Eastern United States (1992)
C. Eagar and M.B. Adams (Eds.)

Volume 97
The Response of Western Forests to Air Pollution (1992)
R.K. Olson, D. Binkley, and M. Böhm (Eds.)

Volume 98
Plankton Regulation Dynamics (1993)
N. Walz (Ed.)

Volume 99
Biodiversity and Ecosystem Function (1993)
E.-D. Schulze and H.A. Mooney (Eds.)

Volume 100
Ecophysiology of Photosynthesis (1994)
E.-D. Schulze and M.M. Caldwell (Eds.)

Volume 101
Effects of Land Use Change on Atmospheric CO_2 Concentrations: South and South East Asia as a Case Study (1993)
V.H. Dale (Ed.)

Volume 102
Coral Reef Ecology (1993)
Y.I. Sorokin

Volume 103
Rocky Shores: Exploitation in Chile and South Africa (1993)
W.R. Siegfried (Ed.)

Volume 104
Long-Term Experiments With Acid Rain in Norwegian Forest Ecosystems (1993)
G. Abrahamsen et al. (Eds.)

Volume 105
Microbial Ecology of Lake Plußsee (1993)
J. Overbeck and R.J. Chrost (Eds.)

Volume 106
Minimum Animal Populations (1994)
H. Remmert (Ed.)

Volume 107
The Role of Fire in Mediterranean-Type Ecosystems (1994)
J.M. Moreno, W.C. Oechel

Volume 108
Ecology and Biogeography of Mediterranean Ecosystems in Chile, California and Australia (1994)
M.T.K. Arroyo, P.H. Zedler, and M.D. Fox (Eds.)

Volume 109
Mediterranean-Type Ecosystems. The Function of Biodiversity (1995)
G.W. Davis and D.M. Richardson (Eds.)

Volume 110
Tropical Montane Cloud Forests (1995)
L.S. Hamilton, J.O. Juvik, and F.N. Scatena (Eds.)

Volume 111
Peatland Forestry. Ecology and Principles (1995)
E. Paavilainen and J. Päivänen

Volume 112
Tropical Forests: Management and Ecology (1995)
A.E. Lugo and C. Lowe (Eds.)

Volume 113
Arctic and Alpine Biodiversity. Patterns, Causes and Ecosystem Consequences (1995)
F.S. Chapin III and C. Körner (Eds.)